Universitext

Universitext

Universitext is a series of textbooks that presents material from a wide variety of mathematical disciplines at master's level and beyond. The books, often well class-tested by their author, may have an informal, personal even experimental approach to their subject matter. Some of the most successful and established books in the series have evolved through several editions, always following the evolution of teaching curricula, to very polished texts.

Thus as research topics trickle down into graduate-level teaching, first textbooks written for new, cutting-edge courses may make their way into *Universitext*.

For further volumes:
http://www.springer.com/series/223

Andries E. Brouwer • Willem H. Haemers

Spectra of Graphs

 Springer

Andries E. Brouwer
Department of Mathematics
Eindhoven University of Technology
Eindhoven
The Netherlands

Willem H. Haemers
Department of Econometrics
and Operations Research
Tilburg University
Tilburg
The Netherlands

ISSN 0172-5939   e-ISSN 2191-6675
ISBN 978-1-4899-9433-2   ISBN 978-1-4614-1939-6 (eBook)
DOI 10.1007/978-1-4614-1939-6
Springer New York Dordrecht Heidelberg London

Mathematics Subject Classification (2010): 05Exx, 05Bxx, 05Cxx, 05Dxx, 15Axx, 51Exx, 94Bxx, 94Cxx

Springer is part of Springer Science+Business Media (www.springer.com)

Preface

Algebraic graph theory is the branch of mathematics that studies graphs by using algebraic properties of associated matrices. In particular, *spectral graph theory* studies the relation between graph properties and the spectrum of the adjacency matrix or Laplace matrix. And the theory of *association schemes* and *coherent configurations* studies the algebra generated by associated matrices.

Spectral graph theory is a useful subject. The founders of Google computed the Perron-Frobenius eigenvector of the web graph and became billionaires. The second-largest eigenvalue of a graph gives information about expansion and randomness properties. The smallest eigenvalue gives information about independence number and chromatic number. Interlacing gives information about substructures. The fact that eigenvalue multiplicities must be integral provides strong restrictions. And the spectrum provides a useful invariant.

This book gives the standard elementary material on spectra in Chapter 1. Important applications of graph spectra involve the largest, second-largest, or smallest eigenvalue, or interlacing, topics that are discussed in Chapters 3 and 4. Afterwards, special topics such as trees, groups and graphs, Euclidean representations, and strongly regular graphs are discussed. Strongly related to strongly regular graphs are regular two-graphs, and Chapter 10 mainly discusses Seidel's work on sets of equiangular lines. Strongly regular graphs form the first nontrivial case of (symmetric) association schemes, and Chapter 11 gives a very brief introduction to this topic and Delsarte's linear programming bound. Chapter 12 very briefly mentions the main facts on distance-regular graphs, including some major developments that have occurred since the monograph [54] was written (proof of the Bannai-Ito conjecture, construction by Van Dam and Koolen of the twisted Grassmann graphs, determination of the connectivity of distance-regular graphs). Instead of working over $\mathbb{R}$, one can work over $\mathbb{F}_p$ or $\mathbb{Z}$ and obtain more detailed information. Chapter 13 considers p-ranks and Smith normal forms. Finally, Chapters 14 and 15 return to the real spectrum and consider when a graph is determined by its spectrum and when it has only few eigenvalues.

In Spring 2006, both authors gave a series of lectures at IPM, the Institute for Studies in Theoretical Physics and Mathematics, in Tehran. The lecture notes were combined and published as an IPM report. Those notes grew into the present text, of which the on-line version still is called ipm.pdf. We aim at researchers, teachers, and graduate students interested in graph spectra. The reader is assumed to be familiar with basic linear algebra and eigenvalues, but we did include a chapter on some more advanced topics in linear algebra, such as the Perron-Frobenius theorem and eigenvalue interlacing. The exercises at the end of the chapters vary from easy but interesting applications of the treated theory to little excursions into related topics.

This book shows the influence of Seidel. For other books on spectral graph theory, see CHUNG [93], CVETKOVIĆ, DOOB & SACHS [115], and CVETKOVIĆ, ROWLINSON & SIMIĆ [120]. For more algebraic graph theory, see BIGGS [30], GODSIL [172], and GODSIL & ROYLE [177]. For association schemes and distance-regular graphs, see BANNAI & ITO [21] and BROUWER, COHEN & NEUMAIER [54].

Amsterdam   *Andries Brouwer*
December 2010   *Willem Haemers*

Contents

Preface v

1 Graph Spectrum 1
  1.1   Matrices associated to a graph 1
  1.2   The spectrum of a graph 2
        1.2.1   Characteristic polynomial 3
  1.3   The spectrum of an undirected graph 3
        1.3.1   Regular graphs 4
        1.3.2   Complements.. 4
        1.3.3   Walks 4
        1.3.4   Diameter 5
        1.3.5   Spanning trees 5
        1.3.6   Bipartite graphs 6
        1.3.7   Connectedness.. 7
  1.4   Spectrum of some graphs 8
        1.4.1   The complete graph 8
        1.4.2   The complete bipartite graph 8
        1.4.3   The cycle 8
        1.4.4   The path 9
        1.4.5   Line graphs 9
        1.4.6   Cartesian products... 10
        1.4.7   Kronecker products and bipartite double 10
        1.4.8   Strong products 11
        1.4.9   Cayley graphs 11
  1.5   Decompositions 11
        1.5.1   Decomposing K_{10} into Petersen graphs 12
        1.5.2   Decomposing K_n into complete bipartite graphs 12
  1.6   Automorphisms 12
  1.7   Algebraic connectivity.. 13
  1.8   Cospectral graphs.. 14
        1.8.1   The 4-cube 14

	1.8.2	Seidel switching	15
	1.8.3	Godsil-McKay switching	16
	1.8.4	Reconstruction......................................	16
1.9	Very small graphs...		16
1.10	Exercises		17

2 Linear Algebra 21
	2.1	Simultaneous diagonalization	21
	2.2	Perron-Frobenius theory	22
	2.3	Equitable partitions	24
		2.3.1 Equitable and almost equitable partitions of graphs	25
	2.4	The Rayleigh quotient	25
	2.5	Interlacing...	26
	2.6	Schur's inequality	28
	2.7	Schur complements	28
	2.8	The Courant-Weyl inequalities	29
	2.9	Gram matrices	29
	2.10	Diagonally dominant matrices	30
		2.10.1 Geršgorin circles	31
	2.11	Projections	31
	2.12	Exercises	32

3 Eigenvalues and Eigenvectors of Graphs 33
	3.1	The largest eigenvalue	33
		3.1.1 Graphs with largest eigenvalue at most 2	34
		3.1.2 Subdividing an edge	35
		3.1.3 The Kelmans operation	36
	3.2	Interlacing...	37
	3.3	Regular graphs	37
	3.4	Bipartite graphs	38
	3.5	Cliques and cocliques	38
		3.5.1 Using weighted adjacency matrices	39
	3.6	Chromatic number	40
		3.6.1 Using weighted adjacency matrices	42
		3.6.2 Rank and chromatic number	42
	3.7	Shannon capacity	42
		3.7.1 Lovász's ϑ-function	44
		3.7.2 The Haemers bound on the Shannon capacity	45
	3.8	Classification of integral cubic graphs	46
		3.8.1 A quotient of the hexagonal grid	47
		3.8.2 Cubic graphs with loops	47
		3.8.3 The classification	47
	3.9	The largest Laplace eigenvalue.............................	50
	3.10	Laplace eigenvalues and degrees	51
	3.11	The Grone-Merris conjecture	53

		3.11.1	Threshold graphs	53
		3.11.2	Proof of the Grone-Merris conjecture	53
	3.12	The Laplacian for hypergraphs		56
		3.12.1	Dominance order	58
	3.13	Applications of eigenvectors		58
		3.13.1	Ranking	59
		3.13.2	Google PageRank	59
		3.13.3	Cutting	60
		3.13.4	Graph drawing	61
		3.13.5	Clustering	61
		3.13.6	Graph isomorphism	62
		3.13.7	Searching an eigenspace	63
	3.14	Stars and star complements		63
	3.15	Exercises		64
4	**The Second-Largest Eigenvalue**			**67**
	4.1	Bounds for the second-largest eigenvalue		67
	4.2	Large regular subgraphs are connected		68
	4.3	Randomness		68
	4.4	Random walks		69
	4.5	Expansion		70
	4.6	Toughness and Hamiltonicity		71
		4.6.1	The Petersen graph is not Hamiltonian	72
	4.7	Diameter bound		72
	4.8	Separation		73
		4.8.1	Bandwidth	74
		4.8.2	Perfect matchings	75
	4.9	Block designs		77
	4.10	Polarities		79
	4.11	Exercises		80
5	**Trees**			**83**
	5.1	Characteristic polynomials of trees		83
	5.2	Eigenvectors and multiplicities		85
	5.3	Sign patterns of eigenvectors of graphs		86
	5.4	Sign patterns of eigenvectors of trees		87
	5.5	The spectral center of a tree		88
	5.6	Integral trees		89
	5.7	Exercises		90
6	**Groups and Graphs**			**93**
	6.1	$\Gamma(G,H,S)$		93
	6.2	Spectrum		93
	6.3	Non-Abelian Cayley graphs		94
	6.4	Covers		95

6.5   Cayley sum graphs... 97
        6.5.1   (3,6)-fullerenes 97
6.6   Exercises 99

7   Topology 101
7.1   Embeddings 101
7.2   Minors 102
7.3   The Colin de Verdière invariant 102
7.4   The Van der Holst-Laurent-Schrijver invariant 103

8   Euclidean Representations 105
8.1   Examples.. 105
8.2   Euclidean representation 105
8.3   Root lattices 106
        8.3.1   Examples 107
        8.3.2   Root lattices.................................... 108
        8.3.3   Classification 109
8.4   The Cameron-Goethals-Seidel-Shult theorem 111
8.5   Further applications 112
8.6   Exercises 113

9   Strongly Regular Graphs 115
9.1   Strongly regular graphs 115
        9.1.1   Simple examples 115
        9.1.2   The Paley graphs................................. 116
        9.1.3   Adjacency matrix 117
        9.1.4   Imprimitive graphs 117
        9.1.5   Parameters 118
        9.1.6   The half case and cyclic strongly regular graphs 118
        9.1.7   Strongly regular graphs without triangles................ 119
        9.1.8   Further parameter restrictions 120
        9.1.9   Strongly regular graphs from permutation groups 121
        9.1.10   Strongly regular graphs from quasisymmetric designs 121
        9.1.11   Symmetric 2-designs from strongly regular graphs 122
        9.1.12   Latin square graphs.............................. 122
        9.1.13   Partial geometries 124
9.2   Strongly regular graphs with eigenvalue -2 124
9.3   Connectivity 125
9.4   Cocliques and colorings 127
9.5   Automorphisms 129
9.6   Generalized quadrangles 129
        9.6.1   Parameters 129
        9.6.2   Constructions of generalized quadrangles 130
        9.6.3   Strongly regular graphs from generalized quadrangles 131
        9.6.4   Generalized quadrangles with lines of size 3 132

	9.7	The $(81, 20, 1, 6)$ strongly regular graph	132
	9.7.1	Descriptions	133
	9.7.2	Uniqueness	134
	9.7.3	Independence and chromatic numbers	135
	9.7.4	Second subconstituent	136
	9.8	Strongly regular graphs and two-weight codes	136
	9.8.1	Codes, graphs, and projective sets	136
	9.8.2	The correspondence between linear codes and subsets of a projective space	137
	9.8.3	The correspondence between projective two-weight codes, subsets of a projective space with two intersection numbers, and affine strongly regular graphs	138
	9.8.4	Duality for affine strongly regular graphs	140
	9.8.5	Cyclotomy	141
	9.9	Table of parameters for strongly regular graphs	143
	9.9.1	Comments	146
	9.10	Exercises	148

10 Regular Two-graphs ... 151
	10.1	Strong graphs	151
	10.2	Two-graphs	152
	10.3	Regular two-graphs	154
	10.3.1	Related strongly regular graphs	155
	10.3.2	The regular two-graph on 276 points	156
	10.3.3	Coherent subsets	156
	10.3.4	Completely regular two-graphs	157
	10.4	Conference matrices	158
	10.5	Hadamard matrices	159
	10.5.1	Constructions	160
	10.6	Equiangular lines	161
	10.6.1	Equiangular lines in $\mathbb{R}^d$ and two-graphs	161
	10.6.2	Bounds on equiangular sets of lines in $\mathbb{R}^d$ or $\mathbb{C}^d$	162
	10.6.3	Bounds on sets of lines with few angles and sets of vectors with few distances	163

11 Association Schemes ... 165
	11.1	Definition	165
	11.2	The Bose-Mesner algebra	166
	11.3	The linear programming bound	168
	11.3.1	Equality	169
	11.3.2	The code-clique theorem	169
	11.3.3	Strengthened LP bounds	170
	11.4	The Krein parameters	170
	11.5	Automorphisms	172
	11.5.1	The Moore graph on 3250 vertices	172

  11.6 P- and Q-polynomial association schemes 173
  11.7   Exercises 175

12 Distance-Regular Graphs 177
  12.1   Parameters 177
  12.2   Spectrum 178
  12.3   Primitivity 178
  12.4   Examples 178
       12.4.1   Hamming graphs 178
       12.4.2   Johnson graphs 179
       12.4.3   Grassmann graphs 180
       12.4.4   Van Dam-Koolen graphs 180
  12.5   Bannai-Ito conjecture 180
  12.6   Connectedness 181
  12.7   Growth 181
  12.8   Degree of eigenvalues 181
  12.9   Moore graphs and generalized polygons 182
  12.10 Euclidean representations 183
  12.11 Extremality 183
  12.12 Exercises 185

13 p-ranks 187
  13.1   Reduction mod p 187
  13.2   The minimal polynomial 188
  13.3   Bounds for the p-rank 188
  13.4   Interesting primes p 189
  13.5   Adding a multiple of J 190
  13.6   Paley graphs 191
  13.7   Strongly regular graphs 192
  13.8   Smith normal form 194
       13.8.1   Smith normal form and spectrum 195
  13.9   Exercises 197

14 Spectral Characterizations 199
  14.1   Generalized adjacency matrices 199
  14.2   Constructing cospectral graphs 200
       14.2.1   Trees 201
       14.2.2   Partial linear spaces 202
       14.2.3   GM switching 202
       14.2.4   Sunada's method 204
  14.3   Enumeration 204
       14.3.1   Lower bounds 204
       14.3.2   Computer results 205
  14.4   DS graphs 206
       14.4.1   Spectrum and structure 206

14.4.2 Some DS graphs 208
14.4.3 Line graphs 210
14.5 Distance-regular graphs 212
14.5.1 Strongly regular DS graphs 213
14.5.2 Distance-regularity from the spectrum 214
14.5.3 Distance-regular DS graphs 215
14.6 The method of Wang and Xu 217
14.7 Exercises 219

15 Graphs with Few Eigenvalues 221
15.1 Regular graphs with four eigenvalues 221
15.2 Three Laplace eigenvalues 223
15.3 Other matrices with at most three eigenvalues 224
15.3.1 Few Seidel eigenvalues............................. 224
15.3.2 Three adjacency eigenvalues 225
15.3.3 Three signless Laplace eigenvalues................... 227
15.4 Exercises 227

References 229

Author Index 243

Subject Index 247

Chapter 1
Graph Spectrum

This chapter presents some simple results on graph spectra. We assume the reader is familiar with elementary linear algebra and graph theory. Throughout, J will denote the all-1 matrix, and $\mathbf{1}$ is the all-1 vector.

1.1 Matrices associated to a graph

Let Γ be a graph without multiple edges. The *adjacency matrix* of Γ is the 0-1 matrix A indexed by the vertex set $V\Gamma$ of Γ, where $A_{xy} = 1$ when there is an edge from x to y in Γ and $A_{xy} = 0$ otherwise. Occasionally we consider multigraphs (possibly with loops), in which case A_{xy} equals the number of edges from x to y.

Let Γ be an undirected graph without loops. The (vertex-edge) *incidence matrix* of Γ is the 0-1 matrix M, with rows indexed by the vertices and columns indexed by the edges, where $M_{xe} = 1$ when vertex x is an endpoint of edge e.

Let Γ be a directed graph without loops. The *directed incidence matrix* of Γ is the matrix N, with rows indexed by the vertices and columns by the edges, where $N_{xe} = -1, 1, 0$ when x is the head of e, the tail of e, or not on e, respectively.

Let Γ be an undirected graph without loops. The *Laplace matrix* of Γ is the matrix L indexed by the vertex set of Γ, with zero row sums, where $L_{xy} = -A_{xy}$ for $x \neq y$. If D is the diagonal matrix, indexed by the vertex set of Γ such that D_{xx} is the degree (valency) of x, then $L = D - A$. The matrix $Q = D + A$ is called the *signless Laplace matrix* of Γ.

An important property of the Laplace matrix L and the signless Laplace matrix Q is that they are positive semidefinite. Indeed, one has $Q = MM^{\top}$ and $L = NN^{\top}$ if M is the incidence matrix of Γ and N the directed incidence matrix of the directed graph obtained by orienting the edges of Γ in an arbitrary way. It follows that for any vector u one has $u^{\top}Lu = \sum_{xy}(u_x - u_y)^2$ and $u^{\top}Qu = \sum_{xy}(u_x + u_y)^2$, where the sum is over the edges of Γ.

1.2 The spectrum of a graph

The (ordinary) *spectrum* of a finite graph Γ is by definition the spectrum of the adjacency matrix A, that is, its set of eigenvalues together with their multiplicities. The *Laplace spectrum* of a finite undirected graph without loops is the spectrum of the Laplace matrix L.

The rows and columns of a matrix of order n are numbered from 1 to n, while A is indexed by the vertices of Γ, so that writing down A requires one to assign some numbering to the vertices. However, the spectrum of the matrix obtained does not depend on the numbering chosen. It is the spectrum of the linear transformation A on the vector space K^X of maps from X into K, where X is the vertex set and K is some field such as $\mathbb{R}$ or $\mathbb{C}$.

The *characteristic polynomial* of Γ is that of A, that is, the polynomial p_A defined by $p_A(\theta) = \det(\theta I - A)$.

Example Let Γ be the path P_3 with three vertices and two edges. Assigning some arbitrary order to the three vertices of Γ, we find that the adjacency matrix A becomes one of

$$\begin{bmatrix} 0 & 1 & 1 \\ 1 & 0 & 0 \\ 1 & 0 & 0 \end{bmatrix} \quad \text{or} \quad \begin{bmatrix} 0 & 1 & 0 \\ 1 & 0 & 1 \\ 0 & 1 & 0 \end{bmatrix} \quad \text{or} \quad \begin{bmatrix} 0 & 0 & 1 \\ 0 & 0 & 1 \\ 1 & 1 & 0 \end{bmatrix}.$$

The characteristic polynomial is $p_A(\theta) = \theta^3 - 2\theta$. The spectrum is $\sqrt{2}, 0, -\sqrt{2}$. The eigenvectors are:

Here, for an eigenvector u, we write u_x as a label at the vertex x. One has $Au = \theta u$ if and only if $\sum_{y \leftarrow x} u_y = \theta u_x$ for all x. The Laplace matrix L of this graph is one of

$$\begin{bmatrix} 2 & -1 & -1 \\ -1 & 1 & 0 \\ -1 & 0 & 1 \end{bmatrix} \quad \text{or} \quad \begin{bmatrix} 1 & -1 & 0 \\ -1 & 2 & -1 \\ 0 & -1 & 1 \end{bmatrix} \quad \text{or} \quad \begin{bmatrix} 1 & 0 & -1 \\ 0 & 1 & -1 \\ -1 & -1 & 2 \end{bmatrix}.$$

Its eigenvalues are 0, 1 and 3. The Laplace eigenvectors are:

One has $Lu = \theta u$ if and only if $\sum_{y \sim x} u_y = (d_x - \theta) u_x$ for all x, where d_x is the degree of the vertex x.

Example Let Γ be the directed triangle with adjacency matrix

$$A = \begin{bmatrix} 0 & 1 & 0 \\ 0 & 0 & 1 \\ 1 & 0 & 0 \end{bmatrix}.$$

Then A has characteristic polynomial $p_A(\theta) = \theta^3 - 1$ and spectrum 1, ω, ω^2, where ω is a primitive cube root of unity.

Example Let Γ be the directed graph with two vertices and a single directed edge. Then $A = \begin{bmatrix} 0 & 1 \\ 0 & 0 \end{bmatrix}$ with $p_A(\theta) = \theta^2$, so A has the eigenvalue 0 with geometric multiplicity (that is, the dimension of the corresponding eigenspace) equal to 1 and algebraic multiplicity (that is, its multiplicity as a root of the polynomial p_A) equal to 2.

1.2.1 Characteristic polynomial

Let Γ be a directed graph on n vertices. For any directed subgraph C of Γ that is a union of directed cycles, let $c(C)$ be its number of cycles. Then the characteristic polynomial $p_A(t) = \det(tI - A)$ of Γ can be expanded as $\sum c_i t^{n-i}$, where $c_i = \sum_C (-1)^{c(C)}$, with C running over all regular directed subgraphs with in- and outdegree 1 on i vertices.

(Indeed, this is just a reformulation of the definition of the determinant as $\det M = \sum_\sigma \text{sgn}(\sigma) M_{1\sigma(1)} \cdots M_{n\sigma(n)}$. Note that when the permutation σ with $n - i$ fixed points is written as a product of nonidentity cycles, its sign is $(-1)^e$, where e is the number of even cycles in this product. Since the number of odd nonidentity cycles is congruent to i (mod 2), we have $\text{sgn}(\sigma) = (-1)^{i+c(\sigma)}$.)

For example, the directed triangle has $c_0 = 1$, $c_3 = -1$. Directed edges that do not occur in directed cycles do not influence the (ordinary) spectrum.

The same description of $p_A(t)$ holds for undirected graphs (with each edge viewed as a pair of opposite directed edges).

Since $\frac{d}{dt} \det(tI - A) = \sum_x \det(tI - A_x)$ where A_x is the submatrix of A obtained by deleting row and column x, it follows that $p_A'(t)$ is the sum of the characteristic polynomials of all single-vertex-deleted subgraphs of Γ.

1.3 The spectrum of an undirected graph

Suppose Γ is undirected and simple with n vertices. Since A is real and symmetric, all its eigenvalues are real. Also, for each eigenvalue θ, its algebraic multiplicity coincides with its geometric multiplicity, so that we may omit the adjective and just speak about "multiplicity". Conjugate algebraic integers have the same multiplicity. Since A has zero diagonal, its trace $\text{tr}\,A$, and hence the sum of the eigenvalues, is zero.

Similarly, L is real and symmetric, so that the Laplace spectrum is real. Moreover, L is positive semidefinite and singular, so we may denote the eigenvalues by

$\mu_1, \ldots, \mu_n$, where $0 = \mu_1 \le \mu_2 \le \ldots \le \mu_n$. The sum of these eigenvalues is $\operatorname{tr} L$, which is twice the number of edges of Γ.

Finally, also Q has real spectrum and nonnegative eigenvalues (but is not necessarily singular). We have $\operatorname{tr} Q = \operatorname{tr} L$.

1.3.1 Regular graphs

A graph Γ is called *regular* of degree (or valency) k when every vertex has precisely k neighbors. So, Γ is regular of degree k precisely when its adjacency matrix A has row sums k, i.e., when $A\mathbf{1} = k\mathbf{1}$ (or $AJ = kJ$).

If Γ is regular of degree k, then for every eigenvalue θ we have $|\theta| \le k$. (One way to see this is by observing that if $|t| > k$ then the matrix $tI - A$ is strictly diagonally dominant, and hence nonsingular, so that t is not an eigenvalue of A.)

If Γ is regular of degree k, then $L = kI - A$. It follows that if Γ has ordinary eigenvalues $k = \theta_1 \ge \ldots \ge \theta_n$ and Laplace eigenvalues $0 = \mu_1 \le \mu_2 \le \ldots \le \mu_n$, then $\theta_i = k - \mu_i$ for $i = 1, \ldots, n$. The eigenvalues of $Q = kI + A$ are $2k, k + \theta_2, \ldots, k + \theta_n$.

1.3.2 Complements

The *complement* $\overline{\Gamma}$ of Γ is the graph with the same vertex set as Γ, where two distinct vertices are adjacent whenever they are nonadjacent in Γ. So, if Γ has adjacency matrix A, then $\overline{\Gamma}$ has adjacency matrix $\overline{A} = J - I - A$ and Laplace matrix $\overline{L} = nI - J - L$.

Because eigenvectors of L are also eigenvectors of J, the eigenvalues of $\overline{L}$ are $0, n - \mu_n, \ldots, n - \mu_2$. (In particular, $\mu_n \le n$.)

If Γ is regular we have a similar result for the ordinary eigenvalues: if Γ is k-regular with eigenvalues $\theta_1 \ge \ldots \ge \theta_n$, then the eigenvalues of the complement are $n - k - 1, -1 - \theta_n, \ldots, -1 - \theta_2$.

1.3.3 Walks

From the spectrum one can read off the number of closed walks of a given length.

Proposition 1.3.1 *Let h be a nonnegative integer. Then $(A^h)_{xy}$ is the number of walks of length h from x to y. In particular, $(A^2)_{xx}$ is the degree of the vertex x, and $\operatorname{tr} A^2$ equals twice the number of edges of Γ; similarly, $\operatorname{tr} A^3$ is six times the number of triangles in Γ.*

1.3.4 Diameter

We saw that all eigenvalues of a single directed edge are zero. For undirected graphs this does not happen.

Proposition 1.3.2 *Let Γ be an undirected graph. All its eigenvalues are zero if and only if Γ has no edges. The same holds for the Laplace eigenvalues and the signless Laplace eigenvalues.*

More generally, we find a lower bound for the diameter:

Proposition 1.3.3 *Let Γ be a connected graph with diameter d. Then Γ has at least $d + 1$ distinct eigenvalues, at least $d + 1$ distinct Laplace eigenvalues, and at least $d + 1$ distinct signless Laplace eigenvalues.*

Proof Let M be any nonnegative symmetric matrix with rows and columns indexed by $V\Gamma$ and such that for distinct vertices x, y we have $M_{xy} > 0$ if and only if $x \sim y$. Let the distinct eigenvalues of M be $\theta_1, \ldots, \theta_t$. Then $(M - \theta_1 I) \cdots (M - \theta_t I) = 0$, so that M^t is a linear combination of $I, M, \ldots, M^{t-1}$. But if $d(x, y) = t$ for two vertices x, y of Γ, then $(M^i)_{xy} = 0$ for $0 \le i \le t - 1$ and $(M^t)_{xy} > 0$, a contradiction. Hence $t > d$. This applies to $M = A$, to $M = nI - L$, and to $M = Q$, where A is the adjacency matrix, L is the Laplace matrix, and Q is the signless Laplace matrix of Γ. $\qquad\square$

Distance-regular graphs, discussed in Chapter 12, have equality here. For an upper bound on the diameter, see §4.7.

1.3.5 Spanning trees

From the Laplace spectrum of a graph one can determine the number of spanning trees (which will be nonzero only if the graph is connected).

Proposition 1.3.4 *Let Γ be an undirected (multi)graph with at least one vertex, and Laplace matrix L with eigenvalues $0 = \mu_1 \le \mu_2 \le \ldots \le \mu_n$. Let ℓ_{xy} be the (x, y)-cofactor of L. Then the number N of spanning trees of Γ equals*

$$N = \ell_{xy} = \det(L + \frac{1}{n^2}J) = \frac{1}{n}\mu_2 \cdots \mu_n \text{ for any } x, y \in V\Gamma.$$

(The (i, j)-cofactor of a matrix M is by definition $(-1)^{i+j} \det M(i, j)$, where $M(i, j)$ is the matrix obtained from M by deleting row i and column j. Note that ℓ_{xy} does not depend on an ordering of the vertices of Γ.)

Proof Let L^S, for $S \subseteq V\Gamma$, denote the matrix obtained from L by deleting the rows and columns indexed by S, so that $\ell_{xx} = \det L^{\{x\}}$. The equality $N = \ell_{xx}$ follows by induction on n, and for fixed $n > 1$ on the number of edges incident with x. Indeed, if $n = 1$ then $\ell_{xx} = 1$. Otherwise, if x has degree 0, then $\ell_{xx} = 0$ since $L^{\{x\}}$ has zero row sums. Finally, if xy is an edge, then deleting this edge from Γ diminishes ℓ_{xx} by $\det L^{\{x,y\}}$, which by induction is the number of spanning trees of Γ with edge xy contracted, which is the number of spanning trees containing the edge xy. This shows $N = \ell_{xx}$.

Now $\det(tI - L) = t \prod_{i=2}^{n}(t - \mu_i)$ and $(-1)^{n-1}\mu_2 \cdots \mu_n$ is the coefficient of t, that is, is $\frac{d}{dt}\det(tI - L)|_{t=0}$. But $\frac{d}{dt}\det(tI - L) = \sum_x \det(tI - L^{\{x\}})$, so $\mu_2 \cdots \mu_n = \sum_x \ell_{xx} = nN$.

Since the sum of the columns of L is zero, so that one column is minus the sum of the other columns, we have $\ell_{xx} = \ell_{xy}$ for any x,y. Finally, the eigenvalues of $L + \frac{1}{n^2}J$ are $\frac{1}{n}$ and $\mu_2, \ldots, \mu_n$, so $\det(L + \frac{1}{n^2}J) = \frac{1}{n}\mu_2 \cdots \mu_n$. $\qquad\square$

For example, the multigraph of valency k on two vertices has Laplace matrix $L = \begin{bmatrix} k & -k \\ -k & k \end{bmatrix}$ so $\mu_1 = 0$, $\mu_2 = 2k$, and $N = \frac{1}{2}.2k = k$.

If we consider the complete graph K_n, then $\mu_2 = \ldots = \mu_n = n$, and therefore K_n has $N = n^{n-2}$ spanning trees. This formula is due to CAYLEY [85]. Proposition 1.3.4 is implicit in KIRCHHOFF [242] and known as the *matrix-tree theorem*. There is a "1-line proof" of the above result using the *Cauchy-Binet formula*.

Proposition 1.3.5 (Cauchy-Binet) *Let A and B be $m \times n$ matrices. Then*

$$\det AB^\top = \sum_S \det A_S \det B_S,$$

where the sum is over the $\binom{n}{m}$ m-subsets S of the set of columns, and A_S (B_S) is the square submatrix of order m of A (resp. B) with columns indexed by S.

Second proof of Proposition 1.3.4 (sketch) Let N_x be the directed incidence matrix of Γ with row x deleted. Then $l_{xx} = \det N_x N_x^\top$. Apply the Cauchy-Binet formula to get l_{xx} as a sum of squares of determinants of size $n-1$. These determinants vanish unless the set S of columns is the set of edges of a spanning tree, in which case the determinant is ± 1. $\qquad\square$

1.3.6 Bipartite graphs

A graph Γ is called *bipartite* when its vertex set can be partitioned into two disjoint parts X_1, X_2 such that all edges of Γ meet both X_1 and X_2. The adjacency matrix of a bipartite graph has the form $A = \begin{bmatrix} 0 & B \\ B^\top & 0 \end{bmatrix}$. It follows that the spectrum of a bipartite graph is symmetric w.r.t. 0: if $\begin{bmatrix} u \\ v \end{bmatrix}$ is an eigenvector with eigenvalue θ, then $\begin{bmatrix} u \\ -v \end{bmatrix}$ is an eigenvector with eigenvalue $-\theta$. (The converse also holds, see Proposition 3.4.1.)

For the ranks one has $\operatorname{rk} A = 2 \operatorname{rk} B$. If $n_i = |X_i|$ ($i = 1,2$) and $n_1 \geq n_2$, then $\operatorname{rk} A \leq 2n_2$, so Γ has eigenvalue 0 with multiplicity at least $n_1 - n_2$.

One cannot, in general, recognize bipartiteness from the Laplace or signless Laplace spectrum. For example, $K_{1,3}$ and $K_1 + K_3$ have the same signless Laplace spectrum and only the former is bipartite. And Figure 14.4 gives an example of a bipartite and a nonbipartite graph with the same Laplace spectrum. However, by Proposition 1.3.10 below, a graph is bipartite precisely when its Laplace spectrum and signless Laplace spectrum coincide.

1.3.7 Connectedness

The spectrum of a disconnected graph is easily found from the spectra of its connected components:

Proposition 1.3.6 *Let Γ be a graph with connected components Γ_i ($1 \leq i \leq s$). Then the spectrum of Γ is the union of the spectra of Γ_i (and multiplicities are added). The same holds for the Laplace spectrum and the signless Laplace spectrum.* $\square$

Proposition 1.3.7 *The multiplicity of 0 as a Laplace eigenvalue of an undirected graph Γ equals the number of connected components of Γ.*

Proof We have to show that a connected graph has Laplace eigenvalue 0 with multiplicity 1. As we saw earlier, $L = NN^\top$, where N is the incidence matrix of an orientation of Γ. Now $Lu = 0$ is equivalent to $N^\top u = 0$ (since $0 = u^\top Lu = ||N^\top u||^2$), that is, for every edge the vector u takes the same value on both endpoints. Since Γ is connected, that means that u is constant. $\square$

Proposition 1.3.8 *Let the undirected graph Γ be regular of valency k. Then k is the largest eigenvalue of Γ, and its multiplicity equals the number of connected components of Γ.*

Proof We have $L = kI - A$. $\square$

One cannot see from the spectrum alone whether a (nonregular) graph is connected: both $K_{1,4}$ and $K_1 + C_4$ have spectrum $2^1, 0^3, (-2)^1$ (we write multiplicities as exponents). And both $\hat{E}_6$ and $K_1 + C_6$ have spectrum $2^1, 1^2, 0, (-1)^2, (-2)^1$.

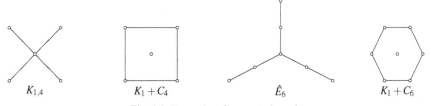

$K_{1,4}$ $K_1 + C_4$ $\hat{E}_6$ $K_1 + C_6$

Fig. 1.1 Two pairs of cospectral graphs

Proposition 1.3.9 *The multiplicity of 0 as a signless Laplace eigenvalue of an undirected graph Γ equals the number of bipartite connected components of Γ.*

Proof Let M be the vertex-edge incidence matrix of Γ, so that $Q = MM^\top$. If $MM^\top u = 0$, then $M^\top u = 0$, so $u_x = -u_y$ for all edges xy, and the support of u is the union of a number of bipartite components of Γ. $\qquad\square$

Proposition 1.3.10 *A graph Γ is bipartite if and only if the Laplace spectrum and the signless Laplace spectrum of Γ are equal.*

Proof If Γ is bipartite, the Laplace matrix L and the signless Laplace matrix Q are similar by a diagonal matrix D with diagonal entries ± 1 (that is, $Q = DLD^{-1}$). Therefore Q and L have the same spectrum. Conversely, if both spectra are the same, then by Propositions 1.3.7 and 1.3.9 the number of connected components equals the number of bipartite components. Hence Γ is bipartite. $\qquad\square$

1.4 Spectrum of some graphs

In this section we discuss some special graphs and their spectra. All graphs in this section are finite, undirected, and simple. Observe that the all-1 matrix J of order n has rank 1, and that the all-1 vector $\mathbf{1}$ is an eigenvector with eigenvalue n, so the spectrum of J is n^1, 0^{n-1}. (Here and throughout, we write multiplicities as exponents where convenient and no confusion seems likely.)

1.4.1 The complete graph

Let Γ be the complete graph K_n on n vertices. Its adjacency matrix is $A = J - I$, and the spectrum is $(n-1)^1$, $(-1)^{n-1}$. The Laplace matrix is $nI - J$, which has spectrum 0^1, n^{n-1}.

1.4.2 The complete bipartite graph

The spectrum of the complete bipartite graph $K_{m,n}$ is $\pm\sqrt{mn}$, 0^{m+n-2}. The Laplace spectrum is 0^1, m^{n-1}, n^{m-1}, $(m+n)^1$.

1.4.3 The cycle

Let Γ be the directed n-cycle D_n. Eigenvectors are $(1, \zeta, \zeta^2, \ldots, \zeta^{n-1})^\top$, where $\zeta^n = 1$, and the corresponding eigenvalue is ζ. Thus, the spectrum consists precisely of the complex n-th roots of unity $e^{2\pi i j/n}$ ($j = 0, \ldots, n-1$).

Now consider the undirected n-cycle C_n. If B is the adjacency matrix of D_n, then $A = B + B^\top$ is the adjacency matrix of C_n. We find the same eigenvectors as before, with eigenvalues $\zeta + \zeta^{-1}$, so that the spectrum consists of the numbers $2\cos(2\pi j/n)$ $(j = 0, \ldots, n-1)$.

This graph is regular of valency 2, so the Laplace spectrum consists of the numbers $2 - 2\cos(2\pi j/n)$ $(j = 0, \ldots, n-1)$.

1.4.4 The path

Let Γ be the undirected path P_n with n vertices. The ordinary spectrum consists of the numbers $2\cos(\pi j/(n+1))$ $(j = 1, \ldots, n)$. The Laplace spectrum is $2 - 2\cos(\pi j/n)$ $(j = 0, \ldots, n-1)$.

The ordinary spectrum follows by looking at C_{2n+2}. If $u(\zeta) = (1, \zeta, \zeta^2, \ldots, \zeta^{2n+1})^\top$ is an eigenvector of C_{2n+2}, where $\zeta^{2n+2} = 1$, then $u(\zeta)$ and $u(\zeta^{-1})$ have the same eigenvalue, $2\cos(\pi j/(n+1))$, and hence so has $u(\zeta) - u(\zeta^{-1})$. This latter vector has two zero coordinates distance $n+1$ apart and (for $\zeta \neq \pm 1$) induces an eigenvector on the two paths obtained by removing the two points where it is zero.

Eigenvectors of L with eigenvalue $2 - \zeta - \zeta^{-1}$ are $(1 + \zeta^{2n-1}, \ldots, \zeta^j + \zeta^{2n-1-j}, \ldots, \zeta^{n-1} + \zeta^n)$, where $\zeta^{2n} = 1$. One can check this directly, or view P_n as the result of folding C_{2n}, where the folding has no fixed vertices. An eigenvector of C_{2n} that is constant on the preimages of the folding yields an eigenvector of P_n with the same eigenvalue.

1.4.5 Line graphs

The *line graph* $L(\Gamma)$ of Γ is the graph with the edge set of Γ as vertex set, where two vertices are adjacent if the corresponding edges of Γ have an endpoint in common. If N is the incidence matrix of Γ, then $N^\top N - 2I$ is the adjacency matrix of $L(\Gamma)$. Since $N^\top N$ is positive semidefinite, the eigenvalues of a line graph are not smaller than -2. We have an explicit formula for the eigenvalues of $L(\Gamma)$ in terms of the signless Laplace eigenvalues of Γ.

Proposition 1.4.1 *Suppose Γ has m edges, and let $\rho_1 \geq \ldots \geq \rho_r$ be the positive signless Laplace eigenvalues of Γ. Then the eigenvalues of $L(\Gamma)$ are $\theta_i = \rho_i - 2$ for $i = 1, \ldots, r$, and $\theta_i = -2$ if $r < i \leq m$.*

Proof The signless Laplace matrix Q of Γ and the adjacency matrix B of $L(\Gamma)$ satisfy $Q = NN^\top$ and $B + 2I = N^\top N$. Because $NN^\top$ and $N^\top N$ have the same nonzero eigenvalues (multiplicities included), the result follows. $\square$

Example Since the path P_n has line graph P_{n-1} and is bipartite, the Laplace and the signless Laplace eigenvalues of P_n are $2 + 2\cos\frac{\pi i}{n}$, $i = 1, \ldots, n$.

Corollary 1.4.2 *If Γ is a k-regular graph ($k \geq 2$) with n vertices, $e = kn/2$ edges, and eigenvalues θ_i ($i = 1, \ldots, n$), then $L(\Gamma)$ is $(2k-2)$-regular with eigenvalues $\theta_i + k - 2$ ($i = 1, \ldots, n$) and $e - n$ times -2.*   □

The line graph of the complete graph K_n ($n \geq 2$) is known as the *triangular graph* $T(n)$. It has spectrum $2(n-2)^1$, $(n-4)^{n-1}$, $(-2)^{n(n-3)/2}$. The line graph of the regular complete bipartite graph $K_{m,m}$ ($m \geq 2$) is known as the *lattice graph* $L_2(m)$. It has spectrum $2(m-1)^1$, $(m-2)^{2m-2}$, $(-2)^{(m-1)^2}$. These two families of graphs, and their complements, are examples of strongly regular graphs, which will be the subject of Chapter 9. The complement of $T(5)$ is the famous *Petersen graph*. It has spectrum $3^1 \, 1^5 \, (-2)^4$.

1.4.6 Cartesian products

Given graphs Γ and Δ with vertex sets V and W, respectively, their *Cartesian product* $\Gamma \square \Delta$ is the graph with vertex set $V \times W$, where $(v, w) \sim (v', w')$ when either $v = v'$ and $w \sim w'$ or $w = w'$ and $v \sim v'$. For the adjacency matrices we have $A_{\Gamma \square \Delta} = A_\Gamma \otimes I + I \otimes A_\Delta$.

If u and v are eigenvectors for Γ and Δ with ordinary or Laplace eigenvalues θ and η, respectively, then the vector w defined by $w_{(x,y)} = u_x v_y$ is an eigenvector of $\Gamma \square \Delta$ with ordinary or Laplace eigenvalue $\theta + \eta$.

For example, $L_2(m) = K_m \square K_m$.

For example, the *hypercube* 2^n, also called Q_n, is the Cartesian product of n factors K_2. The spectrum of K_2 is $1, -1$, and hence the spectrum of 2^n consists of the numbers $n - 2i$ with multiplicity $\binom{n}{i}$ ($i = 0, 1, \ldots, n$).

1.4.7 Kronecker products and bipartite double

Given graphs Γ and Δ with vertex sets V and W, respectively, their *Kronecker product* (or *direct product*, or *conjunction*) $\Gamma \otimes \Delta$ is the graph with vertex set $V \times W$, where $(v, w) \sim (v', w')$ when $v \sim v'$ and $w \sim w'$. The adjacency matrix of $\Gamma \otimes \Delta$ is the Kronecker product of the adjacency matrices of Γ and Δ.

If u and v are eigenvectors for Γ and Δ with eigenvalues θ and η, respectively, then the vector $w = u \otimes v$ (with $w_{(x,y)} = u_x v_y$) is an eigenvector of $\Gamma \otimes \Delta$ with eigenvalue $\theta \eta$. Thus, the spectrum of $\Gamma \otimes \Delta$ consists of the products of the eigenvalues of Γ and Δ.

Given a graph Γ, its *bipartite double* is the graph $\Gamma \otimes K_2$ (with for each vertex x of Γ two vertices x' and x'', and for each edge xy of Γ two edges $x'y''$ and $x''y'$). If Γ is bipartite, its double is just the union of two disjoint copies. If Γ is connected and not bipartite, then its double is connected and bipartite. If Γ has spectrum Φ, then $\Gamma \otimes K_2$ has spectrum $\Phi \cup -\Phi$.

The notation $\Gamma \times \Delta$ is used in the literature both for the Cartesian product and for the Kronecker product of two graphs. We avoid it here.

1.4.8 Strong products

Given graphs Γ and Δ with vertex sets V and W, respectively, their *strong product* $\Gamma \boxtimes \Delta$ is the graph with vertex set $V \times W$, where two distinct vertices (v, w) and (v', w') are adjacent whenever v and v' are equal or adjacent in Γ, and w and w' are equal or adjacent in Δ. If A_Γ and A_Δ are the adjacency matrices of Γ and Δ, then $((A_\Gamma + I) \otimes (A_\Delta + I)) - I$ is the adjacency matrix of $\Gamma \boxtimes \Delta$. It follows that the eigenvalues of $\Gamma \boxtimes \Delta$ are the numbers $(\theta + 1)(\eta + 1) - 1$, where θ and η run through the eigenvalues of Γ and Δ, respectively.

Note that the edge set of the strong product of Γ and Δ is the union of the edge sets of the Cartesian product and the Kronecker product of Γ and Δ.

For example, $K_{m+n} = K_m \boxtimes K_n$.

1.4.9 Cayley graphs

Let G be an Abelian group and $S \subseteq G$. The *Cayley graph* on G with difference set S is the (directed) graph Γ with vertex set G and edge set $E = \{(x, y) \mid y - x \in S\}$. Now Γ is regular with in- and outvalency $|S|$. The graph Γ will be undirected when $S = -S$.

It is easy to compute the spectrum of finite Cayley graphs (on an Abelian group). Let χ be a character of G, that is, a map $\chi : G \to \mathbb{C}^*$ such that $\chi(x + y) = \chi(x)\chi(y)$. Then $\sum_{y \sim x} \chi(y) = (\sum_{s \in S} \chi(s))\chi(x)$, so the vector $(\chi(x))_{x \in G}$ is a right eigenvector of the adjacency matrix A of Γ with eigenvalue $\chi(S) := \sum_{s \in S} \chi(s)$. The $n = |G|$ distinct characters give independent eigenvectors, so one obtains the entire spectrum in this way.

For example, the directed pentagon (with in- and outvalency 1) is a Cayley graph for $G = \mathbb{Z}_5$ and $S = \{1\}$. The characters of G are the maps $i \mapsto \zeta^i$ for some fixed fifth root of unity ζ. Hence the directed pentagon has spectrum $\{\zeta \mid \zeta^5 = 1\}$.

The undirected pentagon (with valency 2) is the Cayley graph for $G = \mathbb{Z}_5$ and $S = \{-1, 1\}$. The spectrum of the pentagon becomes $\{\zeta + \zeta^{-1} \mid \zeta^5 = 1\}$, that is, consists of 2 and $\frac{1}{2}(-1 \pm \sqrt{5})$ (both with multiplicity 2).

1.5 Decompositions

Here we present two nontrivial applications of linear algebra to graph decompositions.

1.5.1 Decomposing K_{10} into Petersen graphs

An amusing application ([35, 310]) is the following. Can the edges of the complete graph K_{10} be colored with three colors such that each color induces a graph isomorphic to the Petersen graph? K_{10} has 45 edges, 9 on each vertex, and the Petersen graph has 15 edges, 3 on each vertex, so at first sight this might seem possible. Let the adjacency matrices of the three color classes be P_1, P_2 and P_3, so that $P_1 + P_2 + P_3 = J - I$. If P_1 and P_2 are Petersen graphs, they both have a 5-dimensional eigenspace for eigenvalue 1, contained in the 9-space $\mathbf{1}^\perp$. Therefore, there is a common 1-eigenvector u and $P_3 u = (J - I)u - P_1 u - P_2 u = -3u$ so that u is an eigenvector for P_3 with eigenvalue -3. But the Petersen graph does not have eigenvalue -3, so the result of removing two edge-disjoint Petersen graphs from K_{10} is not a Petersen graph. (In fact, it follows that P_3 is connected and bipartite.)

1.5.2 Decomposing K_n into complete bipartite graphs

A famous result is the fact that for any edge decomposition of K_n into complete bipartite graphs one needs to use at least $n - 1$ summands. Since K_n has eigenvalue -1 with multiplicity $n - 1$, this follows directly from the following:

Proposition 1.5.1 (H. S. Witsenhausen; GRAHAM & POLLAK [181]) *Suppose a graph Γ with adjacency matrix A has an edge decomposition into r complete bipartite graphs. Then $r \geq n_+(A)$ and $r \geq n_-(A)$, where $n_+(A)$ and $n_-(A)$ are the numbers of positive and negative eigenvalues of A, respectively.*

Proof Let u_i and v_i be the characteristic vectors of both sides of a bipartition of the i-th complete bipartite graph. Then that graph has adjacency matrix $D_i = u_i v_i^\top + v_i u_i^\top$, and $A = \sum D_i$. Let w be a vector orthogonal to all u_i. Then $w^\top A w = 0$ and it follows that w cannot be chosen in the span of eigenvectors of A with positive (negative) eigenvalue. $\square$

1.6 Automorphisms

An *automorphism* of a graph Γ is a permutation π of its point set X such that $x \sim y$ if and only if $\pi(x) \sim \pi(y)$. Given π, we have a linear transformation P_π on V defined by $(P_\pi(u))_x = u_{\pi(x)}$ for $u \in V$, $x \in X$. That π is an automorphism is expressed by $AP_\pi = P_\pi A$. It follows that P_π preserves the eigenspace V_θ for each eigenvalue θ of A.

More generally, if G is a group of automorphisms of Γ, then we find a linear representation of degree $m(\theta) = \dim V_\theta$ of G.

We denote the group of all automorphisms of Γ by Aut Γ. One would expect that when Aut Γ is large then $m(\theta)$ tends to be large, so that Γ has only few distinct eigenvalues. And indeed, the arguments below will show that a transitive group of automorphisms does not go together very well with simple eigenvalues.

Suppose $\dim V_\theta = 1$, say $V_\theta = \langle u \rangle$. Since P_π preserves V_θ, we must have $P_\pi u = \pm u$. So either u is constant on the orbits of π or π has even order, $P_\pi(u) = -u$, and u is constant on the orbits of π^2. For the Perron-Frobenius eigenvector (cf. §2.2) we always have the former case.

Corollary 1.6.1 *If all eigenvalues are simple, then* Aut Γ *is an elementary Abelian 2-group.*

Proof If π has order larger than 2, then there are two distinct vertices x, y in an orbit of π^2, and all eigenvectors have identical x- and y-coordinates, a contradiction. □

Corollary 1.6.2 *Let* Aut Γ *be transitive on X. (Then Γ is regular of degree k, say.)*

 (i) *If $m(\theta) = 1$ for some eigenvalue $\theta \neq k$, then $v = |X|$ is even and $\theta \equiv k$ (mod 2). If* Aut Γ *is, moreover, edge-transitive then Γ is bipartite and $\theta = -k$.*
 (ii) *If $m(\theta) = 1$ for two distinct eigenvalues $\theta \neq k$, then $v \equiv 0$ (mod 4).*
 (iii) *If $m(\theta) = 1$ for all eigenvalues θ, then Γ has at most two vertices.*

Proof (i) Suppose $V_\theta = \langle u \rangle$. Then u induces a partition of X into two equal parts: $X = X_+ \cup X_-$, where $u_x = a$ for $x \in X_+$ and $u_x = -a$ for $x \in X_-$. Now $\theta = k - 2|\Gamma(x) \cap X_-|$ for $x \in X_+$.

(ii) If $m(k) = m(\theta) = m(\theta') = 1$, then we find three pairwise orthogonal (± 1)-vectors, and a partition of X into four equal parts.

(iii) There are not enough integers $\theta \equiv k$ (mod 2) between $-k$ and k. □

For more details, see CVETKOVIĆ, DOOB & SACHS [115], Ch. 5.

1.7 Algebraic connectivity

Let Γ be a graph with at least two vertices. The second-smallest Laplace eigenvalue $\mu_2(\Gamma)$ is called the *algebraic connectivity* of the graph Γ. This concept was introduced by FIEDLER [156]. Now, by Proposition 1.3.7, $\mu_2(\Gamma) \geq 0$, with equality if and only if Γ is disconnected.

The algebraic connectivity is monotone: it does not decrease when edges are added to the graph:

Proposition 1.7.1 *Let Γ and Δ be two edge-disjoint graphs on the same vertex set, and $\Gamma \cup \Delta$ their union. We have $\mu_2(\Gamma \cup \Delta) \geq \mu_2(\Gamma) + \mu_2(\Delta) \geq \mu_2(\Gamma)$.*

Proof Use $\mu_2(\Gamma) = \min_u \{u^\perp L u \mid (u,u) = 1, (u,\mathbf{1}) = 0\}$. □

The algebraic connectivity is a lower bound for the vertex connectivity:

Proposition 1.7.2 *Let Γ be a graph with vertex set X. Suppose $D \subset X$ is a set of vertices such that the subgraph induced by Γ on $X \setminus D$ is disconnected. Then $|D| \geq \mu_2(\Gamma)$.*

Proof   By monotonicity we may assume that Γ contains all edges between D and $X \setminus D$. Now a nonzero vector u that is 0 on D and constant on each component of $X \setminus D$ and satisfies $(u, \mathbf{1}) = 0$, is a Laplace eigenvector with Laplace eigenvalue $|D|$. $\square$

1.8 Cospectral graphs

As noted above (in §1.3.7), there exist pairs of nonisomorphic graphs with the same spectrum. Graphs with the same (adjacency) spectrum are called *cospectral* (or *isospectral*). The two graphs of Figure 1.2 below are nonisomorphic and cospectral. Both graphs are regular, which means that they are also cospectral for the Laplace matrix and any other linear combination of A, I, and J, including the Seidel matrix (see §1.8.2) and the adjacency matrix of the complement.

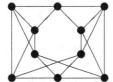 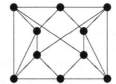

Fig. 1.2 Two cospectral regular graphs
(Spectrum: 4, 1, $(-1)^4$, $\pm\sqrt{5}$, $\frac{1}{2}(1 \pm \sqrt{17})$)

Let us give some more examples and families of examples. A more extensive discussion is found in Chapter 14.

1.8.1 The 4-cube

The hypercube 2^n is determined by its spectrum for $n < 4$, but not for $n \geq 4$. Indeed, there are precisely two graphs with spectrum 4^1, 2^4, 0^6, $(-2)^4$, $(-4)^1$ (HOFFMAN [218]). Consider the two binary codes of word length 4 and dimension 3 given by $C_1 = \mathbf{1}^\perp$ and $C_2 = (0111)^\perp$. Construct a bipartite graph, where one class of the bipartition consists of the pairs $(i, x) \in \{1, 2, 3, 4\} \times \{0, 1\}$ of coordinate position and value, and the other class of the bipartition consists of the code words, and code word u is adjacent to the pairs (i, u_i) for $i \in \{1, 2, 3, 4\}$. For the code C_1 this yields the 4-cube (tesseract), and for C_2 we get its unique cospectral mate.

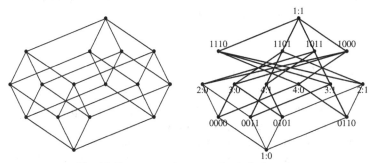

Fig. 1.3 Tesseract and cospectral switched version

1.8.2 Seidel switching

The *Seidel adjacency matrix* of a graph Γ with adjacency matrix A is the matrix S defined by

$$S_{uv} = \begin{cases} 0 \text{ if } u = v \\ -1 \text{ if } u \sim v \\ 1 \text{ if } u \nsim v \end{cases}$$

so that $S = J - I - 2A$. The *Seidel spectrum* of a graph is the spectrum of its Seidel adjacency matrix. For a regular graph on n vertices with valency k and other eigenvalues θ, the Seidel spectrum consists of $n - 1 - 2k$ and the values $-1 - 2\theta$.

Let Γ have vertex set X, and let $Y \subset X$. Let D be the diagonal matrix indexed by X with $D_{xx} = -1$ for $x \in Y$, and $D_{xx} = 1$ otherwise. Then DSD has the same spectrum as S. It is the Seidel adjacency matrix of the graph obtained from Γ by leaving adjacency and nonadjacency inside Y and $X \setminus Y$ as it was, and interchanging adjacency and nonadjacency between Y and $X \setminus Y$. This new graph, Seidel-cospectral with Γ, is said to be obtained by *Seidel switching* with respect to the set of vertices Y.

Being related by Seidel switching is an equivalence relation, and the equivalence classes are called *switching classes*. Here are the three switching classes of graphs with four vertices.

$$\begin{matrix} \bullet\ \bullet \\ \bullet\ \bullet \end{matrix} \sim \cdots \sim \cdots / \begin{matrix} \bullet \\ \bullet \end{matrix}\ \cdots \sim \cdots \sim \cdots \sim \cdots \sim \cdots / \begin{matrix} \bullet \\ \bullet \end{matrix}\ \cdots \sim \cdots \sim \cdots$$

The Seidel matrix of the complementary graph $\overline{\Gamma}$ is $-S$, so a graph and its complement have opposite Seidel eigenvalues.

If two regular graphs of the same valency are Seidel-cospectral, then they are also cospectral.

Figure 1.2 shows an example of two cospectral graphs related by Seidel switching (with respect to the four corners). These graphs are nonisomorphic: they have different local structure.

The Seidel adjacency matrix plays a role in the description of regular two-graphs (see §§10.1–10.3) and equiangular lines (see §10.6).

1.8.3 Godsil-McKay switching

Let Γ be a graph with vertex set X, and let $\{C_1, \ldots, C_t, D\}$ be a partition of X such that $\{C_1, \ldots, C_t\}$ is an equitable partition of $X \setminus D$ (that is, any two vertices in C_i have the same number of neighbors in C_j for all i, j), and for every $x \in D$ and every $i \in \{1, \ldots, t\}$ the vertex x has either 0, $\frac{1}{2}|C_i|$ or $|C_i|$ neighbors in C_i. Construct a new graph Γ' by interchanging adjacency and nonadjacency between $x \in D$ and the vertices in C_i whenever x has $\frac{1}{2}|C_i|$ neighbors in C_i. Then Γ and Γ' are cospectral ([176]).

Indeed, let Q_m be the matrix $\frac{2}{m}J - I$ of order m, so that $Q_m^2 = I$. Let $n_i = |C_i|$. Then the adjacency matrix A' of Γ' is found to be QAQ where Q is the block diagonal matrix with blocks Q_{n_i} $(1 \leq i \leq t)$ and I (of order $|D|$).

The same argument also applies to the complementary graphs, so that also the complements of Γ and Γ' are cospectral. Thus, for example, the second pair of graphs in Figure 1.1 is related by GM switching, and hence has cospectral complements. The first pair does not have cospectral complements and hence does not arise by GM switching.

The 4-cube and its cospectral mate (Figure 1.3) can be obtained from each other by GM switching with respect to the neighborhood of a vertex. Figure 1.2 is also an example of GM switching. Indeed, when two regular graphs of the same degree are related by Seidel switching, the switch is also a case of GM switching.

1.8.4 Reconstruction

The famous Kelly-Ulam conjecture (1941) asks whether a graph Γ can be reconstructed when the (isomorphism types of) the n vertex-deleted graphs $\Gamma \setminus x$ are given. The conjecture is still open (see Bondy [34] for a discussion), but Tutte [339] showed that one can reconstruct the characteristic polynomial of Γ, so any counterexample to the reconstruction conjecture must be a pair of cospectral graphs.

1.9 Very small graphs

Table 1.1 gives various spectra for the graphs on at most four vertices. The columns with heading A, L, Q, S give the spectrum for the adjacency matrix, the Laplace matrix $L = D - A$ (where D is the diagonal matrix of degrees), the signless Laplace matrix $Q = D + A$, and the Seidel matrix $S = J - I - 2A$, respectively.

Label	Picture	A	L	Q	S
0.1					
1.1	•	0	0	0	0
2.1	•—•	$1,-1$	$0,2$	$2,0$	$-1,1$
2.2	• •	$0,0$	$0,0$	$0,0$	$-1,1$
3.1		$2,-1,-1$	$0,3,3$	$4,1,1$	$-2,1,1$
3.2		$\sqrt{2},0,-\sqrt{2}$	$0,1,3$	$3,1,0$	$-1,-1,2$
3.3		$1,0,-1$	$0,0,2$	$2,0,0$	$-2,1,1$
3.4		$0,0,0$	$0,0,0$	$0,0,0$	$-1,-1,2$
4.1		$3,-1,-1,-1$	$0,4,4,4$	$6,2,2,2$	$-3,1,1,1$
4.2		$\rho,0,-1,1-\rho$	$0,2,4,4$	$2+2\tau,2,2,4-2\tau$	$-\sqrt{5},-1,1,\sqrt{5}$
4.3		$2,0,0,-2$	$0,2,2,4$	$4,2,2,0$	$-1,-1,-1,3$
4.4		$\theta_1,\theta_2,-1,\theta_3$	$0,1,3,4$	$2+\rho,2,1,3-\rho$	$-\sqrt{5},-1,1,\sqrt{5}$
4.5		$\sqrt{3},0,0,-\sqrt{3}$	$0,1,1,4$	$4,1,1,0$	$-1,-1,-1,3$
4.6		$\tau,\tau-1,1-\tau,-\tau$	$0,4-\alpha,2,\alpha$	$\alpha,2,4-\alpha,0$	$-\sqrt{5},-1,1,\sqrt{5}$
4.7		$2,0,-1,-1$	$0,0,3,3$	$4,1,1,0$	$-3,1,1,1$
4.8		$\sqrt{2},0,0,-\sqrt{2}$	$0,0,1,3$	$3,1,0,0$	$-\sqrt{5},-1,1,\sqrt{5}$
4.9		$1,1,-1,-1$	$0,0,2,2$	$2,2,0,0$	$-3,1,1,1$
4.10		$1,0,0,-1$	$0,0,0,2$	$2,0,0,0$	$-\sqrt{5},-1,1,\sqrt{5}$
4.11		$0,0,0,0$	$0,0,0,0$	$0,0,0,0$	$-1,-1,-1,3$

Table 1.1 Spectra of very small graphs

Here $\alpha = 2+\sqrt{2}$, $\tau = (1+\sqrt{5})/2$, and $\rho = (1+\sqrt{17})/2$, and $\theta_1 \approx 2.17009$, $\theta_2 \approx 0.31111$, $\theta_3 \approx -1.48119$ are the three roots of $\theta^3 - \theta^2 - 3\theta + 1 = 0$.

1.10 Exercises

Exercise 1.1 Show that no graph has eigenvalue $-1/2$. Show that no undirected graph has eigenvalue $\sqrt{2+\sqrt{5}}$. (Hint: Consider the algebraic conjugates of this number.)

Exercise 1.2 Let Γ be an undirected graph with eigenvalues $\theta_1,\ldots,\theta_n$. Show that for any two vertices a and b of Γ there are constants $c_1,\ldots,c_n$ such that the number of walks of length h from a to b equals $\sum c_i\theta_i^h$ for all h.

Exercise 1.3 Let Γ be a directed graph with constant outdegree $k > 0$ and without directed 2-cycles. Show that Γ has a nonreal eigenvalue.

Exercise 1.4 A *perfect e-error-correcting code* in an undirected graph Γ is a set of vertices C such that each vertex of Γ has distance at most e to precisely one vertex in C. For $e = 1$, this is also known as a *perfect dominating set*. Show that if Γ is regular of degree $k > 0$, and has a perfect dominating set, it has an eigenvalue -1.

Exercise 1.5 (i) Let Γ be a directed graph on n vertices such that there is an h with the property that for any two vertices a and b (distinct or not) there is a unique directed path of length h from a to b. Prove that Γ has constant in-degree and out-degree k, where $n = k^h$, and has spectrum $k^1 \, 0^{n-1}$.

(ii) The *de Bruijn graph* of order m is the directed graph with as vertices the 2^m binary sequences of length m, where there is an arrow from $a_1 \ldots a_m$ to $b_1 \ldots b_m$ when the tail $a_2 \ldots a_m$ of the first equals the head $b_1 \ldots b_{m-1}$ of the second. (For $m = 0$ we take a single vertex with two loops.) Determine the spectrum of the de Bruijn graph.

(iii) A *de Bruijn cycle* of order $m \geq 1$ ([70, 71, 160]) is a circular arrangement of 2^m zeros and ones such that each binary sequence of length m occurs once in this cycle. (In other words, it is a Hamiltonian cycle in the de Bruijn graph of order m, and a Eulerian cycle in the de Bruijn graph of order $m - 1$.) Show that there are precisely $2^{2^{m-1}-m}$ de Bruijn cycles of order m.

Exercise 1.6 ([43, 306]) Let Γ be a *tournament*, that is, a directed graph in which there is precisely one edge between any two distinct vertices, or, in other words, of which the adjacency matrix A satisfies $A^\top + A = J - I$.

(i) Show that all eigenvalues have real part not less than $-1/2$.

(ii) The tournament Γ is called *transitive* if (x,z) is an edge whenever both (x,y) and (y,z) are edges. Show that all eigenvalues of a transitive tournament are zero.

(iii) The tournament Γ is called *regular* when each vertex has the same number of out-arrows. Clearly, when there are n vertices, this number of out-arrows is $(n-1)/2$. Show that all eigenvalues θ have real part at most $(n-1)/2$ and that $\text{Re}(\theta) = (n-1)/2$ occurs if and only if Γ is regular (and then $\theta = (n-1)/2$).

(iv) Show that A either has full rank n or has rank $n - 1$, and that A has full rank when Γ is regular and $n > 1$.

(Hint: For a vector u, consider the expression $\bar{u}^\top (A^\top + A)u$.)

Exercise 1.7 Let Γ be bipartite and consider its line graph $L(\Gamma)$.

(i) Show that Γ admits a directed incidence matrix N such that $N^\top N - 2I$ is the adjacency matrix of $L(\Gamma)$.

(ii) Give a relation between the Laplace eigenvalues of Γ and the ordinary eigenvalues of $L(\Gamma)$.

(iii) Verify this relation in case Γ is the path P_n.

Exercise 1.8 ([102]) Verify (see §1.2.1) that both graphs pictured here have characteristic polynomial $t^4(t^4 - 7t^2 + 9)$, so that these two trees are cospectral.

Note how the coefficients of the characteristic polynomial of a tree count partial matchings (sets of pairwise disjoint edges) in the tree.

Exercise 1.9 ([17]) Verify that both graphs pictured here have characteristic polynomial $(t-1)(t+1)^2(t^3-t^2-5t+1)$ by computing eigenvectors and eigenvalues. Use the observation (§1.6) that the image of an eigenvector under an automorphism is again an eigenvector. In particular, when two vertices x, y are interchanged by an involution (automorphism of order 2), then the eigenspace has a basis consisting of vectors where the x- and y-coordinates are either equal or opposite.

Exercise 1.10 Show that the disjoint union $\Gamma + \Delta$ of two graphs Γ and Δ has characteristic polynomial $p(x) = p_\Gamma(x)p_\Delta(x)$.

Exercise 1.11 If Γ is regular of valency k on n vertices, then show that its complement $\overline{\Gamma}$ has characteristic polynomial

$$p(x) = (-1)^n \frac{x-n+k+1}{x+k+1} p_\Gamma(-x-1).$$

Exercise 1.12 Let the *cone* over a graph Γ be the graph obtained by adding a new vertex and joining that to all vertices of Γ. If Γ is regular of valency k on n vertices, then show that the cone over Γ has characteristic polynomial

$$p(x) = (x^2 - kx - n)p_\Gamma(x)/(x-k).$$

Exercise 1.13 Let the *join* of two graphs Γ and Δ be $\overline{\overline{\Gamma} + \overline{\Delta}}$, the result of joining each vertex of Γ to each vertex of (a disjoint copy of) Δ. If Γ and Δ are regular of valencies k and ℓ, and have m and n vertices, respectively, then the join of Γ and Δ has characteristic polynomial

$$p(x) = ((x-k)(x-\ell) - mn)\frac{p_\Gamma(x)p_\Delta(x)}{(x-k)(x-\ell)}.$$

Exercise 1.14 Let $\Gamma = (V, E)$ be a graph with n vertices and m edges. Construct a new graph Δ with vertex set $V \cup E$ (of size $n + m$), where Γ is the induced subgraph on V and E is a coclique, and each edge $e = xy$ in E is adjacent to its two endpoints x, y in V. Show that if Γ is k-regular, with $k > 1$, then the spectrum of Δ consists of two eigenvalues $\frac{1}{2}(\theta \pm \sqrt{\theta^2 + 4\theta + 4k})$ for each eigenvalue θ of A, together with 0 of multiplicity $m - n$.

Exercise 1.15 Show that the Seidel adjacency matrix S of a graph on n vertices has rank $n-1$ or n. (Hint: $\det S \equiv n-1 \pmod 2$.)

Exercise 1.16 Let Γ be a graph with at least one vertex such that any two distinct vertices have an odd number of common neighbors. Show that Γ has an odd number of vertices. (Hint: Consider $A\mathbf{1}$ and $A^2\mathbf{1} \pmod 2$.)

Chapter 2
Linear Algebra

In this chapter we present some less elementary but relevant results from linear algebra.

2.1 Simultaneous diagonalization

Let V be a complex vector space with finite dimension, and fix a basis $\{e_i \mid i \in I\}$. Then we can define an inner product on V by putting $(x,y) = \sum \bar{x}_i y_i = \bar{x}^\top y$ for $x, y \in V$, $x = \sum x_i e_i$, $y = \sum y_i e_i$, where the bar denotes complex conjugation. If the linear transformation A of V is *Hermitean*, i.e., if $(Ax, y) = (x, Ay)$ for all $x, y \in V$, then all eigenvalues of A are real, and V admits an orthonormal basis of eigenvectors of A.

Proposition 2.1.1 *Suppose $\mathscr{A}$ is a collection of commuting Hermitean linear transformations on V (i.e., $AB = BA$ for $A, B \in \mathscr{A}$). Then V has a basis consisting of common eigenvectors of all $A \in \mathscr{A}$.*

Proof Induction on $\dim V$. If each $A \in \mathscr{A}$ is a multiple of the identity I, then all is clear. Otherwise, let $A \in \mathscr{A}$ not be a multiple of I. If $Au = \theta u$ and $B \in \mathscr{A}$, then $A(Bu) = BAu = \theta Bu$, so B acts as a linear transformation on the eigenspace V_θ for the eigenvalue θ of A. By the induction hypothesis, we can choose a basis consisting of common eigenvectors for each $B \in \mathscr{A}$ in each eigenspace. The union of these bases is the basis of V we were looking for. $\qquad\square$

Given a square matrix A, we can regard A as a linear transformation on a vector space (with fixed basis). Hence the above concepts apply. The matrix A will be Hermitean precisely when $A = \bar{A}^\top$; in particular, a real symmetric matrix is Hermitean.

2.2 Perron-Frobenius theory

Let T be a real $n \times n$ matrix with nonnegative entries. T is called *primitive* if for some k we have $T^k > 0$ and it is called *irreducible* if for all i, j there is a k such that $(T^k)_{ij} > 0$. Here, for a matrix (or vector) A, $A > 0$ (≥ 0) means that all its entries are positive (nonnegative).

The matrix $T = (t_{ij})$ is irreducible if and only if the directed graph Γ_T with vertices $\{1, \ldots, n\}$ and edges (i, j) whenever $t_{ij} > 0$ is strongly connected.

(A directed graph (X, E) is *strongly connected* if for any two vertices x, y there is a directed path from x to y, i.e., there are vertices $x_0 = x, x_1, \ldots, x_m = y$ such that $(x_{i-1}, x_i) \in E$ for $1 \leq i \leq m$.)

Note that if T is irreducible, then $I + T$ is primitive.

The *period* d of an irreducible matrix T is the greatest common divisor of the integers k for which $(T^k)_{ii} > 0$. It is independent of the i chosen.

Theorem 2.2.1 *Let $T \geq 0$ be irreducible. Then there is a (unique) positive real number θ_0 with the following properties:*

(i) *There is a real vector $x_0 > 0$ with $Tx_0 = \theta_0 x_0$.*

(ii) *θ_0 has geometric and algebraic multiplicity 1.*

(iii) *For each eigenvalue θ of T, we have $|\theta| \leq \theta_0$. If T is primitive, then $|\theta| = \theta_0$ implies $\theta = \theta_0$. In general, if T has period d, then T has precisely d eigenvalues θ with $|\theta| = \theta_0$, namely $\theta = \theta_0 e^{2\pi i j/d}$ for $j = 0, 1, \ldots, d-1$. In fact, the entire spectrum of T is invariant under rotation of the complex plane over an angle $2\pi/d$ about the origin.*

(iv) *Any nonnegative left or right eigenvector of T has eigenvalue θ_0. Suppose $t \in \mathbb{R}$, and $x \geq 0$, $x \neq 0$. If $Tx \leq tx$, then $x > 0$ and $t \geq \theta_0$; moreover, $t = \theta_0$ if and only if $Tx = tx$. If $Tx \geq tx$, then $t \leq \theta_0$; moreover, $t = \theta_0$ if and only if $Tx = tx$.*

(v) *If $0 \leq S \leq T$ or if S is a principal minor of T, and S has eigenvalue σ, then $|\sigma| \leq \theta_0$; if $|\sigma| = \theta_0$, then $S = T$.*

(vi) *Given a complex matrix S, let $|S|$ denote the matrix with elements $|S|_{ij} = |S_{ij}|$. If $|S| \leq T$ and S has eigenvalue σ, then $|\sigma| \leq \theta_0$. If equality holds, then $|S| = T$, and there are a diagonal matrix E with diagonal entries of absolute value 1 and a constant c of absolute value 1 such that $S = cETE^{-1}$.*

Proof (i) Let $P = (I+T)^{n-1}$. Then $P > 0$ and $PT = TP$. Let $B = \{x \mid x \geq 0 \text{ and } x \neq 0\}$. Define for $x \in B$:

$$\theta(x) = \max\{t \mid t \in \mathbb{R}, \, tx \leq Tx\} = \min\{\frac{(Tx)_i}{x_i} \mid 1 \leq i \leq n, \, x_i \neq 0\}.$$

Now $\theta(\alpha x) = \theta(x)$ for $\alpha \in \mathbb{R}$, $\alpha > 0$, and ($x \leq y$, $x \neq y$ implies $Px < Py$, so) $\theta(Px) \geq \theta(x)$; in fact, $\theta(Px) > \theta(x)$ unless x is an eigenvector of T. Put $C = \{x \mid x \geq 0 \text{ and } \|x\| = 1\}$. Then, since C is compact and $\theta(.)$ is continuous on $P[C]$ (but not in general on C !), there is an $x_0 \in P[C]$ such that

$$\theta_0 := \sup_{x \in B} \theta(x) = \sup_{x \in C} \theta(x) = \sup_{x \in P[C]} \theta(x) = \theta(x_0).$$

Now $x_0 > 0$ and x_0 is an eigenvector of T, so $Tx_0 = \theta_0 x_0$ and $\theta_0 > 0$.

(ii) For a vector $x = (x_1, \ldots, x_n)^\top$, write $x_+ = (|x_1|, \ldots, |x_n|)^\top$. If $Tx = \theta x$, then by the triangle inequality we have $Tx_+ \geq |\theta| x_+$. For nonzero x this means $|\theta| \leq \theta(x_+) \leq \theta_0$. If, for some vector $z \in B$, we have $Tz \geq \theta_0 z$, then z is an eigenvector of T (otherwise $\theta(Pz) > \theta_0$), and since $0 < Pz = (1 + \theta_0)^{n-1} z$ we have $z > 0$. If x is a real vector such that $Tx = \theta_0 x$, then consider $y = x_0 + \varepsilon x$, where ε is chosen such that $y \geq 0$ but not $y > 0$. By the foregoing, $y \notin B$, so that $y = 0$, and x is a multiple of x_0. If x is a nonreal vector such that $Tx = \theta_0 x$, then both the real and imaginary parts of x are multiples of x_0. This shows that the eigenspace of θ_0 has dimension 1, i.e., that the geometric multiplicity of θ_0 is 1. We shall look at the algebraic multiplicity later.

(iii) We have seen $|\theta| \leq \theta_0$. If $|\theta| = \theta_0$ and $Tx = \theta x$, then $Tx_+ = \theta_0 x_+$ and we have equality in the triangle inequality $|\sum_j t_{ij} x_j| \leq \sum_j t_{ij} |x_j|$. This means that all numbers $t_{ij} x_j$ $(1 \leq j \leq n)$ have the same angular part (argument). If T is primitive, then we can apply this reasoning with T^k instead of T, where $T^k > 0$, and conclude that all x_j have the same angular part. Consequently, in this case x is a multiple of a real vector and may be taken real, nonnegative. Now $Tx = \theta x$ shows that θ is real, and $|\theta| = \theta_0$ so that $\theta = \theta_0$. In the general case, T^d is a direct sum of primitive matrices $T^{(0)}, \ldots, T^{(d-1)}$, and if $x = (x^{(0)}, \ldots, x^{(d-1)})$ is the corresponding decomposition of an eigenvector of T (with eigenvalue θ), then $(x^{(0)}, \zeta x^{(1)}, \ldots, \zeta^{d-1} x^{(d-1)})$ also is an eigenvector of T, with eigenvalue $\zeta \theta$, for any d-th root of unity ζ. (Here we assume that the $T^{(i)}$ are ordered in such a way that in Γ_T the arrows point from the subset corresponding to $T^{(i)}$ to the subset corresponding to $T^{(i+1)}$.) Since T^d has a unique eigenvalue of maximum modulus (let $T_{(i)}^{(i+1)}$ be the (nonsquare) submatrix of T describing the arrows in Γ_T from the subset corresponding to $T^{(i)}$ to the subset corresponding to $T^{(i+1)}$; then $T^{(i)} = \prod_{j=0}^{d-1} T_{(i+j)}^{(i+j+1)}$, and if $T^{(i)} z = \gamma z$, $z > 0$, then $T^{(i-1)} z' = \gamma z'$ where $z' = T_{(i-1)}^{(i)} z \neq 0$, so that all $T^{(i)}$ have the same eigenvalue of maximum modulus), it follows that T has precisely d such eigenvalues.

(iv) Doing the above for left eigenvectors instead of right ones, we find $y_0 > 0$ with $y_0^\top T = \eta_0 y_0^\top$. If $Tx = \theta x$ and $y^\top T = \eta y^\top$, then $\eta y^\top x = y^\top Tx = \theta y^\top x$. It follows that either $\theta = \eta$ or $y^\top x = 0$. Taking $y \in B$, $x = x_0$ or $x \in B$, $y = y_0$ we see that $\theta = \eta$ $(= \theta_0 = \eta_0)$. If $x \in B$ and $Tx \leq tx$, then $t \geq 0$ and $0 < Px \leq (1+t)^{n-1} x$, so $x > 0$. Also $\theta_0 y_0^\top x = y_0^\top Tx \leq ty_0^\top x$, so $\theta_0 \leq t$; in case of equality we have $y_0^\top (Tx - tx) = 0$ and hence $Tx = tx$. For $Tx \geq tx$ the same argument applies.

(v) If $s \neq 0$, $Ss = \sigma s$, then $Ts_+ \geq Ss_+ \geq |\sigma| s_+$, so $|\sigma| \leq \theta_0$. But if $|\sigma| = \theta_0$, then s_+ is an eigenvector of T and $s_+ > 0$ and $(T - S)s_+ = 0$, so $S = T$.

(vi) If $s \neq 0$, $Ss = \sigma s$, then $Ts_+ \geq |S| s_+ \geq |\sigma| s_+$, so $|\sigma| \leq \theta_0$, and if $|\sigma| = \theta_0$, then s_+ is an eigenvector of T and $s_+ > 0$ and $|S| = T$. Equality in $|S| s_+ = |\sigma| s_+$ means that $|\sum S_{ij} s_j| = \sum |S_{ij}|.|s_j|$, so that given i all $S_{ij} s_j$ have the same angular part. Let $E_{ii} = s_i / |s_i|$ and $c = \sigma / |\sigma|$. Then $S_{ij} = cE_{ii} E_{jj}^{-1} |S_{ij}|$.

(vii) Finally, in order to prove that θ_0 is a simple root of χ_T, the characteristic polynomial of T, we have to show that $\frac{d}{d\theta}\chi_T(\theta)$ is nonzero for $\theta = \theta_0$. But $\chi_T(\theta) = \det(\theta I - T)$ and $\frac{d}{d\theta}\chi_T(\theta) = \sum_i \det(\theta I - T_{ii})$, and by (v) we have $\det(\theta I - T_{ii}) > 0$ for $\theta = \theta_0$.   □

Remark In case $T \geq 0$ but T not necessarily irreducible, we can say the following.

(i) The spectral radius θ_0 of T is an eigenvalue, and there are nonnegative left and right eigenvectors corresponding to it.
(ii) If $|S| \leq T$ and S has eigenvalue σ, then $|\sigma| \leq \theta_0$.

(**Proof**   (i) Use continuity arguments; (ii) the old proof still applies.)

For more detail, see the exposition of Perron-Frobenius theory in GANTMACHER [167], Ch. XIII; see also VARGA [341], MARCUS & MINC [269], SENETA [320], Ch. 1, BERMAN & PLEMMONS [26], or HORN & JOHNSON [227], Ch. 8.

2.3 Equitable partitions

Suppose A is a symmetric real matrix whose rows and columns are indexed by $X = \{1,\ldots,n\}$. Let $\{X_1,\ldots,X_m\}$ be a partition of X. The *characteristic matrix* S is the $n \times m$ matrix whose j-th column is the characteristic vector of X_j ($j = 1,\ldots,m$). Define $n_i = |X_i|$ and $K = \mathrm{diag}(n_1,\ldots,n_m)$. Let A be partitioned according to $\{X_1,\ldots,X_m\}$, that is,

$$A = \begin{bmatrix} A_{1,1} & \cdots & A_{1,m} \\ \vdots & & \vdots \\ A_{m,1} & \cdots & A_{m,m} \end{bmatrix},$$

where $A_{i,j}$ denotes the submatrix (block) of A formed by rows in X_i and the columns in X_j. Let $b_{i,j}$ denote the average row sum of $A_{i,j}$. Then the matrix $B = (b_{i,j})$ is called the *quotient matrix* of A w.r.t. the given partition. We easily have

$$KB = S^\top AS, \quad S^\top S = K.$$

If the row sum of each block $A_{i,j}$ is constant then the partition is called *equitable* (or *regular*) and we have $A_{i,j}\mathbf{1} = b_{i,j}\mathbf{1}$ for $i,j = 0,\ldots,d$, so

$$AS = SB.$$

The following result is well-known and useful.

Lemma 2.3.1 *If, for an equitable partition, v is an eigenvector of B for an eigenvalue λ, then Sv is an eigenvector of A for the same eigenvalue λ.*

Proof $Bv = \theta v$ implies $ASv = SBv = \theta Sv$.   □

In the situation of this lemma, the spectrum of A consists of the spectrum of the quotient matrix B (with eigenvectors in the column space of S, i.e., constant on the parts of the partition) together with the eigenvalues belonging to eigenvectors orthogonal to the columns of S (i.e., summing to zero on each part of the partition). These latter eigenvalues remain unchanged if the blocks $A_{i,j}$ are replaced by $A_{i,j} + c_{i,j}J$ for certain constants $c_{i,j}$.

2.3.1 Equitable and almost equitable partitions of graphs

If in the above the matrix A is the adjacency matrix (or the Laplace matrix) of a graph, then an equitable partition of the matrix A is a partition of the vertex set into parts X_i such that each vertex in X_i has the same number $b_{i,j}$ of neighbors in part X_j for any j (or any $j \neq i$). Such partitions are called (almost) equitable partitions of the graph.

For example, the adjacency matrix of the complete bipartite graph $K_{p,q}$ has an equitable partition with $m = 2$. The quotient matrix B equals $\begin{bmatrix} 0 & p \\ q & 0 \end{bmatrix}$ and has eigenvalues $\pm\sqrt{pq}$, which are the nonzero eigenvalues of $K_{p,q}$.

More generally, consider the *join* Γ of two vertex-disjoint graphs Γ_1 and Γ_2, the graph obtained by inserting all possible edges between Γ_1 and Γ_2. If Γ_1 and Γ_2 have n_1 (resp. n_2) vertices and are both regular, say of valency k_1 (resp. k_2), and have spectra Φ_1 (resp. Φ_2), then Γ has spectrum $\Phi = (\Phi_1 \setminus \{k_1\}) \cup (\Phi_2 \setminus \{k_2\}) \cup \{k', k''\}$ where k', k'' are the two eigenvalues of

$$\begin{bmatrix} k_1 & n_2 \\ n_1 & k_2 \end{bmatrix}.$$

Indeed, we have an equitable partition of the adjacency matrix of Γ with the above quotient matrix. The eigenvalues that do not belong to the quotient coincide with those of the disjoint union of Γ_1 and Γ_2.

2.4 The Rayleigh quotient

Let A be a real symmetric matrix and let u be a nonzero vector. The *Rayleigh quotient* of u w.r.t. A is defined as

$$\frac{u^\top A u}{u^\top u}.$$

Let $u_1, \ldots, u_n$ be an orthonormal set of eigenvectors of A, say with $Au_i = \theta_i u_i$, where $\theta_1 \geq \ldots \geq \theta_n$. If $u = \sum \alpha_i u_i$, then $u^\top u = \sum \alpha_i^2$ and $u^\top A u = \sum \alpha_i^2 \theta_i$. It follows that

$$\frac{u^\top A u}{u^\top u} \geq \theta_i \text{ if } u \in \langle u_1, \ldots, u_i \rangle$$

and

$$\frac{u^\top Au}{u^\top u} \leq \theta_i \text{ if } u \in \langle u_1,\ldots,u_{i-1}\rangle^\perp.$$

In both cases, equality implies that u is a θ_i-eigenvector of A. Conversely, one has

Theorem 2.4.1 (Courant-Fischer) *Let W be an i-subspace of V. Then*

$$\theta_i \geq \min_{u\in W, u\neq 0} \frac{u^\top Au}{u^\top u}$$

and

$$\theta_{i+1} \leq \max_{u\in W^\perp, u\neq 0} \frac{u^\top Au}{u^\top u}.$$

Proof   See [227], Theorem 4.2.11. $\Box$

2.5 Interlacing

Consider two sequences of real numbers, $\theta_1 \geq \ldots \geq \theta_n$ and $\eta_1 \geq \ldots \geq \eta_m$, with $m < n$. The second sequence is said to *interlace* the first one whenever

$$\theta_i \geq \eta_i \geq \theta_{n-m+i} \text{ for } i = 1,\ldots,m.$$

The interlacing is *tight* if there exists an integer $k \in [0,m]$ such that

$$\theta_i = \eta_i \text{ for } 1 \leq i \leq k \text{ and } \theta_{n-m+i} = \eta_i \text{ for } k+1 \leq i \leq m.$$

If $m = n-1$, the interlacing inequalities become $\theta_1 \geq \eta_1 \geq \theta_2 \geq \eta_2 \geq \ldots \geq \eta_m \geq \theta_n$, which clarifies the name. GODSIL [172] reserves the name "interlacing" for this particular case and calls it generalized interlacing otherwise.

Theorem 2.5.1 *Let S be a real $n \times m$ matrix such that $S^\top S = I$. Let A be a real symmetric matrix of order n with eigenvalues $\theta_1 \geq \ldots \geq \theta_n$. Define $B = S^\top AS$, and let B have eigenvalues $\eta_1 \geq \ldots \geq \eta_m$ and respective eigenvectors $v_1,\ldots,v_m$.*

  (i) *The eigenvalues of B interlace those of A.*
  (ii) *If $\eta_i = \theta_i$ or $\eta_i = \theta_{n-m+i}$ for some $i \in [1,m]$, then B has a η_i-eigenvector v such that Sv is a η_i-eigenvector of A.*
  (iii) *If, for some integer l, $\eta_i = \theta_i$ for $i = 1,\ldots,l$ (or $\eta_i = \theta_{n-m+i}$ for $i = l,\ldots,m$), then Sv_i is a η_i-eigenvector of A for $i = 1,\ldots,l$ (respectively $i = l,\ldots,m$).*
  (iv) *If the interlacing is tight, then $SB = AS$.*

Proof   Let $u_1,\ldots,u_n$ be an orthonormal set of eigenvectors of the matrix A, where $Au_i = \theta_i u_i$. For each $i \in [1,m]$, take a nonzero vector s_i in

$$\langle v_1,\ldots,v_i\rangle \cap \langle S^\top u_1,\ldots,S^\top u_{i-1}\rangle^\perp. \tag{2.1}$$

Then $Ss_i \in \langle u_1, \ldots, u_{i-1} \rangle^{\perp}$ and hence by Rayleigh's principle,

$$\theta_i \geq \frac{(Ss_i)^{\top} A (Ss_i)}{(Ss_i)^{\top} (Ss_i)} = \frac{s_i^{\top} B s_i}{s_i^{\top} s_i} \geq \eta_i,$$

and similarly (or by applying the above inequality to $-A$ and $-B$) we get $\theta_{n-m+i} \leq \eta_i$, proving (i). If $\theta_i = \eta_i$, then s_i and Ss_i are θ_i-eigenvectors of B and A, respectively, proving (ii). We prove (iii) by induction on l. Assume $Sv_i = u_i$ for $i = 1, \ldots, l - 1$. Then we may take $s_l = v_l$ in (2.1), but in proving (ii) we saw that Ss_l is a θ_l-eigenvector of A. (The statement between parentheses follows by considering $-A$ and $-B$.) Thus we have (iii). Let the interlacing be tight. Then, by (iii), $Sv_1, \ldots, Sv_m$ is an orthonormal set of eigenvectors of A for the eigenvalues $\eta_1, \ldots, \eta_m$. So we have $SBv_i = \eta_i Sv_i = ASv_i$, for $i = 1, \ldots, m$. Since the vectors v_i form a basis, it follows that $SB = AS$. $\qquad\square$

If we take $S = [I\ 0]^{\top}$, then B is just a principal submatrix of A and we obtain:

Corollary 2.5.2 *If B is a principal submatrix of a symmetric matrix A, then the eigenvalues of B interlace the eigenvalues of A.*

The theorem requires the columns of S to be orthonormal. If one has a situation with orthogonal but not necessarily orthonormal vectors, some scaling is required.

Corollary 2.5.3 *Let A be a real symmetric matrix of order n. Let $x_1, \ldots, x_m$ be nonzero orthogonal real vectors of order n. Define a matrix $C = (c_{ij})$ by $c_{ij} = \frac{1}{\|x_i\|^2} x_i^{\top} A x_j$.*

   (i) The eigenvalues of C interlace the eigenvalues of A.

   (ii) If the interlacing is tight, then $Ax_j = \sum c_{ij} x_i$ for all j.

   (iii) Let $x = \sum x_j$. The number $r := \frac{x^{\top} A x}{x^{\top} x}$ lies between the smallest and largest eigenvalue of C. If x is an eigenvector of A with eigenvalue θ, then also C has an eigenvalue θ (for eigenvector $\mathbf{1}$).

Proof Let K be the diagonal matrix with $K_{ii} = \|x_i\|$. Let R be the $n \times m$ matrix with columns x_j, and put $S = RK^{-1}$. Then $S^{\top} S = I$, and the theorem applies with $B = S^{\top} A S = KCK^{-1}$. If interlacing is tight we have $AR = RC$. With $x = \sum x_j = R\mathbf{1}$ and $y = K\mathbf{1}$, we have $\frac{x^{\top} A x}{x^{\top} x} = \frac{y^{\top} B y}{y^{\top} y}$. $\qquad\square$

In particular, this applies when the x_i are the characteristic vectors of a partition (or just a collection of pairwise disjoint subsets).

Corollary 2.5.4 *Let C be the quotient matrix of a symmetric matrix A whose rows and columns are partitioned according to a partitioning $\{X_1, \ldots, X_m\}$.*

   (i) The eigenvalues of C interlace the eigenvalues of A.

   (ii) If the interlacing is tight, then the partition is equitable. $\qquad\square$

Theorem 2.5.1(i) is a classical result; see COURANT & HILBERT [107], Vol. 1, Ch. I. For the special case of a principal submatrix (Corollary 2.5.2), the result even goes back to Cauchy and is therefore often referred to as Cauchy interlacing. Interlacing for the quotient matrix (Corollary 2.5.4) is especially applicable to combinatorial structures (as we shall see). Payne (see, for instance, [291]) has applied the extremal inequalities $\theta_1 \geq \eta_i \geq \theta_n$ to finite geometries several times. He attributes the method to Higman and Sims and therefore calls it the *Higman-Sims technique*.

Remark This theorem generalizes directly to complex Hermitean matrices instead of real symmetric matrices (with conjugate transpose instead of transpose) with virtually the same proof.

For more detailed eigenvalue inequalities, see HAEMERS [197], [199].

2.6 Schur's inequality

Theorem 2.6.1 (SCHUR [307]) *Let A be a real symmetric matrix with eigenvalues $\theta_1 \geq \theta_2 \geq \ldots \geq \theta_n$ and diagonal elements $d_1 \geq d_2 \geq \ldots \geq d_n$. Then $\sum_{i=1}^{t} d_i \leq \sum_{i=1}^{t} \theta_i$ for $1 \leq t \leq n$.*

Proof Let B be the principal submatrix of A obtained by deleting the rows and columns containing $d_{t+1}, \ldots, d_n$. If B has eigenvalues η_i $(1 \leq i \leq t)$, then by interlacing $\sum_{i=1}^{t} d_i = \operatorname{tr} B = \sum_{i=1}^{t} \eta_i \leq \sum_{i=1}^{t} \theta_i$. □

Remark Again "real symmetric" can be replaced by "Hermitean".

2.7 Schur complements

In this section, the square matrix

$$A = \begin{bmatrix} A_{11} & A_{12} \\ A_{21} & A_{22} \end{bmatrix}$$

is a square partitioned matrix (over any field), where A_{11} is nonsingular. The *Schur complement* A/A_{11} of A_{11} in A is the matrix $A_{22} - A_{21}A_{11}^{-1}A_{12}$. The following result is a straightforward but important consequence from the definition.

Theorem 2.7.1 (see [356]) *The Schur complement A/A_{11} satisfies*

(i) $\begin{bmatrix} I & O \\ -A_{21}A_{11}^{-1} & I \end{bmatrix} \begin{bmatrix} A_{11} & A_{12} \\ A_{21} & A_{22} \end{bmatrix} \begin{bmatrix} I & -A_{11}^{-1}A_{12} \\ O & I \end{bmatrix} = \begin{bmatrix} A_{11} & O \\ O & A/A_{11} \end{bmatrix}$,

(ii) $\det(A/A_{11}) = \det A / \det A_{11}$,

(iii) $\operatorname{rk} A = \operatorname{rk} A_{11} + \operatorname{rk}(A/A_{11})$.

Corollary 2.7.2 *If $\operatorname{rk} A = \operatorname{rk} A_{11}$, then $A_{22} = A_{21}A_{11}^{-1}A_{12}$.* □

2.8 The Courant-Weyl inequalities

Denote the eigenvalues of a Hermitean matrix A, arranged in nonincreasing order, by $\lambda_i(A)$.

Theorem 2.8.1 *Let A and B be Hermitean matrices of order n, and let $1 \leq i, j \leq n$.*

   (i) If $i + j - 1 \leq n$, then $\lambda_{i+j-1}(A+B) \leq \lambda_i(A) + \lambda_j(B)$.
   (ii) If $i + j - n \geq 1$, then $\lambda_i(A) + \lambda_j(B) \leq \lambda_{i+j-n}(A+B)$.
   (iii) If B is positive semidefinite, then $\lambda_i(A+B) \geq \lambda_i(A)$.

Proof (i) Let $u_1, \ldots, u_n$ and $v_1, \ldots, v_n$ be orthonormal sets of eigenvectors of A (resp. B) with $Au_i = \lambda_i(A)u_i$ and $Bv_j = \lambda_j(B)v_j$. Let $U = \langle u_h \mid 1 \leq h \leq i-1 \rangle$ and $V = \langle v_h \mid 1 \leq h \leq j-1 \rangle$, and $W = U + V$. For $w \in W^{\perp}$ we have $w^{\top}(A+B)w \leq (\lambda_i(A) + \lambda_j(B))w^{\top}w$. It follows that the space spanned by eigenvectors of $A + B$ with eigenvalue larger than $\lambda_i(A) + \lambda_j(B)$ has dimension at most $i + j - 2$.

   (ii) Apply (i) to $-A$ and $-B$. (iii) Apply the case $j = n$ of (ii). $\qquad\square$

KY FAN [153] shows that $\lambda(A) + \lambda(B)$ dominates $\lambda(A+B)$:

Theorem 2.8.2 *Let A and B be Hermitean matrices of order n. Then, for all t, $0 \leq t \leq n$, we have $\sum_{i=1}^{t} \lambda_i(A+B) \leq \sum_{i=1}^{t} \lambda_i(A) + \sum_{i=1}^{t} \lambda_i(B)$.*

Proof $\sum_{i=1}^{t} \lambda_i(A) = \max \operatorname{tr}(U^*AU)$, where the maximum is over all $n \times t$ matrices U with $U^*U = I$. $\qquad\square$

2.9 Gram matrices

Real symmetric $n \times n$ matrices G are in bijective correspondence with quadratic forms q on $\mathbb{R}^n$ via the relation

$$q(x) = x^{\top} Gx \quad (x \in \mathbb{R}^n).$$

Two quadratic forms q and q' on $\mathbb{R}^n$ are *congruent*, i.e., there is a nonsingular $n \times n$ matrix S such that $q(x) = q'(Sx)$ for all $x \in \mathbb{R}^n$, if and only if their corresponding matrices G and G' satisfy $G = S^{\top}G'S$. Moreover, this occurs for some S if and only if G and G' have the same rank and the same number of nonnegative eigenvalues—this is SYLVESTER [332]'s "law of inertia for quadratic forms", cf. GANTMACHER [167], Vol. 1, Ch. X, §2. We shall now be concerned with matrices that have nonnegative eigenvalues only.

Lemma 2.9.1 *Let G be a real symmetric $n \times n$ matrix. Equivalent are:*

   (i) For all $x \in \mathbb{R}^n$, $x^{\top}Gx \geq 0$.
   (ii) All eigenvalues of G are nonnegative.
   (iii) G can be written as $G = H^{\top}H$, with H an $m \times n$ matrix, where m is the rank of G.

Proof   There is an orthogonal matrix Q and a diagonal matrix D whose nonzero entries are the eigenvalues of G such that $G = Q^\top DQ$. If (ii) holds, then $x^\top Gx = (Qx)^\top D(Qx) \geq 0$ implies (i). Conversely, (ii) follows from (i) by choosing x to be an eigenvector. If $G = H^\top H$ then $x^\top Gx = ||Hx||^2 \geq 0$, so (iii) implies (i). Finally, let $E = D^{1/2}$ be the diagonal matrix that squares to D, and let F be the $m \times n$ matrix obtained from E by dropping the zero rows. Then $G = Q^\top E^\top EQ = Q^\top F^\top FQ = H^\top H$, so that (ii) implies (iii).   □

A symmetric $n \times n$ matrix G satisfying (i) or (ii) is called *positive semidefinite*. It is called *positive definite* when $x^\top Gx = 0$ implies $x = 0$, or, equivalently, when all its eigenvalues are positive. For any collection X of vectors of $\mathbb{R}^m$, we define its *Gram matrix* as the square matrix G indexed by X whose (x, y)-entry G_{xy} is the inner product $(x, y) = x^\top y$. This matrix is always positive semidefinite, and it is definite if and only if the vectors in X are linearly independent. (Indeed, if $n = |X|$, and we use H to denote the $m \times n$ matrix whose columns are the vectors of X, then $G = H^\top H$, and $x^\top Gx = ||Hx||^2 \geq 0$.)

Lemma 2.9.2 *Let N be a real $m \times n$ matrix. Then the matrices $NN^\top$ and $N^\top N$ have the same nonzero eigenvalues (including multiplicities). Moreover, $\mathrm{rk}\,NN^\top = \mathrm{rk}\,N^\top N = \mathrm{rk}\,N$.*

Proof   Let θ be a nonzero eigenvalue of $NN^\top$. The map $u \mapsto N^\top u$ is an isomorphism from the θ-eigenspace of $NN^\top$ onto the θ-eigenspace of $N^\top N$. Indeed, if $NN^\top u = \theta u$, then $N^\top NN^\top u = \theta N^\top u$ and $N^\top u$ is nonzero for nonzero u since $NN^\top u = \theta u$. The final sentence follows since $\mathrm{rk}\,N^\top N \leq \mathrm{rk}\,N$, but if $N^\top Nx = 0$ then $||Nx||^2 = x^\top N^\top Nx = 0$, so $Nx = 0$.   □

2.10 Diagonally dominant matrices

A *diagonally dominant* matrix is a complex matrix B with the property that we have $|b_{ii}| \geq \sum_{j \neq i} |b_{ij}|$ for all i. When all these inequalities are strict, the matrix is called *strictly diagonally dominant*.

Lemma 2.10.1   *(i) A strictly diagonally dominant complex matrix is nonsingular.*

(ii) A symmetric diagonally dominant real matrix with nonnegative diagonal entries is positive semidefinite.

(iii) Let B be a symmetric real matrix with nonnegative row sums and nonpositive off-diagonal entries. Define a graph Γ on the index set of the rows of B, where two distinct indices i, j are adjacent when $b_{i,j} \neq 0$. The multiplicity of the eigenvalue 0 of B equals the number of connected components C of Γ such that all rows $i \in C$ have zero row sum.

Proof   Let $B = (b_{ij})$ be diagonally dominant, and let u be an eigenvector, say, with $Bu = bu$. Let $|u_i|$ be maximal among the $|u_j|$. Then $(b_{ii} - b)u_i = -\sum_{j \neq i} b_{ij}u_j$. In all cases the result follows by comparing the absolute values of both sides.

In order to prove (i), assume that B is singular, and that $Bu = 0$. Take absolute values on both sides. We find $|b_{ii}|.|u_i| \leq \sum_{j \neq i} |b_{ij}|.|u_j| \leq \sum_{j \neq i} |b_{ij}||u_i| < |b_{ii}|.|u_i|$, a contradiction.

For (ii), assume that B has a negative eigenvalue b. Then $(b_{ii} - b).|u_i| \leq |b_{ii}|.|u_i|$, a contradiction.

For (iii), take $b = 0$ again, and see how equality could hold everywhere in $b_{ii}.|u_i| \leq \sum_{j \neq i} |b_{ij}|.|u_j| \leq \sum_{j \neq i} |b_{ij}||u_i| \leq b_{ii}.|u_i|$. We see that u must be constant on the connected components of Γ, and zero where row sums are nonzero.   □

2.10.1 Geršgorin circles

The above can be greatly generalized. Let $B(c,r) = \{z \in \mathbb{C} \mid |z - c| \leq r\}$ be the closed ball in $\mathbb{C}$ with center c and radius r.

Proposition 2.10.2 *Let $A = (a_{ij})$ be a complex matrix of order n, and let λ be an eigenvalue of A. Put $r_i = \sum_{j \neq i} |a_{ij}|$. Then for some i we have $\lambda \in B(a_{ii}, r_i)$. If C is a connected component of $\bigcup_i B(a_{ii}, r_i)$ that contains m of the a_{ii}, then C contains m eigenvalues of A.*

Proof If $Au = \lambda u$, then $(\lambda - a_{ii})u_i = \sum_{j \neq i} a_{ij}u_j$. Let i be an index for which $|u_i|$ is maximal. Then $|\lambda - a_{ii}|.|u_i| \leq \sum_{j \neq i} |a_{ij}|.|u_i|$ so that $\lambda \in B(a_{ii}, r_i)$. For the second part, use that the eigenvalues are continuous functions of the matrix elements. Let $A(\varepsilon)$ be the matrix with the same diagonal as A and with off-diagonal entries εa_{ij}, so that $A = A(1)$. Then $A(0)$ has eigenvalues a_{ii}, and for $0 \leq \varepsilon \leq 1$ the matrix $A(\varepsilon)$ has eigenvalues inside $\bigcup_i B(a_{ii}, r_i)$.   □

This result is due to GERŠGORIN [169]. A book-length treatment was given by VARGA [342].

2.11 Projections

Lemma 2.11.1 *Let $P = \begin{bmatrix} Q & N \\ N^\top & R \end{bmatrix}$ be a real symmetric matrix of order n with two eigenvalues a and b, partitioned with square Q and R. Let Q have h eigenvalues θ_j distinct from a and b. Then R has h eigenvalues $a + b - \theta_j$ distinct from a and b, and $h = m_P(a) - m_Q(a) - m_R(a) = m_P(b) - m_Q(b) - m_R(b)$, where $m_M(\eta)$ denotes the multiplicity of the eigenvalue η of M.*

Proof We may take $a = 1$ and $b = 0$, so that P is a projection and $P^2 = P$. Now if $Qu = \theta u$, then $Rv = (1 - \theta)v$ for $v = N^\top u$ and $NN^\top u = \theta(1 - \theta)u$, so that the eigenvalues of Q and R different from 0 and 1 correspond 1-1. The rest follows by taking traces: $0 = \operatorname{tr} P - \operatorname{tr} Q - \operatorname{tr} R = m_P(1) - m_Q(1) - m_R(1) - h$.   □

2.12 Exercises

Exercise 2.1 Consider a symmetric $n \times n$ matrix A with an equitable partition $\{X_1, \ldots, X_m\}$ of the index set for rows and columns, where all classes have equal size. Let S and B be the characteristic matrix and the quotient matrix of this partition, respectively. Prove that A and $SS^\top$ commute and give an expression for the eigenvalues of $A + \alpha SS^\top$ for $\alpha \in \mathbb{R}$.

Exercise 2.2 Let A be a real symmetric matrix of order n with eigenvalues $\theta_1 \geq \ldots \geq \theta_n$. Let $\{X_1, \ldots, X_m\}$ be a partition of the index set for row and columns of A, and let B be the corresponding quotient matrix, with eigenvalues $\eta_1 \geq \ldots \geq \eta_m$. Show that if $\theta_s = \eta_s$ for some s, then A has a θ_s-eigenvector that is constant on each part X_j.

Exercise 2.3 Let B denote the quotient matrix of a symmetric matrix A whose rows and columns are partitioned according to a partitioning $\{X_1, \ldots, X_m\}$.

(i) Give an example where the eigenvalues of B are a sub(multi)set of the eigenvalues of A and the partition is not equitable.
(ii) Give an example where the partition is equitable and the interlacing is not tight.

Chapter 3
Eigenvalues and Eigenvectors of Graphs

In this chapter, we apply the linear algebra from the previous chapter to graph spectra.

3.1 The largest eigenvalue

The largest eigenvalue of a graph is also known as its *spectral radius* or *index*.

The basic information about the largest eigenvalue of a (possibly directed) graph is provided by the Perron-Frobenius theorem.

Proposition 3.1.1 *Each graph Γ has a real eigenvalue θ_0 with nonnegative real corresponding eigenvector and such that for each eigenvalue θ we have $|\theta| \leq \theta_0$. The value $\theta_0(\Gamma)$ does not increase when vertices or edges are removed from Γ.*

Assume that Γ is strongly connected. Then:

(i) *θ_0 has multiplicity 1.*
(ii) *If Γ is primitive (strongly connected, and such that not all cycles have a length that is a multiple of some integer $d > 1$), then $|\theta| < \theta_0$ for all eigenvalues θ different from θ_0.*
(iii) *The value $\theta_0(\Gamma)$ decreases when vertices or edges are removed from Γ.* $\square$

Now let Γ be undirected. By the Perron-Frobenius theorem and interlacing, we find upper and lower bounds for the largest eigenvalue of a connected graph. (Note that A is irreducible if and only if Γ is connected.)

Proposition 3.1.2 *Let Γ be a connected graph with largest eigenvalue θ_1. If Γ is regular of valency k, then $\theta_1 = k$. Otherwise, we have $k_{\min} < \bar{k} < \theta_1 < k_{\max}$, where $k_{\min}$, $k_{\max}$, and $\bar{k}$ are the minimum, maximum, and average degree, respectively.*

Proof Let **1** be the vector with all entries equal to 1. Then $A\mathbf{1} \leq k_{\max}\mathbf{1}$, and by Theorem 2.2.1(iv) we have $\theta_1 \leq k_{\max}$ with equality if and only if $A\mathbf{1} = \theta_1\mathbf{1}$, that is, if and only if Γ is regular of degree θ_1.

Now consider the partition of the vertex set consisting of a single part. By Corollary 2.5.4, we have $\bar{k} \leq \theta_1$ with equality if and only if Γ is regular. $\qquad\square$

For not necessarily connected graphs, we have $\bar{k} \leq \theta_1 \leq k_{\max}$, and $\bar{k} = \theta_1$ if and only if Γ is regular. If $\theta_1 = k_{\max}$, then we only know that Γ has a regular component with this valency but Γ need not be regular itself.

As was noted already in Proposition 3.1.1, the largest eigenvalue of a connected graph decreases strictly when an edge is removed.

3.1.1 Graphs with largest eigenvalue at most 2

As an example of the application of Theorem 2.2.1, we can mention:

Theorem 3.1.3 (SMITH [327], cf. LEMMENS & SEIDEL [250]). *The only connected graphs having largest eigenvalue 2 are the following graphs (the number of vertices is one more than the index given).*

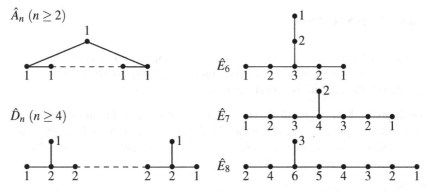

For each graph, the corresponding eigenvector is indicated by the integers at the vertices. Moreover, each connected graph with largest eigenvalue less than 2 is a subgraph of one of the graphs above, i.e., one of the graphs $A_n = P_n$, the path with n vertices $(n \geq 1)$, or

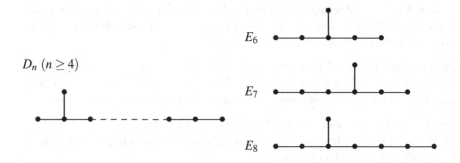

Finally, each connected graph with largest eigenvalue more than 2 *contains one of* $\hat{A}_n$, $\hat{D}_n$, $\hat{E}_6$, $\hat{E}_7$, $\hat{E}_8$ *as a subgraph.*

Proof The vectors indicated are eigenvectors for the eigenvalue 2. Therefore, $\hat{A}_n$, $\hat{D}_n$, and $\hat{E}_m$ ($m = 6, 7, 8$) have largest eigenvalue 2. Any graph containing one of these as an induced proper subgraph has an eigenvalue larger than 2. So, if Γ has largest eigenvalue at most 2 and is not one of $\hat{A}_n$ or $\hat{D}_n$, then Γ is a tree without vertices of degree at least 4 and with at most one vertex of degree 3, and the result easily follows.   □

These graphs occur as the Dynkin diagrams and extended Dynkin diagrams of finite Coxeter groups, cf. [41, 54, 230]. Let us give their eigenvalues:

The eigenvalues of A_n are $2\cos i\pi/(n+1)$ ($i = 1, 2, \ldots, n$).
The eigenvalues of D_n are 0 and $2\cos i\pi/(2n-2)$ ($i = 1, 3, 5, \ldots, 2n-3$).
The eigenvalues of E_6 are $2\cos i\pi/12$ ($i = 1, 4, 5, 7, 8, 11$).
The eigenvalues of E_7 are $2\cos i\pi/18$ ($i = 1, 5, 7, 9, 11, 13, 17$).
The eigenvalues of E_8 are $2\cos i\pi/30$ ($i = 1, 7, 11, 13, 17, 19, 23, 29$).

(Indeed, these eigenvalues are $2\cos(d_i-1)\pi/h$ ($1 \le i \le n$), where h is the Coxeter number and the d_i are the degrees, cf. [54], pp. 84, 308. Note that in all cases the largest eigenvalue is $2\cos\pi/h$.)

The eigenvalues of $\hat{D}_n$ are 2, 0, 0, -2 and $2\cos i\pi/(n-2)$ ($i = 1, \ldots, n-3$).
The eigenvalues of $\hat{E}_6$ are 2, 1, 1, 0, -1, -1, -2.
The eigenvalues of $\hat{E}_7$ are 2, $\sqrt{2}$, 1, 0, 0, -1, $-\sqrt{2}$, -2.
The eigenvalues of $\hat{E}_8$ are 2, τ, 1, τ^{-1}, 0, $-\tau^{-1}$, -1, $-\tau$, -2.

Remark It is possible to go a little bit further and find all graphs with largest eigenvalue at most $\sqrt{2+\sqrt{5}} \approx 2.05817$, cf. BROUWER & NEUMAIER [65]. For the graphs with largest eigenvalue at most $\frac{3}{2}\sqrt{2} \approx 2.12132$, see WOO & NEUMAIER [353] and CIOABĂ et al. [98].

3.1.2 Subdividing an edge

Let Γ be a graph on n vertices, and consider the graph Γ' on $n+1$ vertices obtained from Γ by *subdividing* an edge e (that is, by replacing the edge $e = xy$ by the two edges xz and zy where z is a new vertex). The result below relates the largest eigenvalue of Γ and Γ'.

We say that e lies on an *endpath* if $\Gamma \setminus e$ (the graph on n vertices obtained by removing the edge e from Γ) is disconnected and one of its connected components is a path.

Proposition 3.1.4 (HOFFMAN & SMITH [221]) *Let Γ be a connected graph, and let the graph Γ' be obtained from Γ by subdividing an edge e. Let Γ and Γ' have largest eigenvalues θ and θ', respectively. Then, if e lies on an endpath, we have $\theta' > \theta$, and otherwise $\theta' \le \theta$ with equality only when both equal 2.*

Proof If e lies on an endpath, then Γ is obtained from Γ' by removing a leaf vertex, and $\theta < \theta'$ follows by Proposition 3.1.1. Suppose e is not on an endpath. By Theorem 3.1.3, $\theta \geq 2$. Let A and A' be the adjacency matrices of Γ and Γ', so that $Au = \theta u$ for some vector $u > 0$. We use Theorem 2.2.1 (iv) and conclude $\theta' \leq \theta$ from the existence of a nonzero vector v with $v \geq 0$ and $A'v \leq \theta v$. Such a vector v can be constructed as follows. If z is the new point on the edge $e = xy$, then we can take $v_p = u_p$ for $p \neq z$, and $v_p = \min(u_x, u_y)$, provided that $\theta v_p \geq u_x + u_y$. Suppose not. Without loss of generality, assume $u_x \leq u_y$, so $\theta u_x < u_x + u_y$ and hence $u_x < u_y$. We have $0 \leq \sum_{p \sim x, p \neq y} u_p = \theta u_x - u_y < u_x$. If x has degree 2 in Γ, say $x \sim p, y$, then replace $e = xy$ by $e = px$ to decrease the values of u on the endpoints of e—this does not change Γ'. If x has degree $m > 2$, then define v by $v_x = \theta u_x - u_y$, $v_z = u_x$, and $v_p = u_p$ for $p \neq x, z$. We have to check that $\theta v_x \geq v_x + u_x$, but this follows from $\theta v_x = \theta \sum_{p \sim x, p \neq y} u_p \geq (m-1)u_x \geq 2u_x > v_x + u_x$.   □

3.1.3 The Kelmans operation

As we saw, adding edges causes the largest eigenvalue to increase. The operation described below (due to KELMANS [241]) only moves edges, but also increases θ_1.

Given a graph Γ and two specified vertices u, v, construct a new graph Γ' by replacing the edge vx by a new edge ux for all x such that $v \sim x \nsim u$. The new graph Γ' obtained in this way has the same number of vertices and edges as the old graph, and all vertices different from u, v retain their valency. The vertices u, v are adjacent in Γ' if and only if they are adjacent in Γ. An isomorphic graph is obtained if the roles of u and v are interchanged: if $N(u)$ and $N(v)$ are the sets of neighbors of u, v distinct from u and v, then in the resulting graph the corresponding sets are $N(u) \cup N(v)$ and $N(u) \cap N(v)$.

If $\overline{\Gamma}$ denotes the complementary graph of Γ, then also $\overline{\Gamma'}$ is obtained by a Kelmans operation from $\overline{\Gamma}$.

Proposition 3.1.5 (CSIKVÁRI [110]) *Let Γ be a graph, and let Γ' be obtained from Γ by a Kelmans operation. Then $\theta_1(\Gamma) \leq \theta_1(\Gamma')$. (And hence also $\theta_1(\overline{\Gamma}) \leq \theta_1(\overline{\Gamma'})$.)*

Proof Let A and A' be the adjacency matrices of Γ and Γ', and let $Ax = \theta_1 x$, where $x \geq 0$, $x^\top x = 1$. Without loss of generality, let $x_u \geq x_v$. Then $\theta_1(\Gamma') \geq x^\top A'x = x^\top Ax + 2(x_u - x_v)\sum_{w \in N(v) \setminus N(u)} x_w \geq \theta_1(\Gamma)$.   □

Csikvári continues and uses this to show that $\theta_1(\Gamma) + \theta_1(\overline{\Gamma}) \leq \frac{1}{2}(1 + \sqrt{3})n$.

Earlier, BRUALDI & HOFFMAN [69] had observed that a graph with maximal spectral radius ρ among the graphs with a given number of vertices and edges has a vertex ordering such that if $x \sim y$ and $z \leq x$, $w \leq y$, $z \neq w$, then $z \sim w$. ROWLINSON [303] calls the adjacency matrices of these graphs (ordered this way) *stepwise* and proves that the maximal value of ρ among the graphs on n vertices and e edges is obtained by taking $K_m + (n-m)K_1$, where m is minimal such that $\binom{m}{2} \geq e$, and removing $\binom{m}{2} - e$ edges on a single vertex.

It follows from the above proposition that a graph with maximal $\theta_1(\Gamma) + \theta_1(\overline{\Gamma})$ has a stepwise matrix. It is conjectured that in fact $\theta_1(\Gamma) + \theta_1(\overline{\Gamma}) < \frac{4}{3}n - 1$.

3.2 Interlacing

By the Perron-Frobenius theorem, the largest eigenvalue of a connected graph goes down when one removes an edge or a vertex. Interlacing also gives information about what happens with the other eigenvalues.

The pictures for A and L differ. The eigenvalues for the adjacency matrix A show nice interlacing behavior when one removes a vertex but not when an edge is removed (cf. §1.9). The Laplace eigenvalues behave well in both cases. For A, an eigenvalue can go both up or down when an edge is removed. For L, it cannot increase.

Proposition 3.2.1 *(i) Let Γ be a graph and Δ an induced subgraph. Then the eigenvalues of Δ interlace those of Γ.*

(ii) Let Γ be a graph and let Δ be a subgraph, not necessarily induced, on m vertices. Then the i-th-largest Laplace eigenvalue of Δ is not larger than the i-th-largest Laplace eigenvalue of Γ ($1 \leq i \leq m$), and the i-th-largest signless Laplace eigenvalue of Δ is not larger than the i-th-largest signless Laplace eigenvalue of Γ ($1 \leq i \leq m$).

Proof Part (i) is immediate from Corollary 2.5.2. For part (ii), recall that we have $L = NN^\top$ when N is the directed point-edge incidence matrix obtained by orienting the edges of Γ arbitrarily and that $NN^\top$ and $N^\top N$ have the same nonzero eigenvalues. Removing an edge from Γ corresponds to removing a column from N and leads to a principal submatrix of $N^\top N$, and interlacing holds. Removing an isolated vertex from Γ corresponds to removing a Laplace eigenvalue 0. The same proof applies to the signless Laplace matrix. □

3.3 Regular graphs

It is possible to see from the spectrum whether a graph is regular.

Proposition 3.3.1 *Let Γ be a graph with eigenvalues $k = \theta_1 \geq \theta_2 \geq \ldots \geq \theta_n$. The following are equivalent:*

(i) *Γ is regular (of degree k).*
(ii) *$AJ = kJ$.*
(iii) *$\sum \theta_i^2 = kn$.*

Proof We have seen that (i) and (ii) are equivalent. Also, if Γ is regular of degree k, then $\sum \theta_i^2 = \text{tr} A^2 = kn$. Conversely, if (iii) holds, then $\bar{k} = n^{-1} \sum \theta_i^2 = \theta_1$ and, by Proposition 3.1.2, Γ is regular. □

As we saw above in §1.3.7, it is also possible to see from the spectrum whether a graph is regular and connected. However, for nonregular graphs it is not possible to see from the spectrum whether they are connected.

The following very useful characterization of regular connected graphs was given by HOFFMAN [218].

Proposition 3.3.2 *The graph Γ is regular and connected if and only if there exists a polynomial p such that $J = p(A)$.*

Proof If $J = p(A)$, then J commutes with A and hence Γ is regular (and clearly also connected). Conversely, let Γ be connected and regular. Choose a basis such that the commuting matrices A and J become diagonal. Then A and J become $\mathrm{diag}(k, \theta_2, \ldots, \theta_n)$ and $\mathrm{diag}(n, 0, \ldots, 0)$. Hence, if we put $f(x) = \prod_{i=2}^{n}(x - \theta_i)$, then $J = nf(A)/f(k)$, and $p(x) = nf(x)/f(k)$ satisfies the requirements. □

3.4 Bipartite graphs

Among the connected graphs Γ, those with imprimitive A are precisely the bipartite graphs (and for these, A has period 2). Consequently, from Theorem 2.2.1(iii) we find the following.

Proposition 3.4.1 *(i) A graph Γ is bipartite if and only if, for each eigenvalue θ of Γ, also $-\theta$ is an eigenvalue, with the same multiplicity.*

(ii) If Γ is connected with largest eigenvalue θ_1, then Γ is bipartite if and only if $-\theta_1$ is an eigenvalue of Γ.

Proof For connected graphs all is clear from the Perron-Frobenius theorem. That gives (ii) and (by taking unions) the "only if" part of (i). For the "if" part of (i), let θ_1 be the spectral radius of Γ. Then some connected component of Γ has eigenvalues θ_1 and $-\theta_1$ and hence is bipartite. Removing its contribution to the spectrum of Γ, we see by induction on the number of components that all components are bipartite. □

3.5 Cliques and cocliques

A *clique* in a graph is a set of pairwise adjacent vertices. A *coclique* in a graph is a set of pairwise nonadjacent vertices. The *clique number* $\omega(\Gamma)$ is the size of the largest clique in Γ. The *independence number* $\alpha(\Gamma)$ is the size of the largest coclique in Γ.

Let Γ be a graph on n vertices (undirected, simple, and loopless) having an adjacency matrix A with eigenvalues $\theta_1 \geq \ldots \geq \theta_n$. Both Corollaries 2.5.2 and 2.5.4 lead to a bound for $\alpha(\Gamma)$.

Theorem 3.5.1 $\alpha(\Gamma) \leq n - n_- = |\{i \mid \theta_i \geq 0\}|$ and $\alpha(\Gamma) \leq n - n_+ = |\{i \mid \theta_i \leq 0\}|$.

Proof A has a principal submatrix $B = 0$ of size $\alpha = \alpha(\Gamma)$. Corollary 2.5.2 gives $\theta_\alpha \geq \eta_\alpha = 0$ and $\theta_{n-\alpha-1} \leq \eta_1 = 0$. □

For example, the Higman-Sims graph (see §9.1.7) has spectrum $22^1 \; 2^{77} \; (-8)^{22}$. Each point neighborhood is a coclique of size 22, and equality holds.

Theorem 3.5.2 If Γ is regular of nonzero degree k, then

$$\alpha(\Gamma) \leq n \frac{-\theta_n}{k - \theta_n},$$

and if a coclique C meets this bound, then every vertex not in C is adjacent to precisely $-\theta_n$ vertices of C.

Proof We apply Corollary 2.5.4. The coclique gives rise to a partition of A with quotient matrix

$$B = \begin{bmatrix} 0 & k \\ \frac{k\alpha}{n-\alpha} & k - \frac{k\alpha}{n-\alpha} \end{bmatrix},$$

where $\alpha = \alpha(\Gamma)$. B has eigenvalues $\eta_1 = k = \theta_1$ (the row sum) and $\eta_2 = -k\alpha/(n-\alpha)$ (since trace $B = k + \eta_2$) and so $\theta_n \leq \eta_2$ gives the required inequality. If equality holds, then $\eta_2 = \theta_n$, and since $\eta_1 = \theta_1$, the interlacing is tight and hence the partition is equitable. □

For example, the Petersen graph has spectrum $3^1 \; 1^5 \; (-2)^4$ and its independence number is 4, so equality holds in both bounds.

The first bound is due to CVETKOVIĆ [112]. The second bound is an unpublished result of Hoffman known as the *Hoffman bound* or *ratio bound*. The Hoffman bound was generalized to the nonregular case in [197] as follows.

Proposition 3.5.3 Let Γ have minimum vertex degree δ. Then

$$\alpha(\Gamma) \leq n \frac{-\theta_1 \theta_n}{\delta^2 - \theta_1 \theta_n}.$$ □

3.5.1 Using weighted adjacency matrices

Let us call a real symmetric matrix B a *weighted adjacency matrix* of a graph Γ when B has rows and columns indexed by the vertex set of Γ, has zero diagonal, and satisfies $B_{xy} = 0$ whenever $x \not\sim y$.

The proof of Theorem 3.5.1 applies to B instead of A, and we get

Theorem 3.5.4 $\alpha(\Gamma) \leq n - n_-(B)$ and $\alpha(\Gamma) \leq n - n_+(B)$.

Similarly, the proof of Theorem 3.5.2 remains valid for weighted adjacency matrices B with constant row sums.

Theorem 3.5.5 *Let B be a weighted adjacency matrix of Γ with constant row sums b and smallest eigenvalue s. Then $\alpha(\Gamma) \leq n(-s)/(b-s)$.*

A version that does not mention constant row sums:

Theorem 3.5.5a ([352]) *Let B be a weighted adjacency matrix of Γ such that $I + B - c^{-1}J$ is positive semidefinite. Then $\alpha(\Gamma) \leq c$.*

Proof   We have $0 \leq \chi^{\top}(I + B - c^{-1}J)\chi = |C| - c^{-1}|C|^2$ for the characteristic vector χ of a coclique C. $\square$

3.6 Chromatic number

A *proper vertex coloring* of a graph is an assignment of colors to the vertices so that adjacent vertices get different colors (in other words, a partition of the vertex set into cocliques). The *chromatic number* $\chi(\Gamma)$ is the minimum number of colors of a proper vertex coloring of Γ.

Proposition 3.6.1 (WILF [350]) *Let Γ be connected, with largest eigenvalue θ_1. Then $\chi(\Gamma) \leq 1 + \theta_1$, with equality if and only if Γ is complete or an odd cycle.*

Proof   Put $m = \chi(\Gamma)$. Since Γ cannot be colored with $m-1$ colors, whereas coloring vertices of degree less than $m-1$ is easy, there must be an induced subgraph Δ of Γ with minimum degree at least $m-1$. Now $\theta_1 \geq \theta_1(\Delta) \geq d_{\min}(\Delta) \geq m-1 = \chi(\Gamma) - 1$. If equality holds, then by the Perron-Frobenius theorem $\Gamma = \Delta$ and Δ is regular of degree $m-1$ (by Proposition 3.1.2), and the conclusion follows by Brooks's theorem. $\square$

Since each coclique (color class) has size at most $\alpha(\Gamma)$, we have $\chi(\Gamma) \geq n/\alpha(\Gamma)$ for a graph Γ with n vertices. Thus upper bounds for $\alpha(\Gamma)$ give lower bounds for $\chi(\Gamma)$. For instance, if Γ is regular of degree $k = \theta_1$, then Theorem 3.5.2 implies that $\chi(\Gamma) \geq 1 - \frac{\theta_1}{\theta_n}$. This bound remains valid, however, for nonregular graphs.

Theorem 3.6.2 (HOFFMAN [219]) *If Γ is not edgeless then $\chi(\Gamma) \geq 1 - \dfrac{\theta_1}{\theta_n}$.*

Proof   Put $m = \chi(\Gamma)$. Since Γ is not edgeless, $\theta_n < 0$. Now, by part (i) of the following proposition, $\theta_1 + (m-1)\theta_n \leq \theta_1 + \theta_{n-m+2} + \ldots + \theta_n \leq 0$. $\square$

Proposition 3.6.3 *Put $m = \chi(\Gamma)$. Then*

(i) $\theta_1 + \theta_{n-m+2} + \ldots + \theta_n \leq 0$.
(ii) *If $n > m$, then* $\theta_2 + \ldots + \theta_m + \theta_{n-m+1} \geq 0$.
(iii) *If $n > tm$, then* $\theta_{t+1} + \ldots + \theta_{t+m-1} + \theta_{n-t(m-1)} \geq 0$.

Proof   Let A have orthonormal eigenvectors u_j, so that $Au_j = \theta_j u_j$.

(i) Let $\{X_1, \ldots, X_m\}$ be a partition of Γ into m cocliques, where $m = \chi(\Gamma)$. Let x_j be the pointwise product of u_1 with the characteristic vector of X_j, so that $\sum x_j = u_1$. Now apply Corollary 2.5.3 to the vectors x_j after deleting those that are zero. The matrix C defined there satisfies $C1 = \theta_1 1$, has zero diagonal, and has eigenvalues η_j interlacing those of A. Hence

$$0 = \mathrm{tr}(C) = \eta_1 + \ldots + \eta_m \geq \theta_1 + \theta_{n-m+2} + \ldots + \theta_n.$$

(ii) Put $A' = A - (\theta_1 - \theta_n)u_1 u_1^\top$. Then A' has the same eigenvectors u_j as A, but with eigenvalues $\theta_n, \theta_2, \ldots, \theta_n$. Pick a nonzero vector y in

$$\langle u_{n-m+1}, \ldots, u_n \rangle \cap \langle x_1, \ldots, x_m \rangle^\perp.$$

The two spaces have a nontrivial intersection since the dimensions add up to n and u_1 is orthogonal to both. Let y_j be the pointwise product of y with the characteristic vector of X_j, so that $\sum y_j = y$ and $y_j^\top A' y_j = 0$. Now apply Corollary 2.5.3 to the matrix A' and the vectors y_j after deleting those that are zero. The matrix C defined there has zero diagonal, and smallest eigenvalue smaller than the Rayleigh quotient $\frac{y^\top A y}{y^\top y}$, which by choice of y is at most θ_{n-m+1}. We find

$$0 = \mathrm{tr}(C) = \eta_1 + \ldots + \eta_m \leq \theta_2 + \theta_3 + \ldots + \theta_m + \theta_{n-m+1}.$$

(iii) The proof is as under (ii), but this time we move t (instead of just one) eigenvalues away (by subtracting multiples of $u_j u_j^\top$ for $1 \leq j \leq t$). The vector y must be chosen orthogonal to tm vectors, which can be done inside the $(tm - t + 1)$-space $\langle u_{n-tm+t}, \ldots, u_n \rangle$, assuming that this space is already orthogonal to $u_1, \ldots, u_t$, i.e., assuming that $n > tm$.   □

The above proof of Theorem 3.6.2 using (i) above appeared in [196].

A coloring that meets the bound of Theorem 3.6.2 is called a *Hoffman coloring*. For regular graphs, the color classes of a Hoffman coloring are cocliques that meet Hoffman's coclique bound. So in this case all the color classes have equal size and the corresponding matrix partition is equitable.

In [197], more inequalities of the kind above are given. But the ones mentioned here, especially (i) and (ii), are by far the most useful.

Example The complete multipartite graph $K_{m \times a}$ has chromatic number m and spectrum $(am - a)^1 \, 0^{m(a-1)} \, (-a)^{m-1}$. It has equality in Hoffman's inequality (and hence in (i)) and also in (ii).

Example The graph obtained by removing an edge from K_n has chromatic number $n - 1$ and spectrum $\frac{1}{2}(n - 3 + \sqrt{D})$, 0, $(-1)^{n-3}$, $\frac{1}{2}(n - 3 - \sqrt{D})$, where $D = (n + 1)^2 - 8$, with equality in (i).

Example Consider the generalized octagon of order $(2, 4)$ on 1755 vertices. It has spectrum $10^1 \, 5^{351} \, 1^{650} \, (-3)^{675} \, (-5)^{78}$. It is not 3-chromatic, as one sees by removing the 352 largest eigenvalues, i.e., by applying (iii) with $t = 352$.

The inequality (ii) can be made more explicit if the smallest eigenvalue θ_n has a large multiplicity.

Corollary 3.6.4 *If the eigenvalue θ_n has multiplicity g and $\theta_2 > 0$, then*

$$\chi(\Gamma) \geq \min(1+g, 1 - \frac{\theta_n}{\theta_2}).$$

Proof If $m := \chi(\Gamma) \leq g$, then $\theta_n = \theta_{n-m+1}$, so that $(m-1)\theta_2 + \theta_n \geq 0$.   □

A similar, more explicit form for inequality (iii) follows in the same way.

3.6.1 Using weighted adjacency matrices

If Γ has an m-coloring, then $\Gamma \square K_m$ has an independent set of size n, the number of vertices of Γ. This means that one can use bounds on the size of an independent set to obtain bounds on the chromatic number.

Example Consider the generalized octagon of order $(2,4)$ again. Call it Γ, and call its adjacency matrix A. Now consider the weighted adjacency matrix B of $K_3 \square \Gamma$, where the K_3 is weighted with some number r, where $1 < r < \frac{3}{2}$. For each eigenvalue θ of A, we find eigenvalues $\theta + 2r$ (once) and $\theta - r$ (twice) as eigenvalues of B. Applying Theorem 3.5.4, we see that $\alpha(K_3 \square \Gamma) \leq 3(1+351) + 650 = 1706$, while Γ has 1755 vertices, so Γ is not 3-chromatic.

3.6.2 Rank and chromatic number

The easiest way for A to have low rank is when it has many repeated rows. But then Γ contains large cocliques. People have conjectured that it might be true that $\chi(\Gamma) \leq \mathrm{rk}A$ when $A \neq 0$. A counterexample was given by ALON & SEYMOUR [9], who observed that the complement of the folded 7-cube (on 64 vertices) has chromatic number $\chi = 32$ (indeed, $\alpha = 2$) and rank 29 (indeed, the spectrum of the folded 7-cube is $7^1 \, 3^{21} \, (-1)^{35} \, (-5)^7$).

3.7 Shannon capacity

SHANNON [323] studied the capacity C_0 of the zero-error channel defined by a graph Γ, where a transmission consists of sending a vertex of Γ, and two transmissions can be confused when the corresponding vertices are joined by an edge.

The maximum size of a set of mutually inconfusable messages of length 1 is $\alpha(\Gamma)$, so that one can transmit $\log \alpha(\Gamma)$ bits by sending one vertex. The maximum size of a set of mutually inconfusable messages of length ℓ is the independence number $\alpha(\Gamma^\ell)$, where Γ^ℓ denotes (in this section) the strong product $\Gamma^{\boxtimes \ell}$ of ℓ copies of Γ, that is, the graph on sequences of ℓ vertices from Γ, where two sequences are adjacent when on each coordinate position their elements are equal or adjacent. One can transmit $\log \alpha(\Gamma^\ell)$ bits by sending a sequence of ℓ vertices, and it follows that the channel capacity is $C_0 = \log c(\Gamma)$, where $c(\Gamma) = \sup_{\ell \to \infty} \alpha(\Gamma^\ell)^{1/\ell}$. This value $c(\Gamma)$ is called the *Shannon capacity* of Γ.

For example, for the pentagon, we find $c(\Gamma) \geq \sqrt{5}$, as shown by the 5-coclique 00, 12, 24, 31, 43 in $C_5 \boxtimes C_5$.

Computing $c(\Gamma)$ is a difficult unsolved problem, even for graphs as simple as C_7, the 7-cycle.

Clearly, $\alpha(\Gamma) \leq c(\Gamma) \leq \chi(\overline{\Gamma})$. (Indeed, if $m = \chi(\overline{\Gamma})$, then Γ can be covered by m cliques, and Γ^ℓ can be covered by m^ℓ cliques, and $\alpha(\Gamma^\ell) \leq m^\ell$.) In a few cases this suffices to determine $c(\Gamma)$.

One can sharpen the upper bound to the fractional clique covering number. For example, the vertices of C_5 can be doubly covered by five cliques, so the vertices of C_5^ℓ can be covered 2^ℓ times by 5^ℓ cliques, and $\alpha(C_5^\ell) \leq (5/2)^\ell$, so $c(C_5) \leq 5/2$.

If A is the adjacency matrix of Γ, then $\otimes^\ell(A+I) - I$ is the adjacency matrix of Γ^ℓ.

The Hoffman upper bound for the size of cocliques is also an upper bound for $c(\Gamma)$ (and therefore, when the Hoffman bound holds with equality, also the Shannon capacity is determined).

Proposition 3.7.1 (LOVÁSZ [261]) *Let Γ be regular of valency k. Then*

$$c(\Gamma) \leq n(-\theta_n)/(k - \theta_n).$$

Proof Use the weighted Hoffman bound (Theorem 3.5.5). If $B = A - \theta_n I$, then $\otimes^\ell B - (-\theta_n)^\ell I$ has constant row sums $(k - \theta_n)^\ell - (-\theta_n)^\ell$ and smallest eigenvalue $-(-\theta_n)^\ell$, so that $\alpha(\Gamma^\ell) \leq (n(-\theta_n)/(k - \theta_n))^\ell$.   □

Using $n = 5$, $k = 2$, $\theta_n = (-1 - \sqrt{5})/2$ we find for the pentagon $c(\Gamma) \leq \sqrt{5}$. Hence equality holds.

HAEMERS [194, 195] observed that if B is a matrix indexed by the vertices of Γ and $B_{xx} \neq 0$ for all x, and $B_{xy} = 0$ whenever $x \not\sim y$, then $c(\Gamma) \leq \mathrm{rk}\, B$. Indeed, for such a matrix $\alpha(\Gamma) \leq \mathrm{rk}\, B$ since an independent set determines a submatrix that is zero outside a nonzero diagonal. Now $\otimes^\ell B$ is a suitable matrix for Γ^l, and $\mathrm{rk} \otimes^\ell B = (\mathrm{rk}\, B)^\ell$. The rank here may be taken over any field.

Example The collinearity graph Γ of the generalized quadrangle $GQ(2,4)$ (the complement of the Schläfli graph, cf. §9.6) on 27 vertices has spectrum $10^1\ 1^{20}$ $(-5)^6$. Taking $B = A - I$ shows that $c(\Gamma) \leq 7$. (And $c(\Gamma) \geq \alpha(\Gamma) = 6$.) The complement $\overline{\Gamma}$ has $\alpha(\overline{\Gamma}) = 3$, but this is also the Hoffman bound, so $c(\overline{\Gamma}) = 3$.

ALON [5] proves that $c(\Gamma + \overline{\Gamma}) \geq 2\sqrt{n}$ for all Γ. Combined with the above example, this shows that the Shannon capacity of the disjoint sum of two graphs can be larger than the sum of their Shannon capacities. More detail about the Lovász and Haemers bounds for $c(\Gamma)$ is given in the following sections.

3.7.1 Lovász's ϑ-function

Consider a simple graph Γ of order n, and let $\mathcal{M}_\Gamma$ be the set of real symmetric matrices M indexed by $V\Gamma$ that satisfy $M_{uv} = 1$ when $u = v$ or $u \not\sim v$. The *Lovász parameter* $\vartheta(\Gamma)$ is defined by

$$\vartheta(\Gamma) = \inf_{M \in \mathcal{M}_\Gamma} \theta_1(M),$$

where $\theta_1(M)$ denotes the largest eigenvalue of M. The results below are all due to LOVÁSZ [261].

Lemma 3.7.2 $\vartheta(\Gamma \boxtimes \Delta) \leq \vartheta(\Gamma)\vartheta(\Delta)$. $\qquad\qquad\qquad\qquad\qquad\qquad\qquad\qquad\square$

Proof If $M_\Gamma \in \mathcal{M}_\Gamma$ and $M_\Delta \in \mathcal{M}_\Delta$, then $M_\Gamma \otimes M_\Delta \in \mathcal{M}_{\Gamma \boxtimes \Delta}$. Moreover, $\theta_1(M_\Gamma \otimes M_\Delta) = \theta_1(M_\Gamma)\theta_1(M_\Delta)$. $\qquad\qquad\qquad\qquad\qquad\qquad\qquad\qquad\square$

LOVÁSZ [261] shows that equality holds here.

Theorem 3.7.3 *The Shannon capacity $c(\Gamma)$ satisfies*

$$\alpha(\Gamma) \leq c(\Gamma) \leq \vartheta(\Gamma).$$

Proof Let $M \in \mathcal{M}$. A coclique of size $\alpha(\Gamma)$ corresponds to a principal submatrix J of order $\alpha(\Gamma)$ in M. Interlacing gives $\alpha(\Gamma) = \theta_1(J) \leq \theta_1(M)$, which proves $\alpha(\Gamma) \leq \vartheta(\Gamma)$. By Lemma 3.7.2, we now have $\alpha(\Gamma^\ell) \leq \vartheta(\Gamma^\ell) \leq (\vartheta(\Gamma))^\ell$, and hence $c(\Gamma) \leq \vartheta(\Gamma)$. $\qquad\qquad\qquad\qquad\qquad\qquad\qquad\qquad\square$

The upper bound $\chi(\overline{\Gamma})$ for $c(\Gamma)$ is also an upper bound for $\vartheta(\Gamma)$:

Theorem 3.7.4 ("Sandwich") $\alpha(\Gamma) \leq \vartheta(\Gamma) \leq \chi(\overline{\Gamma})$.

Proof To prove the second inequality, consider a covering of Γ with $\overline{\chi}$ pairwise disjoint cliques. Define $M_{uv} = 1 - \overline{\chi}$ if u and v are distinct vertices in the same clique of the covering, and $M_{uv} = 1$ otherwise. Then $M \in \mathcal{M}_\Gamma$, and $\theta_1(M) = \overline{\chi}$. Indeed, the clique covering gives an equitable partition of M (see §2.3), and the eigenvectors of M orthogonal to the characteristic vectors of the partition have eigenvalue $\overline{\chi}$, while the other eigenvalues are those of the quotient matrix $B = J\Lambda - \overline{\chi}\Lambda + \overline{\chi}I$, where Λ is the diagonal matrix whose diagonal entries are the sizes of the cliques of the covering. Now $\theta_1(B) \leq \overline{\chi}$ because B is similar to $\Lambda^{\frac{1}{2}}J\Lambda^{\frac{1}{2}} - \overline{\chi}\Lambda + \overline{\chi}I$, and $\overline{\chi}\Lambda - \Lambda^{\frac{1}{2}}J\Lambda^{\frac{1}{2}}$ is positive semidefinite since $\overline{\chi}I - J$ is. $\qquad\qquad\qquad\qquad\qquad\qquad\square$

This is an important result: while computing the independence number and the chromatic number of a graph are NP-complete, $\vartheta(\Gamma)$ can be computed to any desired precision in polynomial time (see [184]). In particular, in the cases where $\alpha(\Gamma) = \chi(\overline{\Gamma})$, this value can be found efficiently. For perfect graphs (graphs such that $\alpha(\Delta) = \chi(\overline{\Delta})$ for every induced subgraph Δ) this yields an efficient procedure for actually finding a maximal coclique.

The Hoffman bound for the size of a coclique in a regular graph is also an upper bound for $\vartheta(\Gamma)$ (and therefore, when the Hoffman bound holds with equality, $\vartheta(\Gamma)$ is determined).

Proposition 3.7.5 *Suppose Γ is regular of valency k, with smallest eigenvalue θ_n. Then*

$$\vartheta(\Gamma) \leq \frac{-n\theta_n}{k - \theta_n}.$$

Proof Let A be the adjacency matrix of Γ, and define $M = J - \frac{n}{k-\theta_n}A$. Then $M \in \mathcal{M}_\Gamma$, and $\theta_1(M) = -n\theta_n/(k - \theta_n)$. $\qquad\square$

For example, for the Petersen graph, we have $\alpha(\Gamma) = \vartheta(\Gamma) = 4$. For the pentagon, $c(\Gamma) = \vartheta(\Gamma) = \sqrt{5}$. LOVÁSZ [261] proved that equality holds in the above formula if Γ has an edge-transitive automorphism group. Equality also holds if Γ is strongly regular (see [194]).

Proposition 3.7.6 *One has $\vartheta(\Gamma)\vartheta(\overline{\Gamma}) \geq n$ for a graph Γ of order n. Equality holds if Γ is vertex transitive.* $\qquad\square$

LOVÁSZ [261] gives several equivalent expressions for $\vartheta(\Gamma)$. The following alternative definition uses the set $\mathcal{N}_\Gamma$ of real symmetric matrices N indexed by $V\Gamma$, with the property that N is positive semidefinite, $\operatorname{tr} N = 1$, and $N_{uv} = 0$ if $u \sim v$:

$$\vartheta(\Gamma) = \sup_{N \in \mathcal{N}} \operatorname{tr} NJ.$$

(Note that $\operatorname{tr} NJ$ equals the sum of the entries of N.) Equivalence of the two definitions follows from duality in semidefinite programming. It also follows that the infimum and supremum in the two expressions for $\vartheta(\Gamma)$ are actually a minimum and a maximum.

3.7.2 The Haemers bound on the Shannon capacity

For a graph Γ, let the integer $\eta(\Gamma)$ (the *Haemers invariant*) be the smallest rank of any matrix M (over any field), indexed by $V\Gamma$, which satisfies $M_{uu} \neq 0$ for all u, and $M_{uv} = 0$ if $u \not\sim v$ (see [195]). The following propositions show that this rank parameter has some similarity with $\vartheta(\Gamma)$.

Lemma 3.7.7 $\eta(\Gamma \boxtimes \Delta) \leq \eta(\Gamma)\eta(\Delta)$.

Proof Suppose M_Γ and M_Δ are admissible for Γ and Δ with minimum rank. Then $M_\Gamma \otimes M_\Delta$ is admissible for $\Gamma \boxtimes \Delta$, and $\operatorname{rk}(M_\Gamma \otimes M_\Delta) = \eta(\Gamma)\eta(\Delta)$. $\qquad\square$

Theorem 3.7.8 *The Shannon capacity $c(\Gamma)$ satisfies*

$$\alpha(\Gamma) \leq c(\Gamma) \leq \eta(\Gamma).$$

Proof A coclique in Γ corresponds to a nonsingular diagonal matrix in M. Therefore $\alpha(\Gamma) \leq \operatorname{rk} M$ for every admissible M, so that $\alpha(\Gamma) \leq \eta(\Gamma)$. By Lemma 3.7.7 we have $\alpha(\Gamma^\ell) \leq \eta(\Gamma^\ell) \leq (\eta(\Gamma))^\ell$, and hence $c(\Gamma) \leq \eta(\Gamma)$. $\qquad\square$

Proposition 3.7.9 $\alpha(\Gamma) \leq \eta(\Gamma) \leq \chi(\overline{\Gamma})$.

Proof To prove the second inequality, fix a cover with $\chi(\overline{\Gamma})$ cliques, and take $M_{uv} = 0$ if u and v are in different cliques of the clique cover and $M_{uv} = 1$ otherwise. Then $\mathrm{rk}\,M = \chi(\overline{\Gamma})$. □

In spite of the above similarity, $\eta(\Gamma)$ and $\vartheta(\Gamma)$ are very different. To begin with, $\vartheta(\Gamma)$ need not be an integer, whereas $\eta(\Gamma)$ always is. The computation of $\eta(\Gamma)$ is probably NP-hard ([292]). The two are not related by an inequality: for some graphs $\vartheta(\Gamma) < \eta(\Gamma)$ (for example, $\vartheta(C_5) = \sqrt{5} < 3 = \eta(C_5)$), whereas for other graphs $\eta(\Gamma) < \vartheta(\Gamma)$ (for example, for the collinearity graph Γ of the generalized quadrangle $GQ(2,4)$, we have $\eta(\Gamma) \leq 7 < 9 = \vartheta(\Gamma)$).

Example Consider the graph Γ on the triples from an m-set Σ, adjacent when they meet in precisely one point. Let N be the $m \times \binom{m}{3}$ incidence matrix of symbols and triples. Then $M = N^{\top}N$ is admissible over $\mathbb{F}_2$, so $\eta(\Gamma) \leq \mathrm{rk}_2 M \leq m$. If $4|m$, then consider a partition of Σ into $\frac{1}{4}m$ 4-sets. The triples contained in one of the parts form a coclique of size m, so $\alpha(\Gamma) = c(\Gamma) = \eta(\Gamma) = m$ in this case. Here $\theta(\Gamma) = \frac{m(m-2)(2m-11)}{3(3m-14)}$ (for $m \geq 7$), so that $\theta(\Gamma) > m \geq \eta(\Gamma)$ for $m > 8$. Also $\eta(\overline{\Gamma}) \leq m$, so $\eta(\Gamma)\eta(\overline{\Gamma}) \leq m^2 < \binom{m}{3} = n$ for $m > 8$ ([194]).

3.8 Classification of integral cubic graphs

A graph is called *integral* when all of its eigenvalues are integral. As an application of Proposition 3.3.2, let us classify the cubic graphs (graphs that are regular of valency 3) with integral spectrum. The result is due to BUSSEMAKER & CVETKOVIĆ [75]. See also SCHWENK [309]. There are 13 examples, of which 8 are bipartite.

Case	v	Spectrum	Description
(i)	6	$\pm 3, 0^4$	$K_{3,3}$
(ii)	8	$\pm 3, (\pm 1)^3$	2^3
(iii)	10	$\pm 3, \pm 2, (\pm 1)^2, 0^2$	$K_{2,3}^* \otimes K_2$
(iv)	12	$\pm 3, (\pm 2)^2, \pm 1, 0^4$	$C_6 \,\square\, K_2$
(v)	20	$\pm 3, (\pm 2)^4, (\pm 1)^5$	$\Pi \otimes K_2$
(vi)	20	$\pm 3, (\pm 2)^4, (\pm 1)^5$	$T^* \otimes K_2$
(vii)	24	$\pm 3, (\pm 2)^6, (\pm 1)^3, 0^4$	$\Sigma \otimes K_2$
(viii)	30	$\pm 3, (\pm 2)^9, 0^{10}$	$GQ(2,2)$
(ix)	4	$3, (-1)^3$	K_4
(x)	6	$3, 1, 0^2, (-2)^2$	$K_3 \,\square\, K_2$
(xi)	10	$3, 1^5, (-2)^4$	Π
(xii)	10	$3, 2, 1^3, (-1)^2, (-2)^3$	$(\Pi \otimes K_2)/\sigma$
(xiii)	12	$3, 2^3, 0^2, (-1)^3, (-2)^3$	Σ

3.8.1 A quotient of the hexagonal grid

Let us describe a graph that comes up in the classification. Take a tetrahedron and cut off each corner. Our graph Σ is the 1-skeleton of the resulting polytope, or, equivalently, the result of replacing each vertex of K_4 by a triangle (a $Y - \Delta$ operation). It can also be described as the line graph of the graph obtained from K_4 by subdividing each edge. The bipartite double $\Sigma \otimes K_2$ of Σ is more beautiful (for example, its group is a factor of 6 larger than that of Σ), and can be described as the quotient $\Lambda/6\Lambda$ of the hexagonal grid $\Lambda = \langle a + b\omega \mid a, b \in \mathbb{Z}, \ a + b = 0, 1 \pmod 3 \rangle$ in the complex plane, where $\omega^2 + \omega + 1 = 0$. Now Σ is found e.g. as $\Lambda/\langle 3a + 6b\omega \mid a, b \in \mathbb{Z} \rangle$.

3.8.2 Cubic graphs with loops

For a graph Γ where all vertices have degree 2 or 3, let Γ^* be the cubic graph (with loops) obtained by adding a loop at each vertex of degree 2. Note that the sum of the eigenvalues of Γ^*, the trace of its adjacency matrix, is the number of loops.

The graph $K_{2,3}^*$ has spectrum 3, 1, 1, 0, -2.

Let T be the graph on the singletons and pairs in a 4-set, where adjacency is inclusion. Then T^* has spectrum $3^1, 2^3, 1^2, (-1)^3, (-2)^1$.

3.8.3 The classification

Let Π be the Petersen graph and Σ, T the graphs described above.

We split the result into two propositions, one for the bipartite case and one for the nonbipartite case.

Proposition 3.8.1 *Let Γ be a connected bipartite cubic graph such that all of its eigenvalues are integral. Then Γ is one of eight possible graphs, namely (i) $K_{3,3}$, (ii) 2^3, (iii) $K_{2,3}^* \otimes K_2$, (iv) $C_6 \square K_2$, (v) the Desargues graph (that is, the bipartite double $\Pi \otimes K_2$ of the Petersen graph Π), (vi) T^* (cospectral with the previous), (vii) the bipartite double of Σ, (viii) the point-line incidence graph of the generalized quadrangle of order 2 (that is, the unique 3-regular bipartite graph with diameter 4 and girth 8, also known as Tutte's 8-cage).*

Proof Let Γ have spectrum $(\pm 3)^1 (\pm 2)^a (\pm 1)^b 0^{2c}$ (with multiplicities written as exponents).

The total number of vertices is $v = 2 + 2a + 2b + 2c$. The total number of edges is $\frac{3}{2}v = \frac{1}{2}\operatorname{tr} A^2 = 9 + 4a + b$ (so that $2b + 3c = 6 + a$). The total number of quadrangles is $q = 9 - a - b$, as one finds by computing $\operatorname{tr} A^4 = 15v + 8q = 2(81 + 16a + b)$. The total number of hexagons is $h = 10 + 2b - 2c$, found similarly by computing $\operatorname{tr} A^6 = 87v + 96q + 12h = 2(729 + 64a + b)$.

More in detail, let q_u be the number of quadrangles on the vertex u, and q_{uv} the number of quadrangles on the edge uv, and similarly for h and hexagons. Let uv be an edge. Then $A_{uv} = 1$ and $(A^3)_{uv} = 5 + q_{uv}$ and $(A^5)_{uv} = 29 + 2q_u + 2q_v + 6q_{uv} + h_{uv}$.

The Hoffman polynomial $A(A+3I)(A^2-I)(A^2-4I)$ defines a rank 1 matrix with eigenvalue 720, so $A(A+3I)(A^2-I)(A^2-4I) = \frac{720}{v}J$ and in particular $v|240$ since for an edge xy the xy entry of $\frac{720}{v}J$ must be divisible by 3. This leaves for (a,b,c,v) the possibilities (a) $(0,0,2,6)$, (b) $(0,3,0,8)$, (c) $(1,2,1,10)$, (d) $(2,1,2,12)$, (e) $(3,3,1,16)$, (f) $(4,5,0,20)$, (g) $(5,1,3,20)$, (h) $(6,3,2,24)$, (j) $(9,0,5,30)$.

In case (a) we have $K_{3,3}$, case (i) of the theorem.

In case (b) we have the cube 2^3, case (ii).

In case (c) we have a graph of which the bipartite complement has spectrum $2^2 1^2 0^2 (-1)^2 (-2)^2$ and so is the disjoint union of a 4-cycle and a 6-cycle, case (iii).

In case (d) we have $q = 6$ and $h = 8$. Let uv be an edge, and evaluate $A(A+3I)(A^2-I)(A^2-4I) = 60J$ at the uv position to find $(A^5 - 5A^3 + 4A)_{uv} = 20$ and $2q_u + 2q_v + q_{uv} + h_{uv} = 12$. It follows that uv cannot lie in three or more quadrangles. Suppose u lies in (at least) three quadrangles. Then for each neighbor x of u we have $2q_x + h_{ux} = 4$, so that $q_x = 2$ and $h_u = 0$. The mod 2 sum of two quadrangles on u is not a hexagon, and it follows that we have a $K_{2,3}$ on points u,w,x,y,z (with u,w adjacent to x,y,z). The six quadrangles visible in the $K_{2,3}$ on u,w,x,y,z contribute $6 + 4 + 2 + 0$ to $2q_u + 2q_x + q_{ux} + h_{ux} = 12$, and it follows that there are no further quadrangles or hexagons on these points. So the three further neighbors p,q,r of x,y,z are distinct and have no common neighbors, which is impossible since $v = 12$. So, no vertex is in three or more quadrangles, and hence every vertex u is in precisely two quadrangles. These two quadrangles have an edge uu' in common, and we find an involution interchanging each u and u' and preserving the graph. It follows that we either have $C_6 \square K_2$ (and this has the desired spectrum; it is case (iv)), or a twisted version, but that has only 6 hexagons.

In case (e) we have $v = 16$ vertices. For any vertex x, Hoffman's polynomial yields $(A^6)_{xx} - 5(A^4)_{xx} + 4(A^2)_{xx} = 45$. On the other hand, $(A^{2i})_{xx}$ is odd for each i, since each walk of length $2i$ from x to x can be paired with the reverse walk, so that the parity of $(A^{2i})_{xx}$ is that of the number of self-reverse walks $x...zwz...x$ which is 3^i. Contradiction.

In case (f) we have $v = 20$, $q = 0$, $h = 20$. Since $c = 0$ we can omit the factor A from Hoffman's polynomial and find $(A+3I)(A^2-I)(A^2-4I) = 12J$. If u,w have even distance, then $(A^4 - 5A^2 + 4I)_{uw} = 4$. In particular, if $d(u,w) = 2$ then $9 = (A^4)_{uw} = 7 + h_{uw}$ so that $h_{uw} = 2$: each 2-path uvw lies in two hexagons. If no 3-path $uvwx$ lies in two hexagons then the graph is distance-regular with intersection array $\{3,2,2,1,1; 1,1,2,2,3\}$ (cf. Chapter 12) and hence is the Desargues graph. This is case (v) of the theorem. Now assume that the 3-path $uvwx$ lies in two hexagons, so that there are three paths $u \sim v_i \sim w_i \sim x$ ($i = 1,2,3$). The v_i and w_i need one more neighbor, say $v_i \sim y_i$ and $w_i \sim z_i$ ($i = 1,2,3$). The vertices y_i are distinct since there are no quadrangles, and similarly the z_i are distinct. The vertices y_i and z_j are nonadjacent, otherwise there would be a quadrangle (if $i = j$) or uv_jw_j would be in three hexagons (if $i \neq j$). There remain six more vertices: three that are each adjacent to two vertices y_i, and three that are each adjacent to two vertices z_i. Call them s_i and t_i, where $s_i \sim y_j$ and $t_i \sim z_j$ whenever $i \neq j$. The final part is a matching between the s_i and the t_i. Now the 2-path $v_i w_i z_i$ is in two hexagons, and these must be of the form $t_j z_i w_i v_i y_i s_k$ with $j \neq i \neq k$, and necessarily $j = k$, that is, the graph is uniquely determined. This is case (vi) of the theorem.

In case (g) we have $v = 20$, $q = 3$, $h = 6$. For an edge uv we have $(A^5 - 5A^3 + 4A)_{uv} = 12$, so that $2q_u + 2q_v + q_{uv} + h_{uv} = 4$. But that means that the edge uv cannot be in a quadrangle, contradiction.

In case (h) we have $v = 24$, $q = 0$, $h = 12$. For an edge uv we have $(A^5 - 5A^3 + 4A)_{uv} = 10$, so that $h_{uv} = 2$. It follows that each vertex is in three hexagons, and each 2-path vuw is in a unique hexagon. Now one straightforwardly constructs the unique cubic bipartite graph on 24 vertices without quadrangles and such that each 2-path is in a unique hexagon. Starting from a vertex u, call its neighbors v_i $(i = 1, 2, 3)$, let w_{ij} $(i, j = 1, 2, 3$ and $i \neq j)$ be the six vertices at distance 2, and let x_i $(i = 1, 2, 3)$ be the three vertices opposite u in a hexagon on u, so that the three hexagons on u are $uv_i w_{ij} x_k w_{ji} v_j$ (with distinct i, j, k). Let the third neighbor of w_{ij} be y_{ij}, and let the third neighbor of x_k be z_k. Necessarily $z_k \sim y_{kj}$. Now each vertex y_{ij} still needs a neighbor and there are two more vertices, say $s \sim y_{12}, y_{23}, y_{31}$ and $t \sim y_{13}, y_{21}, y_{32}$. This is case (vii).

In case (j) we have $v = 30$, $q = h = 0$ and we have Tutte's 8-cage. This is case (viii) of the theorem. $\qquad\square$

Proposition 3.8.2 *Let Γ be a connected nonbipartite cubic graph such that all of its eigenvalues are integral. Then Γ is one of five possible graphs, namely (ix) K_4, (x) $K_3 \square K_2$, (xi) the Petersen graph, (xii) the graph on 10 vertices defined by $i \sim (i+1)$ (mod 10), $0 \sim 5$, $1 \sim 3$, $2 \sim 6$, $4 \sim 8$, $7 \sim 9$ (or, equivalently, the graph obtained from $K_{3,3}$ by replacing each of two nonadjacent vertices by a triangle using a $Y - \Delta$ operation), (xiii) Σ.*

Proof Consider $\Gamma \otimes K_2$. It is cubic and has integral eigenvalues, and hence is one of the eight graphs Δ found in the previous proposition. There is an involution σ of $\Delta = \Gamma \otimes K_2$, without fixed edges, that interchanges the two vertices x' and x'' for each vertex x of Γ. Now Γ can be retrieved as Δ / σ.

In cases (i), (iii), (viii) the graph Γ would be cubic on an odd number of vertices, which is impossible.

In case (ii), σ must interchange antipodes, and the quotient $2^3 / \sigma$ is the complete graph K_4. This is case (ix).

In case (iv), $C_6 \square K_2$, σ must interchange antipodes in the same copy of C_6, and the quotient is $K_3 \square K_2$. This is case (x).

In case (v), $\Pi \otimes K_2$, we get the Petersen graph for a σ that interchanges antipodal vertices. This is case (xi). The group is Sym(5).2 and has two conjugacy classes of suitable involutions σ. The second one interchanges x' with $(12)x''$, and its quotient is obtained from Π by replacing the hexagon $13 \sim 24 \sim 15 \sim 23 \sim 14 \sim 25 \sim 13$ by the two triangles $13, 14, 15$ and $23, 24, 25$. This is case (xii).

In case (vi) there is no suitable σ. (An automorphism σ must interchange the two vertices u, x found in the previous proof, since this is the only pair of vertices joined by three 3-paths. But any shortest ux-path is mapped by σ into a different xu-path (since the path has odd length, and σ cannot preserve the middle edge) so that the number of such paths, which is 3, must be even.)

In case (vii) we get Σ. This is case (xiii). (The group of $\Sigma \otimes K_2$ has order 144, six times the order of the group Sym(4) of Σ, and all possible choices of σ are equivalent.) $\square$

Remarks Integral graphs with a small number of vertices have been classified. The number of nonisomorphic connected integral graphs on n vertices, $1 \leq n \leq 11$ is 1, 1, 1, 2, 3, 6, 7, 22, 24, 83, 113, see Sloane EIS sequence #A064731. For integral trees, see §5.6 below.

Most graphs have nonintegral eigenvalues: the integral graphs constitute a fraction of at most $2^{-n/400}$ of all graphs on n vertices ([1]). (Nevertheless, integral graphs are very common, there are far too many to classify.)

Integral graphs (and certain bipartite graphs) occur in quantum information theory in the description of systems with "perfect state transfer", cf. [305, 174].

All Cayley graphs for the elementary Abelian group 2^m are integral.

3.9 The largest Laplace eigenvalue

If $\mu_1 \leq \ldots \leq \mu_n$ are the Laplace eigenvalues of a simple graph Γ, then $0 \leq n - \mu_n \leq \ldots \leq n - \mu_2$ are the Laplace eigenvalues of the complement of Γ (see §1.3.2). Therefore $\mu_n \leq n$ with equality if and only if the complement of Γ is disconnected. If Γ is regular with valency k, we know (by Proposition 3.4.1) that $\mu_n \leq 2k$, with equality if and only if Γ is bipartite. More generally:

Proposition 3.9.1 *Let Γ be a graph with adjacency matrix A (with eigenvalues $\theta_1 \geq \ldots \geq \theta_n$), Laplacian L (with eigenvalues $\mu_1 \leq \ldots \leq \mu_n$), and signless Laplacian Q (with eigenvalues $\rho_1 \geq \ldots \geq \rho_n$). Then*

  (i) (ZHANG & LUO [358])

$$\mu_n \leq \rho_1.$$

   If Γ is connected, then equality holds if and only if Γ is bipartite.
 (ii) *Let d_x be the degree of the vertex x. If Γ has at least one edge, then*

$$\rho_1 \leq \max_{x \sim y} (d_x + d_y).$$

   Equality holds if and only if Γ is regular or bipartite semiregular.
(iii) (YAN [354])

$$2\theta_i \leq \rho_i \ (1 \leq i \leq n).$$

Proof   (i) Apply Theorem 2.2.1 (vi).

(ii) Using Proposition 3.1.2 to bound the largest eigenvalue of $L(\Gamma)$ by its maximum degree $\max_{x \sim y} (d_x + d_y - 2)$, we find $\rho_1 = \theta_1(L(\Gamma)) + 2 \leq \max_{x \sim y} (d_x + d_y)$, with equality if and only if $L(\Gamma)$ is regular so that Γ is regular or bipartite semiregular.

(iii) Since $Q = L + 2A$ and L is positive semidefinite, this follows from the Courant-Weyl inequalities (Theorem 2.8.1 (iii)).   □

Corollary 3.9.2 ([10]) *Let Γ be a graph on n vertices with at least one edge. Then*

$$\mu_n \leq \max_{x \sim y} (d_x + d_y).$$

If Γ is connected, then equality holds if and only if Γ is bipartite regular or semiregular.   □

For bipartite graphs, L and Q have the same spectrum (see Proposition 1.3.10). It follows by the Perron-Frobenius theorem that the largest Laplace eigenvalue of a connected bipartite graph decreases strictly when an edge is removed.

Interlacing provides a lower bound for μ_n:

Proposition 3.9.3 ([186]) *Let Γ be a graph on n vertices with at least one edge, and let d_x be the degree of the vertex x. Then*

$$\mu_n \geq 1 + \max_x d_x.$$

If Γ is connected, then equality holds if and only if $\max_x d_x = n - 1$.

Proof If Γ has a vertex of degree d, then it has a subgraph $K_{1,d}$ (not necessarily induced), and $\mu_n \geq d + 1$. If equality holds, then Γ does not have a strictly larger bipartite subgraph. If Γ is, moreover, connected, then $d = n - 1$.   □

Deriving bounds on μ_n has become an industry—there are many papers, cf. [42, 134, 190, 252, 253, 260, 279, 357].

3.10 Laplace eigenvalues and degrees

The Schur inequality (Theorem 2.6.1) immediately yields an inequality between the sum of the largest m Laplace eigenvalues and the sum of the largest m vertex degrees. GRONE [185] gave a slightly stronger result.

Proposition 3.10.1 *If Γ is connected, with Laplace eigenvalues $v_1 \geq v_2 \geq ... \geq v_n = 0$ and vertex degrees $d_1 \geq d_2 \geq ... \geq d_n > 0$, then for $1 \leq m \leq n - 1$ we have $1 + \sum_{i=1}^{m} d_i \leq \sum_{i=1}^{m} v_i$.*

Proof Let x_i have degree d_i, and put $Z = \{x_1, ..., x_m\}$. Let $N(Z)$ be the set of vertices outside Z with a neighbor in Z. Instead of assuming that Γ is connected, we just use that $N(Z)$ is nonempty. If we delete the vertices outside $Z \cup N(Z)$, then $\sum_{z \in Z} d_z$ does not change, and $\sum_{i=1}^{m} v_i$ does not increase, so we may assume $X = Z \cup N(Z)$. Let R be the quotient matrix of L for the partition $\{\{z\} \mid z \in Z\} \cup \{N(Z)\}$ of X, and let $\lambda_1 \geq ... \geq \lambda_{m+1}$ be the eigenvalues of R. The matrix R has row sums 0, so

$\lambda_{m+1} = 0$. By interlacing (Corollary 2.5.4), we have $\sum_{i=1}^{m} v_i \geq \sum_{i=1}^{m} \lambda_i = \sum_{i=1}^{m+1} \lambda_i = \text{tr} R = \sum_{z \in Z} d_z + R_{m+1,m+1}$, and the desired result follows since $R_{m+1,m+1} \geq 1$. □

Second proof. We prove the following stronger statement:

For any graph Γ (not necessarily connected) and any subset Z of the vertex set X of Γ one has $h + \sum_{z \in Z} d_z \leq \sum_{i=1}^{m} v_i$, where d_z denotes the degree of the vertex z in Γ, and $m = |Z|$, and h is the number of connected components of the graph Γ_Z induced on Z that are not connected components of Γ.

We may assume that Γ is connected, and that Z and $X \setminus Z$ are nonempty. Now h is the number of connected components of Γ_Z.

The partition $\{Z, X \setminus Z\}$ of X induces a partition $L = \begin{bmatrix} B & -C \\ -C^\top & E \end{bmatrix}$. Since Γ is connected, B is nonsingular by Lemma 2.10.1(iii). All entries of B^{-1} are nonnegative. (Write $B = n(I - T)$, where $T \geq 0$, then $B^{-1} = \frac{1}{n}(I + T + T^2 + \dots) \geq 0$. If $h = 1$, then $B^{-1} > 0$.)

Since L is positive semidefinite, we can write $L = MM^\top$, where $M = \begin{bmatrix} P & 0 \\ Q & R \end{bmatrix}$ is a square matrix. Now $B = PP^\top$ and $-C = PQ^\top$. The eigenvalues of $MM^\top$ are the same as those of $M^\top M$, and the latter matrix has submatrix $P^\top P + Q^\top Q$ of order m. By Schur's inequality, we get $\sum_{i=1}^{m} v_i \geq \text{tr}(P^\top P + Q^\top Q) = \sum_{z \in Z} d_z + \text{tr} Q^\top Q$, and it remains to show that $\text{tr} Q^\top Q \geq h$.

Now $Q^\top Q = P^{-1} CC^\top P^{-\top}$, so $\text{tr} Q^\top Q = \text{tr} B^{-1} CC^\top$. We have $B = L_Z + D$, where L_Z is the Laplacian of Γ_Z and D is the diagonal matrix of the row sums of C. Since $CC^\top \geq D$ and $B^{-1} \geq 0$, we have $\text{tr} Q^\top Q \geq \text{tr} B^{-1} D$. If $L_Z u = 0$, then $(L_Z + D)^{-1} Du = u$. Since L_Z has eigenvalue 0 with multiplicity h, the matrix $B^{-1} D$ has eigenvalue 1 with multiplicity h. Since this matrix is positive semidefinite (since $D^{1/2} B^{-1} D^{1/2}$ is), its trace is at least h. □

A lower bound for the individual v_j was conjectured by GUO [191] and proved in BROUWER & HAEMERS [60].

Proposition 3.10.2 *Let Γ be a graph with Laplace eigenvalues $v_1 \geq v_2 \geq \dots \geq v_n = 0$ and with vertex degrees $d_1 \geq d_2 \geq \dots \geq d_n$. Let $1 \leq m \leq n$. If Γ is not $K_m + (n - m)K_1$, then $v_m \geq d_m - m + 2$.* □

We saw the special case $m = 1$ in Proposition 3.9.3. The cases $m = 2$ and $m = 3$ were proved earlier in [251] and [191].

Examples with equality are given by complete graphs K_m with a pending edges at each vertex (where $a > 0$), with Laplace spectrum consisting of 0, $1^{m(a-1)}$, $a+1$, and $\frac{1}{2}(m+a+1 \pm \sqrt{(m+a+1)^2 - 4m})$ with multiplicity $m - 1$ each, so that $v_m = a + 1 = d_m - m + 2$.

Further examples are complete graphs K_m with a pending edges attached at a single vertex. Here $n = m + a$, and the Laplace spectrum consists of $m + a$, m^{m-2}, 1^a, and 0, so that $v_m = 1 = d_m - m + 2$.

Any graph contained in $K_{a,a}$ and containing $K_{2,a}$ has $v_2 = a = d_2$, with equality for $m = 2$.

Any graph on n vertices with $d_1 = n - 1$ has equality for $m = 1$.

More generally, whenever one has an eigenvector u and vertices x, y with $u_x = u_y$, then u remains an eigenvector, with the same eigenvalue, if we add or remove an edge between x and y. Many of the above examples can be modified by adding edges. This leads to many further cases of equality.

3.11 The Grone-Merris conjecture

3.11.1 Threshold graphs

A *threshold graph* is a graph obtained from the graph K_0 by a sequence of operations of the form (i) add an isolated vertex, or (ii) take the complement.

Proposition 3.11.1 *Let Γ be a threshold graph with Laplace eigenvalues (in nonincreasing order) $v_1 \geq v_2 \geq \ldots \geq v_n = 0$. Let d_x be the degree of the vertex x. Then*

$$v_j = \#\{x \mid d_x \geq j\}.$$

Proof Induction on the number of construction steps of type (i) or (ii). □

GRONE & MERRIS [186] conjectured that this is the extreme case, and that for all undirected graphs and all t one has

$$\sum_{j=1}^{t} v_j \leq \sum_{j=1}^{t} \#\{x \mid d_x \geq j\}.$$

For $t = 1$, this is immediate from $v_1 \leq n$. For $t = n$, equality holds. This conjecture was proved in HUA BAI [16], see §3.11.2 below. There is a generalization to higher-dimensional simplicial complexes, see §3.12 below.

A variation on the Grone-Merris conjecture is the following.

Conjecture (Brouwer) *Let Γ be a graph with e edges and Laplace eigenvalues $v_1 \geq v_2 \geq \ldots \geq v_n = 0$. Then for each t we have $\sum_{i=1}^{t} v_i \leq e + \binom{t+1}{2}$.*

It is easy to see (by induction) that this inequality holds for threshold graphs. In [202] it is proved for trees, and for $t = 2$. In [27] it is shown that there is a t such that the t-th inequality of this conjecture is sharper than the t-th Grone-Merris inequality if and only if the graph is nonsplit. In particular, this conjecture holds for split graphs. It also holds for regular graphs.

3.11.2 Proof of the Grone-Merris conjecture

Very recently, HUA BAI [16] proved the Grone-Merris conjecture. We repeat the statement of the theorem.

Theorem 3.11.2 *Let Γ be an undirected graph with Laplace eigenvalues (in non-increasing order) $v_1 \geq v_2 \geq \ldots \geq v_n = 0$. Let d_x be the degree of the vertex x. Then for all t, $0 \leq t \leq n$, we have*

$$\sum_{i=1}^{t} v_i \leq \sum_{i=1}^{t} \#\{x \mid d_x \geq i\}. \tag{3.1}$$

The proof is by reduction to the case of a *split graph*, that is a graph where the vertex set is the disjoint union of a nonempty subset inducing a clique (complete graph), and a nonempty subset inducing a coclique (edgeless graph). Then for split graphs a continuity argument proves the crucial inequality stated in the following lemma.

Lemma 3.11.3 *Let Γ be a split graph with clique of size c and Laplace eigenvalues $v_1 \geq v_2 \geq \ldots \geq v_n = 0$. Let δ be the maximum degree among the vertices in the coclique, so that $\delta \leq c$. If $v_c > c$ or $v_c = c > \delta$, then we have $\sum_{i=1}^{c} v_i \leq \sum_{i=1}^{c} \#\{x \mid d_x \geq i\}$.*

Proof of Theorem 3.11.2 (assuming Lemma 3.11.3). Consider counterexamples to (3.1) with minimal possible t.

Step 1 *If Γ is such a counterexample with minimal number of edges, and x,y are vertices in Γ of degree at most t, then they are nonadjacent.*

Indeed, if $x \sim y$ then let Γ' be the graph obtained from Γ by removing the edge xy. Then $\sum_{i=1}^{t} \#\{x \mid d'_x \geq i\} + 2 = \sum_{i=1}^{t} \#\{x \mid d_x \geq i\}$. The Laplace matrices L and L' of Γ and Γ' satisfy $L = L' + H$ where H has eigenvalues $2, 0^{n-1}$. By Theorem 2.8.2 we have $\sum_{i=1}^{t} v_i \leq \sum_{i=1}^{t} v'_i + 2$, and since Γ' has fewer edges than Γ we find $\sum_{i=1}^{t} v_i \leq \sum_{i=1}^{t} v'_i + 2 \leq \sum_{i=1}^{t} \#\{x \mid d'_x \geq i\} + 2 = \sum_{i=1}^{t} \#\{x \mid d_x \geq i\}$, a contradiction.
□

Step 2 *There is a split counterexample Γ for the same t, with clique size $c := \#\{x \mid d_x \geq t\}$.*

Indeed, we can form a new graph Γ from the Γ of Step 1 by adding edges xy for every pair of nonadjacent vertices x, y, both of degree at least t. Now $\sum_{i=1}^{t} \#\{x \mid d_x \geq i\}$ does not change, and $\sum_{i=1}^{t} v_i$ does not decrease, and the new graph is split with the stated clique size.
□

This will be our graph Γ for the rest of the proof.

Step 3 *A split graph Δ of clique size c satisfies $v_{c+1} \leq c \leq v_{c-1}$.*

Indeed, since Δ contains the complete graph K_c with Laplace spectrum $c^{c-1}, 0$, we see by the Courant-Weyl inequalities (Theorem 2.8.1 (iii)) that $v_{c-1} \geq c$. And since Δ is contained in the complete split graph with clique of size c and coclique of size $n - c$ and all edges in-between, with Laplace spectrum $n^c, c^{n-c-1}, 0$, we have $v_{c+1} \leq c$.
□

Since t was chosen minimal, we have $v_t > \#\{x \mid d_x \geq t\} = c$. The previous step then implies $c \geq t$. If $c = t$, then $v_c > c$ and Lemma 3.11.3 gives a contradiction. So $c > t$.

All vertices in the coclique of Γ have degree at most $t - 1$ and all vertices in the clique have degree at least $c - 1$. So $\#\{x \mid d_x \geq i\} = c$ for $t \leq i \leq c - 1$. From Step 3 we have $v_i \geq v_{c-1} \geq c$ for $t \leq i \leq c - 1$. Since $\sum_{i=1}^t v_i > \sum_{i=1}^t \#\{x \mid d_x \geq i\}$, we also have $\sum_{i=1}^{c-1} v_i > \sum_{i=1}^{c-1} \#\{x \mid d_x \geq i\}$. Now if $v_c \geq c$ we contradict Lemma 3.11.3 (since $\#\{x \mid d_x \geq c\} \leq c$). So $v_c < c$.

Step 4 *The m-th Grone-Merris inequality for a graph Γ is equivalent to the $(n - 1 - m)$-th Grone-Merris inequality for its complement $\overline{\Gamma}$ ($1 \leq m \leq n - 1$).*

Indeed, $\overline{\Gamma}$ has Laplace eigenvalues $\overline{v}_i = n - v_{n-i}$ ($1 \leq i \leq n - 1$) and dual degrees $\#\{x \mid \overline{d}_x \geq i\} = n - \#\{x \mid d_x \geq n - i\}$, and $\sum_{i=1}^n v_i = \sum_{i=1}^n \#\{x \mid d_x \geq i\}$.   □

In our case $\overline{\Gamma}$ is semibipartite with clique size $n - c$, and by the above we have $\overline{v}_{n-c} = n - v_c > n - c$ and $\sum_{i=1}^{n-c} \overline{v}_i > \sum_{i=1}^{n-c} \#\{x \mid \overline{d}_x \geq i\}$. This contradicts Lemma 3.11.3.

This contradiction completes the proof of the Grone-Merris conjecture, except that Lemma 3.11.3 still has to be proved.

3.11.2.1 Proof of Lemma 3.11.3

Let Γ be a split graph with clique of size c and coclique of size $n - c$. The partition of the vertex set induces a partition of the Laplace matrix $L = \begin{bmatrix} K+D & -A \\ -A^\top & E \end{bmatrix}$, where K is the Laplacian of the complete graph K_c and A is the $c \times (n-c)$ adjacency matrix between vertices in the clique and the coclique, and D and E are diagonal matrices with the row and column sums of A.

Step 5 *If $v_c \geq c$, then $\sum_{i=1}^c \#\{x \mid d_x \geq i\} = c^2 + \mathrm{tr}\, D$.*

Indeed, all vertices in the clique have degree at least c, for if some vertex x in the clique had degree $c - 1$, then we could move it to the coclique and find $v_c \leq c - 1$ from Step 3, contrary to the assumption. It follows that $\sum_{i=1}^c \#\{x \mid d_x \geq i\} = \sum_x \min(c, d_x) = c^2 + \mathrm{tr}\, E = c^2 + \mathrm{tr}\, D$.   □

Step 6 *Suppose that the subspace W spanned by the L-eigenvectors belonging to $v_1, \ldots, v_c$ is spanned by the columns of $\begin{bmatrix} I \\ X \end{bmatrix}$. Then $L \begin{bmatrix} I \\ X \end{bmatrix} = \begin{bmatrix} I \\ X \end{bmatrix} Z$ for some matrix Z, and $\sum_{i=1}^c v_i = \mathrm{tr}\, Z$.*

Indeed, if $\begin{bmatrix} U \\ V \end{bmatrix}$ has these eigenvectors as columns, then $L \begin{bmatrix} U \\ V \end{bmatrix} = \begin{bmatrix} U \\ V \end{bmatrix} T$ where T is the diagonal matrix with the eigenvalues. Now $\begin{bmatrix} U \\ V \end{bmatrix} = \begin{bmatrix} I \\ X \end{bmatrix} U$, so that $L \begin{bmatrix} I \\ X \end{bmatrix} = \begin{bmatrix} I \\ X \end{bmatrix} Z$ where $Z = UTU^{-1}$ and $\mathrm{tr}\, Z = \mathrm{tr}\, T = \sum_{i=1}^c v_i$.   □

Suppose we are in the situation of the previous step, and that moreover X is nonpositive. Let δ be the maximum degree among the vertices in the coclique, so that $\delta \leq c$. We have to show that if $v_c > c$ or $v_c = c > \delta$ then $\operatorname{tr} Z \leq c^2 + \operatorname{tr} D$.

Now $L \begin{bmatrix} I \\ X \end{bmatrix} = \begin{bmatrix} K+D-AX \\ -A^\top + EX \end{bmatrix}$, so $Z = K+D-AX$, and $\operatorname{tr} Z = \operatorname{tr}(K+D-AX) = c(c-1) + \operatorname{tr} D - \operatorname{tr}(AX)$, and we need $\operatorname{tr}(AX) \geq -c$. But since $c < n$, the eigenvectors are orthogonal to $\mathbf{1}$ so X has column sums -1. Since X is nonpositive, $\operatorname{tr}(AX) \geq -c$ follows, and we are done.

By interlacing, v_{c+1} is at most the largest eigenvalue of E, that is δ, which by hypothesis is smaller than v_c. Hence the subpace of vectors $\begin{bmatrix} 0 \\ * \end{bmatrix}$ meets W trivially, so that W has a basis of the required form. Only nonpositivity of X remains, and the following lemma completes the proof.

Lemma 3.11.4 *If $v_c > \delta$, then the invariant subspace W spanned by the L-eigenvectors for v_i, $1 \leq i \leq c$, is spanned by the columns of $\begin{bmatrix} I \\ X \end{bmatrix}$, where X is nonpositive.*

Proof We argue by continuity, viewing $L = L(A)$ and $X = X(A)$ as functions of the real-valued matrix A, where $0 \leq A \leq J$. (Now D has the row sums of A, and E has the column sums, and δ is the largest element of the diagonal matrix E.) We write J for the $c \times (n-c)$ all-1 matrix, and J_c for the all-1 matrix of order c, so that $JX = -J_c$.

Our hypothesis $v_c > \delta$ holds for all matrices $L^{(\alpha)} := L(\alpha A + (1-\alpha)J) = \alpha L + (1-\alpha)L^{(0)}$, for $0 \leq \alpha \leq 1$. Indeed, let $L^{(\alpha)}$ have eigenvalues $v_i^{(\alpha)}$, so that $v_c^{(0)} = n$ and $v_{c+1}^{(0)} = v_{n-1}^{(0)} = c$. The matrix $L^{(\alpha)}$ has lower left-hand corner $\alpha E + (1-\alpha)cI$ so that $\delta^{(\alpha)} = \alpha\delta + (1-\alpha)c$. The c-space W is orthogonal to $\mathbf{1}$, so that $v_c^{(\alpha)} \geq \alpha v_c + (1-\alpha)c$ (by Theorem 2.4.1), and hence $v_c^{(\alpha)} > \delta^{(\alpha)}$ for $0 < \alpha \leq 1$, and also for $\alpha = 0$, since $v_c^{(0)} = n$ and $\delta^{(0)} = c$. It follows that $v_c^{(\alpha)} > v_{c+1}^{(\alpha)}$ for $0 \leq \alpha \leq 1$.

As we already used, $L(J)$ has spectrum n^c, c^{n-c-1}, 0, and one checks that $X(J) = -\frac{1}{n-c}J^\top < 0$, as desired. Above we found the condition $XZ = -A^\top + EX$ on X, that is, $X(K+D-AX) - EX + A^\top = 0$, that is, $X(K+J_c+D) = XJ_c + XAX + EX - A^\top = -X(J-A)X + EX - A^\top$. It follows, since $K+J_c+D$ is a positive diagonal matrix, that if $X \leq 0$ and $A > 0$, then $X < 0$. The matrix $X(A)$ depends continuously on A (in the region where $v_{c+1} < v_c$) and is strictly negative when $A > 0$. Then it is nonpositive when $A \geq 0$. $\square$

3.12 The Laplacian for hypergraphs

Let a *simplicial complex* on a finite set S be a collection $\mathscr{C}$ of subsets of S (called simplices) that is an order ideal for inclusion, that is, is such that if $A \in \mathscr{C}$ and $B \subseteq A$ then also $B \in \mathscr{C}$. Let the *dimension* of a simplex A be one less than its cardinality, and let the dimension of a simplicial complex be the maximum of the dimensions of its simplices. Given a simplicial complex $\mathscr{C}$, let $\mathscr{C}_i$ (for $i \geq -1$) be the vector space (over any field) that has the simplices of dimension i as basis. Order the simplices arbitrarily (say, using some order on S) and define $\partial_i : \mathscr{C}_i \to \mathscr{C}_{i-1}$ by $\partial_i s_0 \ldots s_i = \sum_j (-1)^j s_0 \ldots \widehat{s_j} \ldots s_i$. Then $\partial_{i-1} \partial_i = 0$ for all $i \geq 0$.

Let N_i be the matrix of ∂_i on the standard basis, and put $L_i = N_{i+1} N_{i+1}^\top$ and $L_i' = N_i^\top N_i$. The matrices L_i generalize the Laplacian. Indeed, in the case of a 1-dimensional simplicial complex (that is, a graph) the ordinary Laplace matrix is just L_0, and L_0' is the all-1 matrix J.

Since $\partial_i \partial_{i+1} = 0$ we have $L_i L_i' = L_i' L_i = 0$, generalizing $LJ = JL = 0$.

We have $\operatorname{tr} L_{i-1} = \operatorname{tr} L_i' = (i+1)|\mathscr{C}_i|$. This generalizes the facts that $\operatorname{tr} L$ is twice the number of edges, and $\operatorname{tr} J$ the number of vertices.

In case the underlying field is $\mathbb{R}$, we have the direct sum decomposition $\mathscr{C}_i = \operatorname{im} N_{i+1} \oplus \ker(L_i + L_i') \oplus \operatorname{im} N_i^\top$. (Because then $M^\top M x = 0$ if and only if $Mx = 0$.) Now $\ker N_i = \operatorname{im} N_{i+1} \oplus \ker(L_i + L_i')$ so that the i-th reduced homology group is $H_i(\mathscr{C}) := \ker N_i / \operatorname{im} N_{i+1} \cong \ker(L_i + L_i')$.

Example *The spectrum of L_{m-2} for a simplicial complex containing all m-subsets of an n-set (the complete m-uniform hypergraph) consists of the eigenvalue n with multiplicity $\binom{n-1}{m-1}$ and all further eigenvalues are 0.*

Indeed, we may regard simplices $s_0 \ldots s_{m-1}$ as elements $s_0 \wedge \ldots \wedge s_{m-1}$ of an exterior algebra. Then the expression $s_0 \ldots s_{m-1}$ is defined regardless of the order of the factors, and also when factors are repeated. Now $N_i t_0 \ldots t_i = \sum_j (-1)^j t_0 \ldots \widehat{t_j} \ldots t_i$ and for the complete $(i+2)$-uniform hypergraph we have $N_i^\top t_0 \ldots t_i = \sum_t t t_0 \ldots t_i$, so that $L_i + L_i' = nI$. It follows that $N_{i+1}^\top N_{i+1} N_{i+1}^\top = n N_{i+1}^\top$, and L_i has eigenvalues 0 and n. The multiplicities follow by taking the trace.

DUVAL & REINER [149] generalized the Grone-Merris conjecture. Given an m-uniform hypergraph $\mathscr{H}$, let d_x be the number of edges containing the vertex x. Let the spectrum of $\mathscr{H}$ be that of the matrix L_{m-2} for the simplicial complex consisting of all subsets of edges of $\mathscr{H}$.

Conjecture *Let the m-uniform hypergraph $\mathscr{H}$ have degrees d_x and Laplace eigenvalues v_i, ordered such that $v_1 \geq v_2 \geq \ldots \geq 0$. Then for all t we have*

$$\sum_{j=1}^t v_j \leq \sum_{j=1}^t \#\{x \mid d_x \geq j\}.$$

Equality for all t holds if and only if $\mathscr{H}$ is invariant under downshifting.

The part about "downshifting" means the following: Put a total order on the vertices of $\mathscr{H}$ in such a way that if $x \leq y$ then $d_x \geq d_y$. Now $\mathscr{H}$ is said to be invariant under downshifting if whenever $\{x_1, \ldots, x_m\}$ is an edge of $\mathscr{H}$, and $\{y_1, \ldots, y_m\}$ is an m-set with $y_i \leq x_i$ for all i, then also $\{y_1, \ldots, y_m\}$ is an edge of $\mathscr{H}$. If this holds for one total order, then it holds for any total order that is compatible with the degrees.

For $m = 2$ this is precisely the Grone-Merris conjecture. (And the graphs that are invariant for downshifting are precisely the threshold graphs.) The "if" part of the equality case is a theorem:

Theorem 3.12.1 (DUVAL & REINER [149]) *If $\mathcal{H}$ is an m-uniform hypergraph with degrees d_x and Laplace eigenvalues v_i with $v_1 \geq v_2 \geq \ldots \geq 0$, and $\mathcal{H}$ is invariant for downshifting, then $v_j = \#\{x \mid d_x \geq j\}$ for all t.*

In particular it follows that hypergraphs invariant for downshifting have integral Laplace spectrum.

For example, the complete m-uniform hypergraph on an underlying set of size n has degrees $\binom{n-1}{m-1}$ so that $v_j = n$ for $1 \leq j \leq \binom{n-1}{m-1}$ and $v_j = 0$ for $\binom{n-1}{m-1} < j \leq \binom{n}{m}$, as we already found earlier.

3.12.1 Dominance order

The conjecture and the theorem can be formulated more elegantly in terms of *dominance order*. Let $\mathbf{a} = (a_i)$ and $\mathbf{b} = (b_i)$ be two finite nonincreasing sequences of nonnegative real numbers. We say that $\mathbf{b}$ *dominates* $\mathbf{a}$, and write $\mathbf{a} \trianglelefteq \mathbf{b}$, when $\sum_{i=1}^t a_i \leq \sum_{i=1}^t b_i$ for all t, and $\sum_{i=1}^\infty a_i = \sum_{i=1}^\infty b_i$, where missing elements are taken to be zero.

For example, in this notation Schur's inequality (Theorem 2.6.1) says that $\mathbf{d} \trianglelefteq \theta$ if $\mathbf{d}$ is the sequence of diagonal elements and θ the sequence of eigenvalues of a real symmetric matrix.

If $\mathbf{a} = (a_j)$ is a finite nonincreasing sequence of nonnegative integers, then $\mathbf{a}^\top$ denotes the sequence $(a_j^\top)$ with $a_j^\top = \#\{i \mid a_i \geq j\}$. If $\mathbf{a}$ is represented by a Ferrers diagram, then $\mathbf{a}^\top$ is represented by the transposed diagram.

For example, the Duval-Reiner conjecture says that $v \trianglelefteq \mathbf{d}^\top$.

If $\mathbf{a}$ and $\mathbf{b}$ are two nonincreasing sequences, then let $\mathbf{a} \cup \mathbf{b}$ denote the (multiset) union of both sequences, with elements sorted in nonincreasing order.

Lemma 3.12.2
  (i) $\mathbf{a}^{\top\top} = \mathbf{a}$,
  (ii) $(\mathbf{a} \cup \mathbf{b})^\top = \mathbf{a}^\top + \mathbf{b}^\top$ *and* $(\mathbf{a} + \mathbf{b})^\top = \mathbf{a}^\top \cup \mathbf{b}^\top$,
  (iii) $\mathbf{a} \trianglelefteq \mathbf{b}$ *if and only if* $\mathbf{b}^\top \trianglelefteq \mathbf{a}^\top$.   □

3.13 Applications of eigenvectors

Sometimes it is not the eigenvalue but the eigenvector that is needed. We very briefly sketch some of the applications.

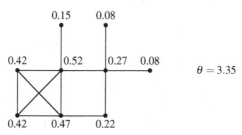

Fig. 3.1 Graph with Perron-Frobenius eigenvector

3.13.1 Ranking

In a network, important people have many connections. One would like to pick out the vertices of highest degree and call them the most important. But it is not just the number of neighbors. Important people have connections to many other important people. If one models this and says that up to some constant of proportionality one's importance is the sum of the importances of one's neighbors in the graph, then the vector giving the importance of each vertex becomes an eigenvector of the graph, necessarily the Perron-Frobenius eigenvector (cf. Figure 3.1) if importance cannot be negative. The constant of proportionality is then the largest eigenvalue.

3.13.2 Google PageRank

Google uses a similar scheme to compute the PageRank [46] of web pages. The authors described (in 1998) the algorithm as follows:

Suppose pages x_1, ..., x_m are the pages that link to a page y. Let page x_i have d_i outgoing links. Then the PageRank of y is given by

$$PR(y) = 1 - \alpha + \alpha \sum_i \frac{PR(x_i)}{d_i}.$$

The PageRanks form a probability distribution: $\sum_x PR(x) = 1$. The vector of PageRanks can be calculated using a simple iterative algorithm, and corresponds to the principal eigenvector of the normalized link matrix of the web. A PageRank for 26 million web pages can be computed in a few hours on a medium size workstation. A suitable value for α is $\alpha = 0.85$.

In other words, let Γ be the directed graph on n vertices consisting of all web pages found, with an arrow from x to y when page x contains a hyperlink to page y. Let A be the adjacency matrix of Γ (with $A_{xy} = 1$ if there is a link from x to y). Let D be the diagonal matrix of outdegrees, so that the scaled matrix $S = D^{-1}A$ has row sums 1, and construct the positive linear combination $M = \frac{1-\alpha}{n}J + \alpha S$ with $0 < \alpha < 1$. Since $M > 0$, the matrix M has a unique positive left eigenvector u, normed so that $\sum u_x = 1$. Now $M\mathbf{1} = \mathbf{1}$ and hence $uM = u$. The PageRank of the web page x is the value u_x.

A small detail is the question of what to do when page x does not have outgoing edges, so that row x in A is zero. One possibility is to do nothing (and take $D_{xx} = 1$). Then u will have eigenvalue less than 1.

The vector u is found by starting with an approximation (or just any positive vector) u_0 and then computing the limit of the sequence $u_i = u_0 M^i$. That is easy: the matrix M is enormous, but A is sparse: on average a web page does not have more than a dozen links. The constant α regulates the speed of convergence: convergence is determined by the 2nd largest eigenvalue, which is bounded by α ([212]). It is reported that 50 to 100 iterations suffice. A nonzero α guarantees that the matrix is irreducible. An α much less than 1 guarantees quick convergence. But an α close to 1 is better at preserving the information in A. Intuitively, u_x represents the expectation of finding oneself at page x after many steps, where each step consists of either (with probability α) clicking on a random link on the current page, or (with probability $1 - \alpha$) picking a random internet page. Note that the precise value of u_x is unimportant—only the ordering among the values u_x is used.

There are many papers (and even books) discussing Google's PageRank. See e.g. [72], [28].

3.13.3 Cutting

Often the cheapest way to cut a connected graph into two pieces is by partitioning it into a single vertex (of minimal valency) and the rest. But in the area of clustering (see also below) one typically wants relatively large pieces. Here the second Laplace eigenvector helps. Without going into any detail, let us try the same example as above in Figure 3.2 below.

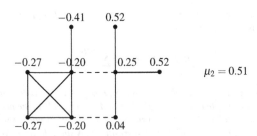

Fig. 3.2 Graph with 2nd Laplace eigenvector

We see that cutting the edges where the second Laplace eigenvector changes sign is fairly successful in this case. See also §3.13.5 below.

3.13.4 Graph drawing

Often, a reasonable way to draw a connected graph is to take Laplace eigenvectors u and v for the 2nd and 3rd smallest Laplace eigenvalues, and draw the vertex x at the point with coordinates (u_x, v_x). See, e.g., [246].

One can justify this as follows. Let the *energy* of an embedding $\rho : \Gamma \to \mathbb{R}^m$ be the sum of the squared edge lengths $\sum_e ||\rho(x) - \rho(y)||^2$ where the sum is over all edges $e = xy$. Let R be the $m \times n$ matrix of which the columns are the vertex images $\rho(x)$. Then the energy of ρ equals $RLR^\top$. For graph drawing one would like to minimize the energy, given some normalization so that not all vertices are mapped close to the origin or close to some lower-dimensional subspace of $\mathbb{R}^m$. PISANSKI & SHAWE-TAYLOR [295] propose to require $R\mathbf{1} = 0$ and $RR^\top = I$, so that the origin is the center of mass, and $||R^\top v||^2 = ||v||^2$ for all vectors $v \in \mathbb{R}^m$: no vector is almost perpendicular to the entire drawing. In this situation the minimum energy is $\sum_{i=2}^{m+1} \mu_i$, and this minimum is achieved when the row space of R contains the Laplace eigenvectors of $\mu_2, \ldots, \mu_{m+1}$. The authors also discuss variations of this setup.

3.13.5 Clustering

Given a large data set, one often wants to cluster it. If the data is given as a set of vectors in some Euclidean space $\mathbb{R}^m$, then a popular clustering algorithm is *k-means*:

Given a set X of N vectors in $\mathbb{R}^m$ and a number k, find a partition of X into k subsets $X_1, \ldots, X_k$ such that $\sum_{i=1}^k \sum_{x \in X_i} ||x - c_i||^2$ is as small as possible, where $c_i = (1/|X_i|) \sum_{x \in X_i} x$ is the centroid of X_i.

The usual algorithm uses an iterative approach. First choose the k vectors c_i in some way, arbitrary or not. Then take X_i to be the subset of X consisting of the vectors closer to c_i than to the other c_j (breaking ties arbitrarily). Then compute new vectors c_i as the centroids of the sets X_i, and repeat. In common practical situations this algorithm converges quickly, but one can construct examples where this takes exponential time. The final partition found need not be optimal, but since the algorithm is fast, it can be repeated a number of times with different starting points c_i.

Now if the data is given as a graph, one can compute eigenvectors $u_1, \ldots, u_m$ for the m smallest eigenvalues $\mu_1, \ldots, \mu_m$ of the Laplace matrix L, and assign to the vertex x the vector $(u_i(x))_i$ and apply a vector space clustering algorithm such as k-means to the resulting vectors.

This is reasonable. For example, if the graph is disconnected with c connected components, then the first c eigenvalues of L are zero, and the first c eigenvectors are (linear combinations of) the characteristic functions of the connected components.

This approach also works when one has more detailed information—not adjacent/nonadjacent but a (nonnegative) similarity or closeness measure. (One uses an edge-weighted graph, with $A_{xy} = w(x,y)$ and $d_x = \sum_y w(x,y)$ and D the diagonal matrix with $D_{xx} = d_x$, and $L = D - A$. Again L is positive semidefinite, with $u^\top L u = \sum w(x,y)(u(x) - u(y))^2$. The multiplicity of the eigenvalue 0 is the number of connected components of the underlying graph where points x, y are adjacent when $w(x,y) > 0$.)

Especially important is the special case where one searches for the cheapest cut of the graph into two relatively large pieces. If the graph is connected, then map the vertices into $\mathbb{R}^1$ using $x \mapsto u(x)$, where u is the eigenvector for the second smallest eigenvalue of L, and then use 2-means to cluster the resulting points. Compare §1.7 on the algebraic connectivity of a graph.

Several matrices related to the Laplacian have been used in this context. It seems useful to normalize the matrix, and to retain the property that if the graph is disconnected the characteristic functions of components are eigenvectors. Suitable matrices are for example $D^{-1}L = I - D^{-1}A$ and the symmetric version $L_{\mathrm{norm}} = D^{-1/2}LD^{-1/2} = I - D^{-1/2}AD^{-1/2}$ known as the *normalized Laplacian*.

There is a large body of literature on clustering in general and spectral clustering in particular. A few references are [189, 266, 324, 330, 351].

3.13.6 Graph isomorphism

No polynomial algorithm for graph isomorphism is known. But for graphs of bounded eigenvalue multiplicity, graph isomorphism can be decided in polynomial time (BABAI, GRIGORYEV & MOUNT [14]).

The graph isomorphism problem (on graphs with n vertices) can be reduced to the problem of finding the size of the automorphism group of a graph (on at most $2n$ vertices) (MATHON [273]). Suppose graphs Γ, Δ given. If they are not connected, replace them by their complements. Now in order to determine whether $\Gamma \cong \Delta$, it suffices to compute $|\mathrm{Aut}(\Gamma)|$, $|\mathrm{Aut}(\Delta)|$, $|\mathrm{Aut}(\Gamma + \Delta)|$. The graphs are isomorphic when the third number is larger than the product of the former two.

Let Γ be a graph with vertex set X of size n. We want to find $|\mathrm{Aut}(\Gamma)|$ in polynomial time. One cannot test all $n!$ permutations of X, so one needs enough structure on Γ to restrict the number of potential automorphisms.

Let V be the real vector space with basis X and natural inner product. Let G be a group of automorphisms of Γ. Then the elements of G can be regarded as orthogonal linear transformations of V, permuting its basis, and the eigenspaces of Γ are G-invariant. Let Y be a G-invariant subset of X, and let S be a G-invariant subspace of V. Let $\mathrm{proj}_S : V \to S$ be the orthogonal projection of V onto S. This commutes with the action of G, and it follows that the partition of Y into fibers $Y_s = \{y \in Y \mid \mathrm{proj}_S(y) = s\}$ (with $s \in S$) forms a system of imprimitivity for G.

If S has dimension m, then a maximal independent subset B of $\mathrm{proj}_S(Y)$ has size at most m. Since the induced action $\bar{G}$ of G on the fibers is determined by the images of the elements in B, we have $|\bar{G}| \leq n^m$, and the potential elements of $\bar{G}$ can be enumerated in polynomial time. The elements that permute $\text{proj}_S(Y)$ will form an explicitly known overgroup $\hat{G}$ of $\bar{G}$.

This basic step can be used recursively, or used for the sum S of a number of eigenspaces. For details, see [14].

FÜRER [165, 166] shows that there are combinatorial ways to distinguish pairs of vertices in a graph that are at least as strong as the known approaches using spectrum and eigenspaces. His approach moreover has the advantage that high precision numerical mathematics (to distinguish different eigenvalues) is avoided.

3.13.7 Searching an eigenspace

There exists a unique strongly regular graph[1] with parameters $(v,k,\lambda,\mu) = (162,56, 10,24)$ found as the second subconstituent of the McLaughlin graph. Its vertex set can be split into two halves such that each half induces a strongly regular graph with parameters $(v,k,\lambda,\mu) = (81,20,1,6)$ (cf. §9.7). How many such splits are there? Can we find them all?

In this and many similar situations one can search an eigenspace. The first graph has spectrum $56^1\, 2^{140}\, (-16)^{21}$ and a split gives an eigenvector with eigenvalue -16 if we take the vector that is 1 on the subgraph and -1 on the rest.

It is easy to construct an explicit basis (u_i) for the 21-dimensional eigenspace, where the j-th coordinate of u_i is δ_{ij}. Construct the 2^{21} eigenvectors that are ± 1 on the first 21 coordinates and inspect the remaining coordinates. If all are ± 1, one has found a split into two regular graphs of valency 20. In this particular case there are 224 such subgraphs, 112 splits, and all subgraphs occurring are strongly regular with the abovementioned parameters.

3.14 Stars and star complements

Consider a graph Γ with vertex set X. By interlacing, the multiplicity of any given eigenvalue changes by at most 1 if we remove a vertex. But there is always a vertex such that removing it actually decreases the multiplicity. And that means that if θ is an eigenvalue of multiplicity m we can find a *star subset* for θ, that is, a subset S of X of size m such that $\Gamma \setminus S$ does not have eigenvalue θ. Now $X \setminus S$ is called a *star complement*.

Why precisely can we decrease the multiplicity? Let u be a θ-eigenvector of A, so that $(\theta I - A)u = 0$, and let x be a vertex with $u_x \neq 0$. Then removing x from Γ decreases the multiplicity of θ.

---

[1] For strongly regular graphs, see Chapter 9. No properties are used except that the substructure of interest corresponds to an eigenvector of recognizable shape.

Indeed, removing x is equivalent to the two actions: (i) forcing $u_x = 0$ for eigenvectors u, and (ii) omitting the condition $\sum_{y \sim x} u_y = \theta u_x$ (row x of the matrix equation $(\theta I - A)u = 0$) for eigenvectors u. Since A is symmetric, the column dependency $(\theta I - A)u = 0$ given by u is also a row dependency, and row x is dependent on the remaining rows, so that (ii) doesn't make a difference. But (i) does, as the vector u shows. So the multiplicity goes down.

This argument shows that the star sets for θ are precisely the sets S of size m such that no θ-eigenvector vanishes on all of S. Also, that any subgraph without eigenvalue θ is contained in a star complement.

Proposition 3.14.1 ([150, 117]) *Let Γ be a graph with eigenvalue θ of multiplicity m. Let S be a subset of the vertex set X of Γ, and let the partition $\{S, X \setminus S\}$ of X induce a partition $A = \begin{bmatrix} B & C \\ C^{\top} & D \end{bmatrix}$ of the adjacency matrix A. If S is a star set for θ (i.e., if $|S| = m$ and D does not have eigenvalue θ), then $B - \theta I = C(D - \theta I)^{-1}C^{\top}$.*

Proof The row space of $A - \theta I$ has rank $n - m$. If S is a star set, then this row space is spanned by the rows of $[C^{\top} \; D - \theta I]$. Alternatively, apply Corollary 2.7.2 to $A - \theta I$. □

This proposition says that the edges inside a star set are determined by the rest of the graph (and the value θ). Especially when m is large, this may be useful.

Stars and star complements have been used to study exceptional graphs with smallest eigenvalue not less than -2, see, e.g., [116, 118, 119]. (One starts with the observation that if θ is the smallest eigenvalue of a graph, then a star complement has smallest eigenvalue larger than θ. But all graphs with smallest eigenvalue larger than -2 are explicitly known.) Several graphs and classes of graphs have been characterized by graph complement. See, e.g., [234, 118].

A *star partition* is a partition of X into star sets S_θ for θ, where θ runs through the eigenvalues of Γ. It was shown in [117] that every graph has a star partition.

3.15 Exercises

Exercise 3.1 Let Γ be an undirected graph with smallest eigenvalue -1. Show that Γ is the disjoint union of complete graphs.

Exercise 3.2 Consider a graph with largest eigenvalue θ_1 and maximum valency $k_{\max}$. Use interlacing to show that $\theta_1 \geq \sqrt{k_{\max}}$. When does equality hold?

Exercise 3.3 Let Γ be a k-regular graph with n vertices and eigenvalues $k = \theta_1 \geq \ldots \geq \theta_n$. Let Γ' be an induced subgraph of Γ with n' vertices and average degree k'.

(i) Prove that $\theta_2 \geq \frac{nk' - n'k}{n - n'} \geq \theta_n$.
(ii) What can be said in case of equality (on either side)?
(iii) Deduce Hoffman's bound (Theorem 3.5.2) from the above inequality.

Exercise 3.4 Deduce Proposition 3.6.3(iii) (the part that says $(m-1)\theta_{t+1} + \theta_{n-t(m-1)} \geq 0$) from Theorem 3.5.4.

Exercise 3.5 ([151]) Let the Ramsey number $R(k_1, k_2)$ be the smallest integer r such that for each coloring of the edges K_r with two colors c_1, c_2 there is a subgraph of size k_i of which all edges have the same color c_i for $i = 1$ or $i = 2$. Show that $\alpha(\Gamma \boxtimes \Delta) \leq R(\alpha(\Gamma) + 1, \alpha(\Delta) + 1) - 1$.

Exercise 3.6 Show that the Lovász parameter $\vartheta(\Gamma)$ is the minimum possible value of s such that there exists a Euclidean representation of Γ that assigns a unit vector in $\mathbb{R}^n$ to each vertex, where the images of any two nonadjacent vertices have inner product $-1/(s-1)$.

Exercise 3.7 Let an *orthonormal labeling* of a graph Γ be the assignment of a unit vector u_x (in some $\mathbb{R}^m$) to each vertex x, where the $u_x^\top u_y = 0$ whenever $x \not\sim y$. Show that $\vartheta(\Gamma) = \min_c \max_x (c^\top u_x)^{-2}$, where the minimum is over all unit vectors c, and the maximum over all vertices x.
  (Hint: Consider the matrix M with $M_{xx} = 1$ and $M_{xy} = 1 - \dfrac{u_x^\top u_y}{(c^\top u_x)(c^\top u_y)}$.)

Exercise 3.8 Show that $\vartheta(\Gamma) \leq d(\Gamma) \leq \chi(\overline{\Gamma})$, where $d(\Gamma)$ is the smallest d such that Γ has an orthonormal labeling in $\mathbb{R}^d$. (Hint: Consider the new orthonormal labeling in $\mathbb{R}^{d^2}$ given by the vectors $u_x \otimes u_x$, and take $c = d^{-1/2} \sum_i e_i \otimes e_i$.)

Exercise 3.9 (cf. [2, 152]) Let $\mathscr{K}_\Gamma$ denote the class of real symmetric matrices M indexed by $V\Gamma$ such that $M_{uv} = 0$ if $u \not\sim v$, and $M_{uv} \neq 0$ if $u \sim v$ (nothing is required for the diagonal of M). The parameter

$$mr(\Gamma) = \min_{M \in \mathscr{K}_\Gamma} \mathrm{rk}\, M.$$

is called the *minimum rank* of Γ. Show that

(i) $mr(K_n) = 1$ and that $mr(\Gamma) \leq n - 1$ with equality if Γ is the path P_n.
(ii) $mr(\Delta) \leq mr(\Gamma)$ if Δ is an induced subgraph of Γ.
(iii) $mr(L(K_n)) = n - 2$.
(iv) $mr(L(\Gamma)) \leq n - 2$ for every line graph $L(\Gamma)$ of a graph Γ of order n, with equality if Γ has a Hamilton path.

Exercise 3.10 ([192, 274]) The *energy* $E(\Gamma)$ of a graph Γ, as defined by Gutman, is $\sum_i |\theta_i|$, the sum of the absolute values of the eigenvalues of the adjacency matrix A. Show that if Γ has n vertices and m edges, then

$$\sqrt{2m + n(n-1)|\det A|^{2/n}} \leq E(\Gamma) \leq \sqrt{2mn}.$$

(Hint: Use the arithmetic-geometric mean inequality and Cauchy-Schwarz.)

Exercise 3.11 ([245]) (i) Let Γ be a graph on n vertices with m edges, so that its average valency is $\bar{k} = 2m/n$. If $\bar{k} \geq 1$ then

$$E(\Gamma) \le \bar{k} + \sqrt{\bar{k}(n-\bar{k})(n-1)}$$

with equality if and only if Γ is mK_2, or K_n, or a strongly regular graph with parameters (n,k,λ,μ), where $\lambda = \mu = k(k-1)/(n-1)$.

(Hint: Use Cauchy-Schwarz.)

(ii) Let Γ be a graph on n vertices. Then

$$E(\Gamma) \le \frac{1}{2}n(1+\sqrt{n})$$

with equality if and only if Γ is a strongly regular graph with parameters (n,k,λ,μ), where $k = (n+\sqrt{n})/2$ and $\lambda = \mu = (n+2\sqrt{n})/4$. There are infinitely many examples with equality.

Exercise 3.12 Prove the conjecture from §3.11.1 for regular graphs.

(Hint: Use Cauchy-Schwarz.)

Exercise 3.13 Suppose the vertex set of a graph Γ is partitioned into m classes of equal size $\ell = n/m$. Let $0 = \mu_1 \le \mu_2 \le \cdots \le \mu_n$ be the Laplace eigenvalues of Γ, and let e denote the total number of edges with endpoints in different classes of the partition. Prove that

$$\ell \sum_{i=2}^{m} \mu_i \le 2e \le \ell \sum_{i=n-m+2}^{n} \mu_i,$$

and in particular

$$\ell(m-1)\mu_2 \le 2e \le \ell(m-1)\mu_n.$$

What can be said in case of equality on either side of both formulas?

Chapter 4
The Second-Largest Eigenvalue

There is a tremendous amount of literature about the second-largest eigenvalue of a regular graph. If the gap between the largest and second-largest eigenvalues is large, then the graph has good connectivity, expansion and randomness properties. (About connectivity, see also §1.7.)

4.1 Bounds for the second-largest eigenvalue

In this connection it is of interest how large this gap can become. Theorems by Alon-Boppana and Serre say that for k-regular graphs on n points, where k is fixed and n tends to infinity, θ_2 cannot be much smaller than $2\sqrt{k-1}$, and that in fact a positive fraction of all eigenvalues is not much smaller.

Proposition 4.1.1 (Alon-Boppana, see ALON [4]) *If $k \geq 3$ then for k-regular graphs on n vertices one has*

$$\theta_2 \geq 2\sqrt{k-1}\,(1 - O(\frac{\log(k-1)}{\log n})).$$

Proposition 4.1.2 (SERRE [322]) *Fix $k \geq 1$. For each $\varepsilon > 0$, there exists a positive constant $c = c(\varepsilon, k)$ such that for any k-regular graph Γ on n vertices, the number of eigenvalues of Γ larger than $(2 - \varepsilon)\sqrt{k-1}$ is at least cn.*

Quenell gives (weaker) explicit bounds:

Proposition 4.1.3 ([298]) *Let Γ be a finite graph with diameter d and minimal degree $k \geq 3$. Then for $2 \leq m \leq 1 + d/4$, the m-th eigenvalue of the adjacency matrix A of Γ satisfies $\theta_m > 2\sqrt{k-1}\cos(\frac{\pi}{r+1})$, where $r = \lfloor d/(2m-2) \rfloor$.*

ALON [4] conjectured, and FRIEDMAN [163] proved that large random k-regular graphs have second-largest eigenvalue smaller than $2\sqrt{k-1} + \varepsilon$ (for fixed k, $\varepsilon > 0$ and n sufficiently large). Friedman remarks that numerical experiments seem to indicate that random k-regular graphs in fact satisfy $\theta_2 < 2\sqrt{k-1}$.

A connected k-regular graph is called a *Ramanujan graph* when $|\theta| \leq 2\sqrt{k-1}$ for all eigenvalues $\theta \neq k$. (This notion was introduced in [265].) It is not difficult to find such graphs. For example, complete graphs, or Paley graphs, will do. Highly nontrivial was the construction of infinite sequences of Ramanujan graphs with given, constant, valency k and size n tending to infinity. LUBOTZKY, PHILLIPS & SARNAK [265] and MARGULIS [271] constructed for each prime $p \equiv 1 \pmod 4$ an infinite series of Ramanujan graphs with valency $k = p+1$.

In the general case where n is not assumed to be large, a trivial estimate using $\text{tr} A^2 = kn$ shows that $\lambda^2 \geq k(n-k)/(n-1)$ where $\lambda = \max_{2 \leq i \leq n} |\theta_i|$. This holds with equality for complete graphs, and is close to the truth for Paley graphs (which have $k \approx \frac{1}{2}n$, $\lambda \approx \frac{1}{2}\sqrt{n}$).

4.2 Large regular subgraphs are connected

We note the following trivial but useful result.

Proposition 4.2.1 *Let Γ be a graph with second-largest eigenvalue θ_2. Let Δ be a nonempty regular induced subgraph with largest eigenvalue $\rho > \theta_2$. Then Δ is connected.*

Proof The multiplicity of the eigenvalue ρ of Δ is the number of connected components of Δ, and by interlacing this is 1. $\qquad\square$

4.3 Randomness

Let Γ be a regular graph of valency k on n vertices, and assume that (for some real constant λ) we have $|\theta| \leq \lambda$ for all eigenvalues $\theta \neq k$. The ratio λ/k determines randomness and expansion properties of Γ: the smaller λ/k, the more random, and the better expander Γ is.

For example, the following proposition says that most points have approximately the expected number of neighbors in a given subset of the vertex set. Here $\Gamma(x)$ denotes the set of neighbors of the vertex x in the graph Γ.

Proposition 4.3.1 *Let R be a subset of size r of the vertex set X of Γ. Then*

$$\sum_{x \in X} (|\Gamma(x) \cap R| - \frac{kr}{n})^2 \leq \frac{r(n-r)}{n} \lambda^2.$$

Proof Apply interlacing to A^2 and the partition $\{R, X \setminus R\}$ of X. The sum of all entries of the matrix A^2 in the (R,R)-block equals the number of paths $y \sim x \sim z$, with $y,z \in R$ and $x \in X$, that is, $\sum_x (|\Gamma(x) \cap R|)^2$. $\qquad\square$

Rather similarly, the following proposition, a version of the *expander mixing lemma* from ALON & CHUNG [7], says that there are about the expected number of edges between two subsets.

Proposition 4.3.2 *Let S and T be two subsets of the vertex set of Γ, of sizes s and t, respectively. Let $e(S,T)$ be the number of ordered edges xy with $x \in S$ and $y \in T$. Then*

$$|e(S,T) - \frac{kst}{n}| \leq \lambda \sqrt{st(1 - \frac{s}{n})(1 - \frac{t}{n})} \leq \lambda \sqrt{st}.$$

Proof Write the characteristic vectors χ_S and χ_T of the sets S and T as a linear combination of a set of orthonormal eigenvectors of A: $\chi_S = \sum \alpha_i u_i$ and $\chi_T = \sum \beta_i u_i$, where $Au_i = \theta_i u_i$. Then $e(S,T) = \chi_S^\top A \chi_T = \sum \alpha_i \beta_i \theta_i$. We have $\alpha_1 = s/\sqrt{n}$ and $\beta_1 = t/\sqrt{n}$ and $\theta_1 = k$. Now $|e(S,T) - \frac{kst}{n}| = |\sum_{i>1} \alpha_i \beta_i \theta_i| \leq \lambda \sum_{i>1} |\alpha_i \beta_i|$ and $\sum_{i>1} \alpha_i^2 \leq (\chi_S, \chi_S) - s^2/n = s(n-s)/n$, and $\sum_{i>1} \beta_i^2 \leq t(n-t)/n$, so that $|e(S,T) - \frac{kst}{n}| \leq \lambda \sqrt{st(n-s)(n-t)}/n$. $\qquad \square$

If S and T are equal or complementary, this says that

$$|e(S,T) - \frac{kst}{n}| \leq \lambda \frac{s(n-s)}{n}.$$

In particular, the average valency k_S of an induced subgraph S of size s satisfies $|k_S - \frac{ks}{n}| \leq \lambda \frac{n-s}{n}$. For example, the Hoffman-Singleton graph has $\theta_2 = 2$, $\theta_n = -3$, so $\lambda = 3$ and we find equality for subgraphs $\overline{K_{15}}$ ($s = 15$, $k_S = 0$), $10K_2$ ($s = 20$, $k_S = 1$) and $5C_5$ ($s = 25$, $k_S = 2$).

4.4 Random walks

Let Γ be a connected graph, possibly with loops, with n vertices and m edges, where $m > 0$. Let A be its adjacency matrix, and D the diagonal matrix of vertex degrees (so that $D_{xx} = d_x$ is the degree of the vertex x and $A\mathbf{1} = D\mathbf{1}$). A *random walk* on Γ is a sequence of vertices $x_0, \ldots, x_{t-1}, x_t, \ldots$ starting at some vertex x_0, where at the t-th step the vertex x_t is chosen at random among the neighbors of x_{t-1}.

Given an initial probability distribution $p = (p_x)_x$ over the vertices, we have after t steps a distribution $(AD^{-1})^t p$ for the 'current vertex'. It follows that the distribution $\frac{1}{2m} D\mathbf{1} = (d_x/2m)_x$ is stationary. Conversely, a stationary distribution p satisfies $AD^{-1}p = p$, hence $LD^{-1}p = 0$, and since Γ is connected, p is uniquely determined. We see that if v is any vertex, the expected time T_v for a random walk starting at v to return to v is given by $T_v = 2m/d_v$, and if vw is any edge, the expected time T between two traversals of this edge in the same direction is given by $T = d_v T_v = 2m$.

Now suppose that Γ is regular of degree k. If $\lambda < k$, that is, if Γ is not bipartite, then an arbitrary initial distribution p converges to the stationary distribution. Indeed, $||(\frac{1}{k}A)^t p - \frac{1}{n}\mathbf{1}||^2 < (\lambda/k)^{2t}$ (as one sees in the usual way: writing $p = \sum \alpha_i u_i$, where $Au_i = \theta_i u_i$). This shows that the mixing rate is determined by λ/k. Similar things hold for nonregular Γ.

(Thus, if λ/k is small, then any initial distribution p converges quickly to the stationary distribution. It follows that Γ has good connectivity and expansion properties.)

Fix a vertex v, and let q_x be the expected number of steps for a random walk starting at x to reach v (the *access time* of v from x), where $q_v = 0$. For $x \neq v$ we have $q_x = 1 + \sum_{y \sim x} \frac{1}{d_x} q_y$, so that $(I - D^{-1}A)q = \mathbf{1} - T_v e_v$, and q is determined by $Lq = D\mathbf{1} - 2me_v$ and $q_v = 0$. This leads to explicit formulas for the access time.

This is a large subject. See Lovász [262] for a survey.

4.5 Expansion

An *expander* is a (preferably sparse) graph with the property that the number of points at distance at most 1 from any given (not too large) set is at least a fixed constant (larger than 1) times the size of the given set. Expanders became famous because of their role in sorting networks (cf. AJTAI, KOMLÓS & SZEMERÉDI [3]) and have since found many other applications. Proposition 4.3.1 already implies that there cannot be too many vertices without neighbors in a given subset of the vertex set. A better bound was given by Tanner in order to show that generalized polygons are good expanders.

Proposition 4.5.1 (cf. TANNER [334]) *Let Γ be connected and regular of degree k, and let $|\theta| \leq \lambda$ for all eigenvalues $\theta \neq k$ of Γ. Let R be a set of r vertices of Γ and let $\Gamma(R)$ be the set of vertices adjacent to some point of R. Then*

$$\frac{|\Gamma(R)|}{n} \geq \frac{\rho}{\rho + \frac{\lambda^2}{k^2}(1-\rho)}$$

where $\rho = r/n$.

Proof Let χ be the characteristic vector of R. Write it as a linear combination of a set of orthonormal eigenvectors of A: $\chi = \sum \alpha_i u_i$ where $Au_i = \theta_i u_i$. Then $A\chi = \sum \alpha_i \theta_i u_i$ and $(A\chi, A\chi) = \sum \alpha_i^2 \theta_i^2$, so that $\|A\chi\|^2 \leq \alpha_0^2(\theta_0^2 - \lambda^2) + \lambda^2 \sum \alpha_i^2 = (\chi, u_0)^2(k^2 - \lambda^2) + \lambda^2(\chi, \chi) = \frac{r^2}{n}(k^2 - \lambda^2) + r\lambda^2$. Now let ψ be the characteristic vector of $\Gamma(R)$. Then $k^2 r^2 = (A\chi, \mathbf{1})^2 = (A\chi, \psi)^2 \leq \|A\chi\|^2 \|\psi\|^2 \leq |\Gamma(R)|.(\frac{r^2}{n}(k^2 - \lambda^2) + r\lambda^2)$, proving our claim. $\square$

The above used two-sided bounds on the eigenvalues different from the valency. It suffices to bound θ_2. Let the *edge expansion constant* $h(\Gamma)$ (a.k.a. *isoperimetric constant* or *Cheeger number*) of a graph Γ be the minimum of $e(S,T)/|S|$ where the minimum is taken over all partitions $\{S,T\}$ of the vertex set with $|S| \leq |T|$, and where $e(S,T)$ is the number of edges meeting both S and T. We have

Proposition 4.5.2 ([280]) *Let Γ be regular of degree k, not K_n with $n \leq 3$. Then*

$$\frac{1}{2}(k - \theta_2) \le h(\Gamma) \le \sqrt{k^2 - \theta_2^2}.$$

Proof For the lower bound, apply interlacing to A and a partition $\{S, T\}$ of the vertex set, with $s = |S|$ and $t = |T|$. Put $e = e(S, T)$. One finds $ne/st \ge k - \theta_2$, so that $e/s \ge (t/n)(k - \theta_2) \ge \frac{1}{2}(k - \theta_2)$. For the upper bound, consider a nonnegative vector w indexed by the point set X of Γ, with support of size at most $\frac{1}{2}n$. If w_x takes t different nonzero values $a_1 > \dots > a_t > 0$, then let $S_i = \{x \mid w_x \ge a_i\}$ $(1 \le i \le t)$, and let $m_i = |S_i \setminus S_{i-1}|$ (with $S_0 = \emptyset$). Let $h = h(\Gamma)$. Now

$$h \sum_x w_x \le \sum_{x \sim y} |w_x - w_y|.$$

Indeed, all S_i have size at most $\frac{1}{2}n$, so at least $h|S_i|$ edges stick out of S_i, and these contribute at least $h(m_1 + \dots + m_i)(a_i - a_{i+1})$ to $\sum_{x \sim y} |w_x - w_y|$ (with $a_{t+1} = 0$). The total contribution is at least $h \sum_i m_i a_i = h \sum_x w_x$.

Let u be an eigenvector of A with $Au = \theta_2 u$. We may assume that $u_x > 0$ for at most $\frac{1}{2}n$ points x (otherwise replace u by $-u$). Define a vector v by $v_x = \max(u_x, 0)$. Since $(Av)_x = \sum_{y \sim x} v_y \ge \sum_{y \sim x} u_y = (Au)_x = \theta_2 u_x = \theta_2 v_x$ if $v_x > 0$, we have $v^\top Av = \sum_x v_x (Av)_x \ge \theta_2 \sum_x v_x^2$.

Note that $\sum_{x \sim y} (v_x \pm v_y)^2 = k \sum_x v_x^2 \pm v^\top Av$.

Apply the above to the nonnegative vector w given by $w_x = v_x^2$. We find $h \sum_x v_x^2 \le \sum_{x \sim y} |v_x^2 - v_y^2| \le (\sum_{x \sim y} (v_x - v_y)^2 \cdot \sum_{x \sim y} (v_x + v_y)^2)^{1/2} = ((k \sum_x v_x^2)^2 - (v^\top Av)^2)^{1/2} \le (\sum_x v_x^2)\sqrt{k^2 - \theta_2^2}$, assuming $\theta_2 \ge 0$. □

For similar results for not necessarily regular graphs, see §4.8.

4.6 Toughness and Hamiltonicity

As application of the above ideas, one can give bounds for the toughness of a graph in terms of the eigenvalues.

A connected, noncomplete graph Γ is called t-*tough* if one has $|S| \ge tc$ for every disconnecting set of vertices S such that the graph induced on its complement has $c \ge 2$ connected components. The *toughness* $\tau(\Gamma)$ of a graph Γ is the largest t such that Γ is t-tough. For example, the Petersen graph has toughness $4/3$.

This concept was introduced by CHVÁTAL [94], who hoped that t-tough graphs would be *Hamiltonian* (i.e., have a circuit passing through all vertices) for sufficiently large t. People tried to prove this for $t = 2$, the famous "2-tough conjecture", but examples were given in [23] of t-tough non-Hamiltonian graphs for all $t < 9/4$. Whether a larger bound on τ suffices is still open.

Still, being tough seems to help. In [22] it was shown that a t-tough graph Γ on $n \ge 3$ vertices with minimum degree δ is Hamiltonian when $(t + 1)(\delta + 1) > n$.

Proposition 4.6.1 ([50]) *Let Γ be a connected noncomplete regular graph of valency k and let $|\theta| \leq \lambda$ for all eigenvalues $\theta \neq k$. Then $\tau(\Gamma) > k/\lambda - 2$.*

This proposition gives the right bound, in the sense that there are infinitely many graphs with $\tau(\Gamma) \leq k/\lambda$. The constant 2 can be improved a little. The result can be refined by separating out the smallest and the second-largest eigenvalue. The main tool in the proof is Proposition 4.3.1.

See also the remarks following Theorem 9.3.2.

KRIVELEVICH & SUDAKOV [247] show that, when n is large enough, a graph on n vertices, regular of degree $k = \theta_1$, and with second-largest eigenvalue θ_2 satisfying

$$\frac{\theta_2}{\theta_1} < \frac{(\log \log n)^2}{1000 \log n \log \log \log n}$$

is Hamiltonian. PYBER [297] shows that it follows that every sufficiently large strongly regular graph is Hamiltonian.

4.6.1 The Petersen graph is not Hamiltonian

An amusing application of interlacing (cf. [281, 214]) shows that the Petersen graph is not Hamiltonian. Indeed, a Hamilton circuit in the Petersen graph would give an induced C_{10} in its line graph. Now the line graph of the Petersen graph has spectrum $4^1\, 2^5\, (-1)^4\, (-2)^5$ and by interlacing the seventh eigenvalue $2\cos \frac{3}{5}\pi = (1 - \sqrt{5})/2$ of C_{10} should be at most -1, a contradiction.

4.7 Diameter bound

CHUNG [92] gave the following diameter bound.

Proposition 4.7.1 *Let Γ be a connected noncomplete graph on $n \geq 2$ vertices, regular of valency k, and with diameter d. Let $|\theta| \leq \lambda$ for all eigenvalues $\theta \neq k$. Then*

$$d \leq \left\lceil \frac{\log(n-1)}{\log(k/\lambda)} \right\rceil.$$

Proof The graph Γ has diameter at most m when $A^m > 0$. Let A have orthonormal eigenvectors u_i with $Au_i = \theta_i u_i$. Then $A = \sum_i \theta_i u_i^\top u_i$. Take $u_1 = \frac{1}{\sqrt{n}}\mathbf{1}$. Now $(A^m)_{xy} = \sum_i \theta_i^m (u_i^\top u_i)_{xy} \geq \frac{k^m}{n} - \lambda^m \sum_{i>1} |(u_i)_x| \cdot |(u_i)_y|$ and $\sum_{i>1} |(u_i)_x| \cdot |(u_i)_y| \leq (\sum_{i>1} |(u_i)_x|^2)^{1/2}(\sum_{i>1} |(u_i)_y|^2)^{1/2} = (1 - |(u_1)_x|^2)^{1/2}(1 - |(u_1)_y|^2)^{1/2} = 1 - \frac{1}{n}$, so that $(A^m)_{xy} > 0$ if $k^m > (n-1)\lambda^m$. □

4.8 Separation

Let Γ be a graph with Laplace matrix L and Laplace eigenvalues $0 = \mu_1 \leq \ldots \leq \mu_n$. The Laplace matrix of a subgraph Γ' of Γ is not a submatrix of L, unless Γ' is a component. So the interlacing techniques of §2.5 do not work in such a straightforward manner here. But we can obtain results if we consider off-diagonal submatrices of L.

Proposition 4.8.1 *Let X and Y be disjoint sets of vertices of Γ such that there is no edge between X and Y. Then*

$$\frac{|X||Y|}{(n-|X|)(n-|Y|)} \leq \left(\frac{\mu_n - \mu_2}{\mu_n + \mu_2}\right)^2.$$

Proof Put $\mu = \frac{1}{2}(\mu_n + \mu_2)$ and define a matrix A of order $2n$ by

$$A = \begin{bmatrix} 0 & L - \mu I \\ L - \mu I & 0 \end{bmatrix}.$$

Let A have eigenvalues $\theta_1 \geq \ldots \geq \theta_{2n}$. Then $\theta_{2n+1-i} = -\theta_i$ ($1 \leq i \leq 2n$) and $\theta_1 = \mu$ and $\theta_2 = \frac{1}{2}(\mu_n - \mu_2)$. The sets X and Y give rise to a partitioning of A (with rows and columns indexed by $Y, \overline{Y}, \overline{X}, X$) with quotient matrix

$$B = \begin{bmatrix} 0 & 0 & -\mu & 0 \\ 0 & 0 & -\mu + \mu\frac{|X|}{n-|Y|} & -\mu\frac{|X|}{n-|Y|} \\ -\mu\frac{|Y|}{n-|X|} & -\mu + \mu\frac{|Y|}{n-|X|} & 0 & 0 \\ 0 & -\mu & 0 & 0 \end{bmatrix}.$$

Let B have eigenvalues $\eta_1 \geq \ldots \geq \eta_4$. Then $\eta_1 = \theta_1 = \mu$ and $\eta_4 = \theta_{2n} = -\mu$, and $\eta_1 \eta_2 \eta_3 \eta_4 = \det B = \mu^4 \frac{|X||Y|}{(n-|X|)(n-|Y|)} > 0$. Using interlacing, we find

$$\mu^2 \frac{|X||Y|}{(n-|X|)(n-|Y|)} = -\eta_2\eta_3 \leq -\theta_2\theta_{2n-1} = (\tfrac{1}{2}(\mu_n - \mu_2))^2,$$

which gives the required inequality. $\qquad\square$

One can rewrite Tanner's inequality (applied with $R = X$, $\Gamma(R) = V\Gamma \setminus Y$) in the form $|X||Y|/(n-|X|)(n-|Y|) \leq (\lambda/k)^2$ where $\lambda = \max(\theta_2, -\theta_n)$, and this is slightly weaker than the above, equivalent only when $\theta_n = -\theta_2$.

The vertex sets X and Y with the above property are sometimes called *disconnected vertex sets*. In the complementary graph, X and Y become sets such that all edges between X and Y are present. Such a pair is called a *biclique*.

For applications another form is sometimes handy:

Corollary 4.8.2 *Let Γ be a connected graph on n vertices, and let X and Y be disjoint sets of vertices such that there is no edge between X and Y. Then*

$$\frac{|X||Y|}{n(n-|X|-|Y|)} \le \frac{(\mu_n - \mu_2)^2}{4\mu_2\mu_n}.$$

Proof Let K be the constant for which Proposition 4.8.1 says $|X||Y| \le K(n - |X|)(n - |Y|)$. Then $|X||Y|(1-K) \le n(n-|X|-|Y|)K$. $\square$

The above proposition gives bounds on vertex connectivity. For edge connectivity one has

Proposition 4.8.3 (ALON & MILMAN [8]) *Let A and B be subsets of VΓ such that each point of A has distance at least ρ to each point of B. Let F be the set of edges that do not have both ends in A or both in B. Then*

$$|F| \ge \rho^2\mu_2 \frac{|A||B|}{|A|+|B|}.$$

For $\rho = 1$ this yields:

Corollary 4.8.4 *Let Γ be a graph on n vertices, A a subset of VΓ, and F the set of edges with one end in A and one end outside A. Then*

$$|F| \ge \mu_2|A|(1 - \frac{|A|}{n}).$$

Let χ be the characteristic vector of A. Then equality holds if and only if $\chi - \frac{|A|}{n}\mathbf{1}$ is a Laplace eigenvector with eigenvalue μ_2.

Proof Let u_i be an orthonormal system of Laplace eigenvectors, so that $Lu_i = \mu_i u_i$. Take $u_1 = \frac{1}{\sqrt{n}}\mathbf{1}$. Let $\chi = \sum \alpha_i u_i$. Now $|A| = (\chi, \chi) = \sum \alpha_i^2$ and $\alpha_1 = (\chi, u_1) = \frac{1}{\sqrt{n}}|A|$. We find $|F| = \sum_{a \in A, b \notin A, a \sim b} 1 = \sum_{x \sim y}(\chi_x - \chi_y)^2 = \chi^\top L\chi = \sum \alpha_i^2 \mu_i \ge (\sum_{i>1} \alpha_i^2)\mu_2$. $\square$

This is best possible in many situations.

Example The Hoffman-Singleton graph has Laplace spectrum $0^1 5^{28} 10^{21}$, and we find $|F| \ge |A||B|/10$. This holds with equality for the 10-40 split into a Petersen subgraph and its complement.

4.8.1 Bandwidth

A direct consequence of Proposition 4.8.1 is an inequality of HELMBERG et al. [213] concerning the bandwidth of a graph. A symmetric matrix M is said to have bandwidth w if $(M)_{i,j} = 0$ for all i, j satisfying $|i - j| > w$. The bandwidth $w(\Gamma)$ of a graph Γ is the smallest possible bandwidth for its adjacency matrix (or Laplace matrix). This number (and the vertex order realizing it) is of interest for some combinatorial optimization problems.

Theorem 4.8.5 *Suppose Γ is not edgeless and define $b = \left\lceil n\frac{\mu_2}{\mu_n} \right\rceil$. Then*

$$w(\Gamma) \geq \begin{cases} b & \text{if } n-b \text{ is even,} \\ b-1 & \text{if } n-b \text{ is odd.} \end{cases}$$

Proof Order the vertices of Γ such that L has bandwidth $w = w(\Gamma)$. If $n-w$ is even, let X be the first $\frac{1}{2}(n-w)$ vertices and let Y be the last $\frac{1}{2}(n-w)$ vertices. Then Proposition 4.8.1 applies and thus we find the first inequality. If $n-w$ is odd, take for X and Y the first and last $\frac{1}{2}(n-w-1)$ vertices and the second inequality follows. If b and w have different parity, then $w-b \geq 1$ and so the better inequality holds. □

In case $n-w$ is odd, the bound can be improved a little by applying Proposition 4.8.1 with $|X| = \frac{1}{2}(n-w+1)$ and $|Y| = \frac{1}{2}(n-w-1)$. It is clear that the result remains valid if we consider graphs with weighted edges.

4.8.2 Perfect matchings

A more recent application of Proposition 4.8.1 is the following sufficient condition for existence of a perfect matching (a *perfect matching* in a graph is a subset of the edges such that every vertex of the graph is incident with exactly one edge of the subset).

Theorem 4.8.6 ([59]) *Let Γ be a graph with n vertices, and Laplace eigenvalues $0 = \mu_1 \leq \mu_2 \leq \ldots \leq \mu_n$. If n is even and $\mu_n \leq 2\mu_2$, then Γ has a perfect matching.*

Except for Proposition 4.8.1, we need two more tools for the proof. The first one is Tutte's famous characterization of graphs with a perfect matching. The second one is an elementary observation.

Theorem 4.8.7 (TUTTE [338]) *A graph $\Gamma = (V,E)$ has no perfect matching if and only if there exists a subset $S \subset V$ such that the subgraph of Γ induced by $V \setminus S$ has more than $|S|$ odd components.*

Lemma 4.8.8 *Let $x_1 \ldots x_n$ be n positive integers such that $\sum_{i=1}^{n} x_i = k \leq 2n-1$. Then, for every integer ℓ satisfying $0 \leq \ell \leq k$, there exists an $I \subset \{1,\ldots,n\}$ such that $\sum_{i \in I} x_i = \ell$.*

Proof Induction on n. The case $n = 1$ is trivial. If $n \geq 2$, assume $x_1 \geq \ldots \geq x_n$. Then $n-1 \leq k - x_1 \leq 2(n-1) - 1$ and we apply the induction hypothesis to $\sum_{i=2}^{n} x_i = k - x_1$ with the same ℓ if $\ell \leq n-1$, and $\ell - x_1$ otherwise. □

Proof of Theorem 4.8.6. Assume $\Gamma = (V,E)$ has no perfect matching. By Tutte's theorem, there exists a set $S \subset V$ of size s (say) such that the subgraph Γ' of Γ induced by $V \setminus S$ has $q > s$ odd components. But since n is even, $s + q$ is even, and hence $q \geq s+2$.

First assume $n \leq 3s + 3$. Then Γ' has at most $2s + 3$ vertices and at least $s + 2$ components. By Lemma 4.8.8, Γ' and hence Γ has a pair of disconnected vertex sets X and Y with $|X| = \lfloor \frac{1}{2}(n-s) \rfloor$ and $|Y| = \lceil \frac{1}{2}(n-s) \rceil$. Now Proposition 4.8.1 implies

$$\left(\frac{\mu_n - \mu_2}{\mu_n + \mu_2} \right)^2 \geq \frac{|X| \cdot |Y|}{ns + |X| \cdot |Y|} = \frac{(n-s)^2 - \varepsilon}{(n+s)^2 - \varepsilon},$$

where $\varepsilon = 0$ if $n-s$ is even and $\varepsilon = 1$ if $n-s$ is odd. Using $n \geq 2s+2$ we obtain

$$\frac{\mu_n - \mu_2}{\mu_n + \mu_2} > \frac{n-s-1}{n+s} \geq \frac{s+1}{3s+2} > \frac{1}{3}.$$

Hence $2\mu_2 < \mu_n$.

Next assume $n \geq 3s+4$. Now Γ', and hence Γ, has a pair of disconnected vertex sets X and Y with $|X| + |Y| = n - s$ and $\min\{|X|, |Y|\} \geq s+1$, so $|X| \cdot |Y| \geq (s+1)(n-2s-1) > ns - 2s^2$. Now Proposition 4.8.1 implies

$$\left(\frac{\mu_n - \mu_2}{\mu_n + \mu_2} \right)^2 \geq \frac{|X| \cdot |Y|}{ns + |X| \cdot |Y|} \geq \frac{ns - 2s^2}{2ns - 2s^2} = \frac{1}{2} - \frac{s}{2n - 2s} > \frac{1}{4},$$

since $n \geq 3s+4$. So

$$\frac{\mu_n - \mu_2}{\mu_n + \mu_2} > \frac{1}{2} > \frac{1}{3},$$

and again $2\mu_2 < \mu_n$.   □

The complete bipartite graphs $K_{l,m}$ with $l \leq m$ have Laplace eigenvalues $\mu_2 = m$ and $\mu_n = n = l + m$. This shows that $2\mu_2$ can get arbitrarily close to μ_n for graphs with n even and no perfect matching.

If the graph is regular, the result can be improved considerably.

Theorem 4.8.9 ([59, 96]) *A connected k-regular graph on n vertices, where n is even, with (ordinary) eigenvalues $k = \theta_1 \geq \theta_2 \ldots \geq \theta_n$, which satisfies*

$$\theta_3 \leq \begin{cases} k - 1 + \frac{3}{k+1} & \text{if } k \text{ is even,} \\ k - 1 + \frac{4}{k+2} & \text{if } k \text{ is odd,} \end{cases}$$

has a perfect matching.

Proof. Let $\Gamma = (V, E)$ be a k-regular graph with $n = |V|$ even and no perfect matching. By Theorem 4.8.7 there exists a set $S \subset V$ of size s such that $V \setminus S$ induces a subgraph with $q \geq s+2$ odd components $\Gamma_1, \Gamma_2, \ldots, \Gamma_q$ (say). Let t_i denote the number of edges in Γ between S and Γ_i, and let n_i be the number of vertices of Γ_i. Then clearly $\sum_{i=1}^q t_i \leq ks$, $s \geq 1$, and $t_i \geq 1$ (since Γ is connected). Hence $t_i < k$ and $n_i > 1$ for at least three values of i, say $i = 1, 2$, and 3. Let ℓ_i denote the largest eigenvalue of Γ_i, and assume $\ell_1 \geq \ell_2 \geq \ell_3$. Then eigenvalue interlacing applied to the subgraph induced by the union of Γ_1, Γ_2, and Γ_3 gives $\ell_i \leq \theta_i$ for $i = 1, 2, 3$.

Consider Γ_3 with n_3 vertices and e_3 edges (say). Then $2e_3 = kn_3 - t_3 \leq n_3(n_3 - 1)$. We saw that $t_3 < k$ and $n_3 > 1$, hence $k < n_3$. Moreover, the average degree $\bar{d}_3$ of Γ_3 equals $2e_3/n_3 = k - t_3/n_3$. Because n_3 is odd and $kn_3 - t_3$ is even, k and t_3 have the same parity, therefore $t_3 < k$ implies $t_3 \leq k - 2$. Also, $k < n_3$ implies $k \leq n_3 - 1$ if k

is even, and $k \leq n_3 - 2$ if k is odd. Hence

$$\bar{d}_3 \geq \begin{cases} k - \frac{k-2}{k+1} & \text{if } k \text{ is even,} \\ k - \frac{k-2}{k+2} & \text{if } k \text{ is odd.} \end{cases}$$

Note that $t_3 < n_3$ implies that Γ_3 cannot be regular. Next we use the fact that the largest adjacency eigenvalue of a graph is bounded from below by the average degree with equality if and only if the graph is regular (Proposition 3.1.2). Thus $\bar{d}_3 < \ell_3$. We saw that $\ell_3 \leq \theta_3$, which finishes the proof.   □

From the above it is clear that n even and $\theta_2 \leq k - 1$ implies existence of a perfect matching. In terms of the Laplace matrix this translates into:

Corollary 4.8.10 *A regular graph with an even number of vertices and algebraic connectivity at least 1 has a perfect matching.*

But we can say more. The Laplace matrix of a disjoint union of $n/2$ edges has eigenvalues 0 and 2. By the Courant-Weyl inequalities (Theorem 2.8.1), this implies that deletion of the edges of a perfect matching of a graph Γ reduces the eigenvalues of the Laplace matrix of Γ by at most 2. Hence:

Corollary 4.8.11 *A regular graph with an even number of vertices and algebraic connectivity μ_2 has at least $\lfloor (\mu_2 + 1)/2 \rfloor$ pairwise disjoint perfect matchings.*

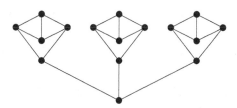

Fig. 4.1 A 3-regular graph with no perfect matching

CIOABĂ, GREGORY & HAEMERS [97] have improved the sufficient condition for a perfect matching from Theorem 4.8.9 to $\theta_3 < g_k$, where $g_3 = 2.85577...$ (the largest root of $x^3 - x^2 - 6x + 2$), $g_k = (k - 2 + \sqrt{k^2 + 12})/2$ if $k \geq 4$ and even, and $g_k = (k - 3 + \sqrt{(k+1)^2 + 16})/2$ if $k \geq 5$ and odd. They also prove that this bound is best possible by giving examples of k-regular graphs with n even and $\lambda_3 = g_k$ that have no perfect matching. The example for $k = 3$ is presented in Figure 4.1.

4.9 Block designs

In case we have a nonsymmetric matrix N (say), we can still use interlacing by considering the matrix

$$A = \begin{bmatrix} 0 & N \\ N^\top & 0 \end{bmatrix}.$$

We find results in terms of the eigenvalues of A, which now satisfy $\theta_i = -\theta_{n-i+1}$ for $i = 1, \ldots, n$. The positive eigenvalues of A are the singular values of N, they are also the square roots of the nonzero eigenvalues of $NN^\top$ (and of $N^\top N$).

Suppose N is the 0-1 incidence matrix of an incidence structure (P, B) with point set P (rows) and block set B (columns). Then we consider the so-called *incidence graph* Γ of (P, B), which is the bipartite graph with vertex set $P \cup B$, where two vertices are adjacent if they correspond to an incident point-block pair. An edge of Γ is called a *flag* of (P, B).

An incidence structure (P, B) is called a t-(v, k, λ) design if $|P| = v$, all blocks are incident with k points, and for every t-set of points there are exactly λ blocks incident with all t points. For example, (P, B) is a 1-(v, k, r) design precisely when N has constant column sums k (i.e., $N^\top \mathbf{1} = k\mathbf{1}$) and constant row sums r (i.e., $N\mathbf{1} = r\mathbf{1}$), in other words, when Γ is semiregular with degrees k and r. Moreover, (P, B) is a 2-(v, k, λ) design if and only if $N^\top \mathbf{1} = k\mathbf{1}$ and $NN^\top = \lambda J + (r - \lambda)I$. Note that, for $t \geq 1$, a t-design is also a $(t-1)$-design. In particular, a 2-(v, k, λ) design is also a 1-(v, k, r) design with $r = \lambda(v-1)/(k-1)$. A *Steiner system* $S(t, k, v)$ is a t-(v, k, λ) design with $\lambda = 1$.

Theorem 4.9.1 *Let (P, B) be a 1-(v, k, r) design with b blocks and let (P', B') be a substructure with m' flags. Define $b = |B|$, $v' = |P'|$ and $b' = |B'|$. Then*

$$\left(m' \frac{v}{v'} - b'k \right)\left(m' \frac{b}{b'} - v'r \right) \leq \theta_2^2 (v - v')(b - b').$$

Equality implies that all four substructures induced by P' or $V \setminus V'$ and B' or $B \setminus B'$ form a 1-design (possibly degenerate).

Proof We apply Corollary 2.5.4. The substructure (P', B') gives rise to a partition of A with the quotient matrix

$$B = \begin{bmatrix} 0 & 0 & \frac{m'}{v'} & r - \frac{m'}{v'} \\ 0 & 0 & \frac{b'k-m'}{v-v'} & r - \frac{b'k-m'}{v-v'} \\ \frac{m'}{b'} & k - \frac{m'}{b'} & 0 & 0 \\ \frac{v'r-m'}{b-b'} & k - \frac{v'r-m'}{b-b'} & 0 & 0 \end{bmatrix}.$$

We easily have $\theta_1 = -\theta_n = \eta_1 = -\eta_4 = \sqrt{rk}$ and

$$\det(B) = rk \left(\frac{m' \frac{v}{v'} - b'k}{v - v'} \right)\left(\frac{m' \frac{b}{b'} - v'r}{b - b'} \right).$$

Interlacing gives

$$\frac{\det(B)}{rk} = -\eta_2 \eta_3 \leq -\theta_2 \theta_{n-1} = \theta_2^2,$$

which proves the first statement. If equality holds, then $\theta_1 = \eta_1$, $\theta_2 = \eta_2$, $\theta_{n-1} = \eta_3$, and $\theta_n = \eta_4$, so we have tight interlacing, which implies the second statement. □

The above result becomes especially useful if we can express θ_2 in terms of the design parameters. For instance, if (P,B) is a 2-(v,k,λ) design, then $\theta_2^2 = r - \lambda = \lambda\frac{v-k}{k-1}$ (see the exercises) and if (P,B) is a generalized quadrangle of order (s,t), then $\lambda_2^2 = s+t$ (see §9.6). Let us consider two special cases. (A 2-design (P,B) with $|P| = |B|$ is called *symmetric*.)

Corollary 4.9.2 *If a symmetric 2-(v,k,λ) design (P,B) has a symmetric 2-(v',k',λ') subdesign (P',B') (possibly degenerate), then*

$$(k'v - kv')^2 \le (k - \lambda)(v - v')^2.$$

If equality holds, then the subdesign $(P',B\setminus B')$ is a 2-$(v',v'(k-k')/(v-v'),\lambda - \lambda')$ design (possibly degenerate).

Proof In Theorem 4.9.1, take $b = v$, $r = k$, $b' = v'$, $m' = v'k'$, and $\theta_2^2 = k - \lambda$. □

Corollary 4.9.3 *Let X be a subset of the points and let Y be a subset of the blocks of a 2-(v,k,λ) design (P,B) such that no point of X is incident with a block of Y. Then*

$$kr|X||Y| \le (r - \lambda)(v - |X|)(b - |Y|).$$

If equality holds, then the substructure $(X,B') = (X,B\setminus Y)$ is a 2-design.

Proof Take $m' = 0$, $v' = |X|$, $b' = |Y|$ and $\theta_2^2 = r - \lambda$. Now Theorem 4.9.1 gives the inequality and that (X,B') is a 1-design. But then (X,B') is a 2-design, because (P,B) is. □

An example of a subdesign of a symmetric design is the incidence structure formed by the absolute point and lines of a polarity in a projective plane of order q. This gives a (degenerate) 2-$(v',1,0)$ design in a 2-$(q^2 + q + 1, q + 1, 1)$ design. The bound gives $v' \le q\sqrt{q} + 1$. (See also the following section.) The 2-$(q\sqrt{q}+1, q+1, 1)$ design that is obtained in case of equality is called a *unital*. Other examples of symmetric designs that meet the bound can be found in HAEMERS & SHRIKHANDE [205] or JUNGNICKEL [237]. Wilbrink used Theorem 4.9.1 to shorten the proof of Feit's result on the number of points and blocks fixed by an automorphism group of a symmetric design (see [67]). The inequality of the second corollary is tight for hyperovals and (more generally) maximal arcs in finite projective planes.

4.10 Polarities

A *projective plane* is a point-line geometry such that any two points are on a unique line, and any two lines meet in a unique point. It is said to be of order q when all lines have $q+1$ points and all points are on $q+1$ lines. A projective plane of order q has q^2+q+1 points and as many lines.

A *polarity* of a point-block incidence structure is a map of order 2 interchanging points and blocks and preserving incidence. An *absolute point* is a point incident with its image under the polarity.

Suppose we have a projective plane of order q with a polarity σ. The polarity enables us to write the point-line incidence matrix N as a symmetric matrix, and then the number of absolute points is $\operatorname{tr} N$. By definition, we have $N^2 = NN^\top = J + qI$, which has one eigenvalue equal to $(q+1)^2$ and all other eigenvalues equal to q. That means that N has spectrum $(q+1)^1$, $\sqrt{q}^m$, $-\sqrt{q}^n$, for certain integers m,n, where this time exponents indicate multiplicities. The number of absolute points equals $a = q+1+(m-n)\sqrt{q}$. It follows that if q is not a square then $m=n$ and there are precisely $q+1$ absolute points. If q is a square, and p is a prime dividing q, then $a \equiv 1 \pmod{p}$ so that a is nonzero.

(This is false in the infinite case: the polarity sending the point (p,q,r) to the line $pX + qY + rZ = 0$ has no absolute points over $\mathbb{R}$.)

With slightly more effort one finds bounds for the number of absolute points:

Proposition 4.10.1 *A polarity of a projective plane of order q has at least $q+1$ and at most $q\sqrt{q}+1$ absolute points.*

Proof Suppose z is a nonabsolute point. Now σ induces a map τ on the line z^σ defined for $y \in z^\sigma$ by: y^τ is the common point of y^σ and z^σ. Now $\tau^2 = 1$, and $y^\tau = y$ precisely when y is absolute. This shows that the number of absolute points on a nonabsolute line is $q+1 \pmod 2$.

Now if q is odd, then take an absolute point x. This observation says that each line on x different from x^σ contains another absolute point, for a total of at least $q+1$. On the other hand, if q is even, then each nonabsolute line contains an absolute point, so $q^2+q+1-a \le aq$ and $a \ge q+1$.

For the upper bound, use interlacing: partition the matrix N into absolute / non-absolute points/lines and find the matrix of average row sums $\begin{bmatrix} 1 & q \\ \frac{aq}{v-a} & q+1-\frac{aq}{v-a} \end{bmatrix}$, where $v = q^2+q+1$, with eigenvalues $q+1$ and $1-\frac{aq}{v-a}$. Now interlacing yields $1-\frac{aq}{v-a} \ge -\sqrt{q}$, that is, $a \le q\sqrt{q}+1$, just like we found in the previous section. $\square$

The essential part of the proof of the lower bound was to show that there is at least one absolute point, and this used an eigenvalue argument.

4.11 Exercises

Exercise 4.1 Prove the following bipartite version of Proposition 4.5.1. Let Γ be a connected and bipartite graph, semiregular with degrees k and l. Let $|\theta| \le \lambda$ for every eigenvalue $\theta \ne \pm\sqrt{kl}$. If R is a subset of the set K of vertices of degree k, and $\rho = |R|/|K|$, then

$$\frac{|\Gamma(R)|}{|R|} \geq \frac{k^2}{\rho(kl - \lambda^2) + \lambda^2}.$$

(This is the result from TANNER [334].)

Exercise 4.2 (i) Determine the isoperimetric number $h(K_n)$.
(ii) Using Proposition 4.5.2, show that the n-cube has $h(Q_n) = 1$.

Exercise 4.3 An (ℓ, m)-biclique in a graph Γ is a complete bipartite subgraph $K_{\ell,m}$ of Γ (not necessarily induced). Let $0 = \mu_1 \leq \ldots \leq \mu_n$ be the Laplace eigenvalues of Γ. Show that $\ell m/(n - \ell)(n - m) \leq ((\mu_n - \mu_2)/(2n - \mu_2 - \mu_n))^2$ if Γ is noncomplete and contains an (ℓ, m)-biclique.

Exercise 4.4 Let A be the incidence graph of a 2-(v, k, λ) design with b blocks and r blocks incident with each point. Express the spectrum of A in the design parameters v, k, λ, b and r.

Exercise 4.5 Let (P, B) be a 2-(v, k, λ) design, and suppose that some block is repeated ℓ times (i.e., ℓ blocks are incident with exactly the same set of k points). Prove that $b \geq \ell v$. (This is *Mann's inequality*.)

Chapter 5
Trees

Trees have a simpler structure than general graphs, and we can prove stronger results. For example, interlacing tells us that the multiplicity of an eigenvalue decreases by at most one when a vertex is removed. For trees Godsil's lemma gives the same conclusion also when a path is removed.

5.1 Characteristic polynomials of trees

For a graph Γ with adjacency matrix A, let $\phi_\Gamma(t) := \det(tI - A)$ be its characteristic polynomial.

Note that since the characteristic polynomial of the disjoint union of two graphs is the product of their characteristic polynomials, results for trees immediately yield results for forests as well.

It will be useful to agree that $\phi_{T\setminus x,y} = 0$ if $x = y$.

Proposition 5.1.1 *Let T be a tree, and for $x, y \in T$, let P_{xy} be the unique path joining x and y in T.*

(i) Let $e = xy$ be an edge in T that separates T into two subtrees A and B, with $x \in A$ and $y \in B$. Then

$$\phi_T = \phi_A \phi_B - \phi_{A\setminus x} \phi_{B\setminus y}.$$

(ii) Let x be a vertex of T. Then

$$\phi_T(t) = t\phi_{T\setminus x}(t) - \sum_{y \sim x} \phi_{T\setminus \{x,y\}}(t).$$

(iii) Let x be a vertex of T. Then

$$\phi_{T\setminus x}(t)\phi_T(s) - \phi_{T\setminus x}(s)\phi_T(t) = (s-t)\sum_{y \in T} \phi_{T\setminus P_{xy}}(s)\phi_{T\setminus P_{xy}}(t).$$

(iv) Let x be a vertex of T. Then

$$\phi_{T\setminus x}\phi'_T - \phi'_{T\setminus x}\phi_T = \sum_{y\in T}\phi^2_{T\setminus P_{xy}}.$$

(v) Let x,y be vertices of T. Then

$$\phi_{T\setminus x}\phi_{T\setminus y} - \phi_{T\setminus x,y}\phi_T = \phi^2_{T\setminus P_{xy}}.$$

(vi) Let x,y,z be vertices of T, where $z \in P_{xy}$. Then

$$\phi_{T\setminus x,y,z}\phi_T = \phi_{T\setminus x}\phi_{T\setminus y,z} - \phi_{T\setminus z}\phi_{T\setminus x,y} + \phi_{T\setminus y}\phi_{T\setminus x,z}.$$

(vii) We have $\phi'_T = \sum_{x\in T}\phi_{T\setminus x}$.

(viii) Let T have n vertices and c_m matchings of size m. Then

$$\phi_T(t) = \sum_m(-1)^m c_m t^{n-2m}.$$

Proof   Part (i) follows by expansion of the defining determinant. It can also be phrased as $\phi_T = \phi_{T\setminus e} - \phi_{T\setminus\{x,y\}}$. Part (ii) follows by applying (i) to all edges on x. Note that $\phi_{\{x\}}(t) = t$. Part (iii) follows from (ii) by induction on the size of T. Expand $\phi_T(s)$ and $\phi_T(t)$ on the left-hand side using (ii), and then use induction. Part (iv) is immediate from (iii). Part (vii) follows by taking the derivative of the defining determinant. Part (viii) is a reformulation of the description in §1.2.1. Note that the only directed cycles in a tree are those of length 2. Part (v) is true if $T = P_{xy}$, and the general case follows from part (vi) and induction: the statement remains true when a subtree S is attached via an edge e at a vertex $z \in P_{xy}$. Finally, part (vi) follows from: *if $\Gamma\setminus z = A+B$, then $\phi_\Gamma = \phi_{A\cup z}\phi_B + \phi_A\phi_{B\cup z} - \phi_A\phi_{\{z\}}\phi_B$, where of course $\phi_{\{z\}}(t) = t$.* □

Theorem 5.1.2 ("Godsil's lemma", [173]) *Let T be a tree and θ an eigenvalue of multiplicity $m > 1$. Let P be a path in T. Then θ is an eigenvalue of $T\setminus P$ with multiplicity at least $m-1$.*

Proof   By parts (iv) and (vii) of the above Proposition we have

$$\phi'_T(t)^2 - \phi''_T(t)\phi_T(t) = \sum_{x,y\in T}\phi_{T\setminus P_{xy}}(t)^2.$$

Now θ is a root of multiplicity at least $2m-2$ of the left-hand side, and hence also of each of the terms on the right-hand side.   □

As an application of Godsil's lemma, consider a tree T with e distinct eigenvalues and maximum possible diameter $e-1$. Let P be a path of length $e-1$ (that is, with e vertices) in T. Then $T\setminus P$ has a spectrum that is independent of the choice of P: for each eigenvalue θ with multiplicity m of T, the forest $T\setminus P$ has eigenvalue θ with multiplicity $m-1$ (and it has no other eigenvalues).

In particular, all eigenvalues of a path have multiplicity 1.

Note that going from T to $T\setminus x$ changes multiplicities by at most 1: they go up or down by at most 1. Godsil's lemma is one-sided: going from T to $T\setminus P$, the multiplicities go down by at most 1, but they may well go up by more. For example, if one joins the centers x, y of two copies of $K_{1,m}$ by an edge, one obtains a tree T that has 0 as an eigenvalue of multiplicity $2m - 2$. For $P = xy$, the forest $T \setminus P$ has eigenvalue 0 with multiplicity $2m$.

5.2 Eigenvectors and multiplicities

For trees we have rather precise information about eigenvectors and eigenvalue multiplicities (FIEDLER [157]).

Lemma 5.2.1 *Let T be a tree with eigenvalue θ, and let $Z = Z_T(\theta)$ be the set of vertices in T where all θ-eigenvectors vanish. If for some vertex $t \in T$ some component S of $T \setminus t$ has eigenvalue θ (in particular, if some θ-eigenvector of T vanishes at t), then $Z \neq \emptyset$.*

Proof Consider proper subtrees S of T with eigenvalue θ and with a single edge st joining some vertex $s \in S$ with some vertex $t \in T \setminus S$, and pick a minimal one. If $|S| = 1$, then $\theta = 0$, and $t \in Z$. Assume $|S| > 1$. If a θ-eigenvector u of S is the restriction to S of a θ-eigenvector v of T, then v vanishes in t. So, if some θ-eigenvector v of T does not vanish at t, then u and $v|_S$ are not dependent, and some linear combination vanishes in s and is a θ-eigenvector of $S \setminus s$, contradicting the minimality of S. This shows that $t \in Z$.   □

Note that it is not true that the hypothesis of the lemma implies that $t \in Z$. For example, consider the tree T of type D_6 given by $1 \sim 2 \sim 3 \sim 4 \sim 5, 6$. It has $Z(0) = \{2, 4\}$, and the component $S = \{4, 5, 6\}$ of $T \setminus 3$ has eigenvalue 0, but $3 \notin Z(0)$.

Proposition 5.2.2 *Consider a tree T with eigenvalue θ, and let $Z = Z(\theta)$ be the set of vertices in T where all θ-eigenvectors vanish. Let $Z_0 = Z_0(\theta)$ be the set of vertices in Z that have a neighbor in $T \setminus Z$.*

(i) Let S be a connected component of $T \setminus Z$. Then S has eigenvalue θ with multiplicity 1. If u is a θ-eigenvector of S, then u is nowhere zero.

(ii) Let $T \setminus Z$ have c connected components, and let $d = |Z_0|$. Then θ has multiplicity $c - d$.

The components of $T \setminus Z(\theta)$ are called the *eigenvalue components* of T for θ.

Proof (i) Suppose θ is eigenvalue of T with multiplicity greater than 1. Then some θ-eigenvector has a zero coordinate and Lemma 5.2.1 shows that $Z \neq \emptyset$.

If S is a connected component of $T \setminus Z$ then it has eigenvalue θ (otherwise $S \subseteq Z$, a contradiction). Apply Lemma 5.2.1 to S instead of T to find that if some θ-eigenvector of S vanishes on a point of S, then there is a point $s \in S$ where all of its θ-eigenvectors vanish. But the restriction to S of a θ-eigenvector of T is a θ-eigenvector of S, so $s \in Z$, contradiction.

(ii) Each point of Z_0 imposes a linear condition, and since T is a tree, these conditions are independent. □

We see that if the multiplicity of θ is not 1, then Z contains a vertex of degree at least 3. In particular, $Z \neq \emptyset$, and hence $Z_0 \neq \emptyset$. Deleting a vertex in Z_0 from T increases the multiplicity of θ.

In particular, we see again that all eigenvalues of a path have multiplicity 1.

5.3 Sign patterns of eigenvectors of graphs

For a path, the i-th-largest eigenvalue has multiplicity 1 and an eigenvector with $i-1$ sign changes, that is, i areas of constant sign. It is possible to generalize this observation to more general graphs. One obtains discrete analogues of Courant's nodal domain theorem. See also [135].

Given a real vector u, let the *support* $\operatorname{supp} u$ be the set $\{i|u_i \neq 0\}$. For $*$ one of $<, >, \leq, \geq$, we also write $\operatorname{supp}^* u$ for $\{i|u_i * 0\}$. Let $N(u)$ (resp. $N^*(u)$) be the number of connected components C of the subgraph induced by $\operatorname{supp} u$ (resp. $\operatorname{supp}^* u$) such that u does not vanish identically on C.

Proposition 5.3.1 *Let Γ be a graph with eigenvalues $\theta_1 \geq \ldots \geq \theta_n$, and let u be an eigenvector with eigenvalue $\theta = \theta_j = \theta_{j+m-1}$ of multiplicity m. Let Δ be the subgraph of Γ induced by $\operatorname{supp} u$, with eigenvalues $\eta_1 \geq \ldots \geq \eta_t$. Then*

(i)
$$N^>(u) + N^<(u) \leq \#\{i \mid \eta_i \geq \theta\} \leq j+m-1,$$

(ii)
$$N^>(u) + N^<(u) - N(u) \leq \#\{i \mid \eta_i > \theta\} \leq j-1, \text{ and}$$

(iii) if Γ has c connected components, then

$$N^\geq(u) + N^\leq(u) \leq j+c-1.$$

Proof For a subset S of the vertex set of Γ, let I_S be the diagonal matrix with ones on the positions indexed by elements of S and zeros elsewhere.

Let C run through the connected components of $\operatorname{supp}^> u$ and $\operatorname{supp}^< u$ (resp. $\operatorname{supp}^\geq u$ and $\operatorname{supp}^\leq u$). Put $u_C = I_C u$. Then the space $U := \langle u_C \mid C \rangle$ has dimension $N^>(u) + N^<(u)$ (resp. $N^\geq(u) + N^\leq(u)$).

Let A be the adjacency matrix of Δ (resp. Γ). Define a real symmetric matrix B by $B_{CD} = u_C^\top (A - \theta I) u_D$. Then B has zero row sums and nonpositive off-diagonal entries, so B is positive semidefinite. It follows that for $y \in U$ we have $y^\top (A - \theta I)y \geq 0$. This means that U intersects the space spanned by the eigenvectors of $A - \theta I$ with negative eigenvalue in 0.

For (i), $N^>(u) + N^<(u) \leq \#\{i \mid \eta_i \geq \theta\}$ follows.

The vectors $y \in U$ with $y^\top (A - \theta I)y = 0$ correspond to eigenvectors with eigenvalue 0 of B, and by Lemma 2.10.1 there are $N(u)$ (resp. c) such independent eigenvectors. This proves (ii) (resp. (iii)). □

Remarks (i) For $j = 1$, the results follow from the Perron-Frobenius theorem. (If Γ is connected, then the eigenvector for θ_1 is nowhere zero and has constant sign.)

(ii) The only thing used about A is that its off-diagonal elements are nonnegative, and zero for nonadjacent pairs of vertices. For example, the conclusions also hold for $-L$.

Examples (a) Let Γ be connected and bipartite, and let θ be the smallest eigenvalue of Γ. The corresponding eigenvector u has different signs on the two sides of the bipartition, so $\text{supp}^> u$ and $\text{supp}^< u$ are the two sides of the bipartition, $N^>(u) + N^<(u) = n$ and $N(u) = 1$. We have equality in (i)–(iii).

(b) Let Γ be the star $K_{1,s}$. The spectrum is $\sqrt{s}^1$, 0^{s-1}, $(-\sqrt{s})^1$. Let u be an eigenvector with eigenvalue $\theta = 0$ that has t nonzero coordinates. (Then $2 \leq t \leq s$.) Now $N^>(u) + N^<(u) = N(u) = t$ and $N^{\geq}(u) + N^{\leq}(u) = 2$, and for $t = s$ equality holds in (i)–(iii).

(c) Let Γ be the Petersen graph. It has spectrum 3^1, 1^5, $(-2)^4$. Let u be an eigenvector with eigenvalue $\theta = 1$ that vanishes on four points, so that $\text{supp}\, u$ induces $3K_2$ with spectrum 1^3, $(-1)^3$. We find $N^>(u) + N^<(u) = N(u) = 3$ and $N^{\geq}(u) + N^{\leq}(u) = 2$, again equality in (i)–(iii).

(d) Let Γ be the path P_n. The eigenvalues are $\theta_k = 2\cos(k\pi/(n+1))$ for $k = 1, \ldots, n$. The eigenvector u corresponding to θ_k has $k - 1$ sign changes, so that $N^>(u) + N^<(u) = k$. If $\gcd(k, n+1) = 1$ then u has no zero entries, so that $N(u) = 1$. Now we have equality in (i)–(iii). If $\gcd(k, n+1) = r$, then u has $r - 1$ zero entries, so that $N(u) = r$. Also, the eigenvalue θ_k is the k/r-th of each component of $\text{supp}\, u$, so $\#\{i \mid \eta_i \geq \theta\} = k$ and $\#\{i \mid \eta_i > \theta\} = k - r$, with equality in (i) and the first inequality of (ii).

Remark It is not true that $N(u) \leq m$ if m is the multiplicity of θ for Γ. For example, in case (b) above we have $N(u) = s$ and $m = s - 1$. (And in case (c) the opposite happens: $N(u) = 3$ and $m = 5$.)

Proposition 5.3.2 *Let Γ be a connected graph with second-largest eigenvalue θ_2. Let u be a θ_2-eigenvalue with minimal support. Then $N^>(u) = N^<(u) = 1$.*

Proof By the Perron-Frobenius theorem, only θ_1 has an eigenvector (say z) with constant sign, so $N^>(u)$ and $N^<(u)$ are both nonzero. If C and D are two connected components of $\text{supp}^> u$, and we put $u_C = I_C u$, etc., as before, then a suitable linear combination y of u_C and u_D is orthogonal to z and has Rayleigh quotient at least θ_2, so that y is a θ_2-eigenvector with support strictly contained in that of u. □

This proposition will play a role in the discussion of the Colin de Verdière parameter (in the proof of Proposition 7.3.3). Remark (ii) above also applies here.

5.4 Sign patterns of eigenvectors of trees

For trees we have more precise information.

Proposition 5.4.1 *Let T be a tree with eigenvalue θ, and put $Z = Z(\theta)$. Let $T \setminus Z$ have eigenvalues $\eta_1 \geq \ldots \geq \eta_m$. Let $g = \#\{i \mid \eta_i \geq \theta\}$ and $h = \#\{i \mid \eta_i > \theta\}$. Let u be a θ-eigenvector of T. Then $N^>(u) + N^<(u) = g$ and $N^>(u) + N^<(u) - N(u) = h$.*

Proof Since $N()$ and g and h are additive over connected components, we may assume that Z is empty. Now by Proposition 5.2.2(i), θ has multiplicity 1 and u is nowhere 0. Let T have n vertices, and let there be p edges xy with $u_x u_y > 0$ and q edges xy with $u_x u_y < 0$. Then $p + q = n - 1$. Since T is bipartite, also $-\theta$ is an eigenvalue, and an eigenvector v for $-\theta$ is obtained by switching the sign of u on one bipartite class. By Proposition 5.3.1, we have $q = N^>(u) + N^<(u) - 1 \leq h$ and $p = N^>(v) + N^<(v) - 1 \leq n - h - 1$, that is $q \geq h$, and hence equality holds everywhere. ◻

Let a *sign change* for an eigenvector u of T be an edge $e = xy$ such that $u_x u_y < 0$.

Proposition 5.4.2 *Let T be a tree with j-th eigenvalue θ. If u is an eigenvector for θ with s sign changes, and $d = |Z_0(\theta)|$, then $d + s \leq j - 1$.*

Proof Let $T \setminus Z$ have c connected components, and let u be identically zero on c_0 of these. Then $s + c - c_0 = N^>(u) + N^<(u)$. Let $\theta = \theta_j = \theta_{j+m-1}$, where $m = c - d$ is the multiplicity of θ. By Proposition 5.3.1(i), we have $s + c - c_0 \leq j + m - 1$, that is, $d + s - c_0 \leq j - 1$. But we can make c_0 zero by adding a small multiple of some θ-eigenvector that is nonzero on all of $T \setminus Z$. ◻

Example For $T = E_6$ all eigenvalues have multiplicity 1, and $N^>(u) + N^<(u)$ takes the values 1, 2, 3, 4, 4, 6 for the six eigenvectors u. The sign patterns are:

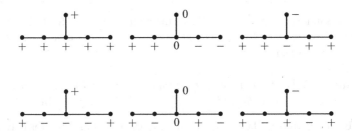

We see that a small perturbation that would make u nonzero everywhere would give the two zeros in the second eigenvector the same sign, but the two zeros in the fifth eigenvector different signs (because $\theta_2 > 0$ and $\theta_5 < 0$), and for the perturbed vector u' we would find 0, 1, 2, 3, 4, 5 sign changes.

5.5 The spectral center of a tree

There are various combinatorial concepts of "center" for trees. One has the center/bicenter and the centroid/bicentroid. Here we define a concept of center using spectral methods. Closely related results can be found in NEUMAIER [286].

Proposition 5.5.1 *Let T be a tree (with at least two vertices) with second-largest eigenvalue λ. Then there is a unique minimal subtree Y of T such that no connected component of $T \setminus Y$ has largest eigenvalue larger than λ. If $Z(\lambda) \neq \emptyset$ (and in particular if λ has multiplicity larger than 1) then $Y = Z_0(\lambda)$ and $|Y| = 1$. Otherwise $|Y| = 2$, and Y contains the endpoints of the edge on which the unique λ-eigenvector changes sign. In this latter case, all connected components of $T \setminus Y$ have largest eigenvalue strictly smaller than λ.*

We call the set Y the *spectral center* of T.

Proof If for some vertex y all connected components of $T \setminus y$ have largest eigenvalue at most λ, then pick $Y = \{y\}$. Otherwise, for each vertex y of T there is a unique neighbor y' in the unique component of $T \setminus y$ that has largest eigenvalue more than λ. Since T is finite, we must have $y'' = y$ for some vertex y. Now pick $Y = \{y, y'\}$. Clearly, Y has the stated property and is minimal.

Put $Z = Z(\lambda)$. If $Z = \emptyset$, then λ has multiplicity 1 and by Proposition 5.4.2 there is a unique edge $e = pq$ such that the unique λ-eigenvector has different signs on p and q. Both components of $T \setminus e$ have largest eigenvalue strictly larger than λ (e.g., by Theorem 2.2.1 (iv)), so that Y must contain both endpoints of e.

If $Z \neq \emptyset$, then all eigenvalue components for λ have eigenvalue λ, and any strictly larger subgraph has a strictly larger eigenvalue, so Y must contain $Z_0 := Z_0(\lambda)$. By Proposition 5.4.2 we have $|Z_0| = 1$, say $Z_0 = \{y\}$. If Y is not equal to $\{y\}$, then Y also contains y'. This proves uniqueness.

Suppose that $Z_0 = \{y\}$. If $T \setminus Z$ has c connected components, then λ has multiplicity c in $T \setminus y$ and $c - 1$ in T. Since T has precisely c eigenvalues $\theta \geq \lambda$, by interlacing $T \setminus y$ has at most c such eigenvalues, so that $T \setminus y$ has no eigenvalues larger than λ. This shows that $|Y| = 1$ when Z is nonempty.

Finally, suppose that $Y = \{y, y'\}$ and that $T \setminus Y$ has largest eigenvalue λ. By Lemma 5.2.1, $Z \neq \emptyset$, a contradiction. $\qquad\square$

Example If T is the path P_n with n vertices, then $\lambda = 2\cos 2\pi/(n+1)$. If $n = 2m+1$ is odd, then Y consists of the middle vertex, and $T \setminus Y$ is the union of two paths P_m with largest eigenvalue $\lambda = 2\cos \pi/(m+1)$. If $n = 2m$ is even, then Y consists of the middle two vertices, and $T \setminus Y$ is the union of two paths P_{m-1} with largest eigenvalue $2\cos \pi/m < \lambda$.

5.6 Integral trees

An *integral tree* is a tree with only integral eigenvalues. Such trees are rare. A list of all integral trees on at most 50 vertices can be found in [52].

A funny result is

Proposition 5.6.1 (WATANABE [347]) *An integral tree cannot have a perfect matching, that is, must have an eigenvalue 0, unless it is K_2.*

Proof The constant term of the characteristic polynomial of a tree is, up to sign, the number of perfect matchings. It is also the product of all eigenvalues. If it is nonzero, then it is 1, since the union of two distinct perfect matchings contains a cycle. But then all eigenvalues are ± 1 and P_3 is not an induced subgraph, so we have K_2.   □

This result can be extended a little. Let $SK_{1,m}$ be the tree on $2m+1$ vertices obtained by subdividing all edges of $K_{1,m}$. The spectrum is $\pm\sqrt{m+1}\,(\pm 1)^{m-1}\,0$.

Proposition 5.6.2 (BROUWER [52]) *If an integral tree has eigenvalue 0 with multiplicity 1, then it is* $SK_{1,m}$, *where* $m = t^2 - 1$ *for some integer* $t \geq 1$.   □

For a long time it has been an open question whether there exist integral trees of arbitrarily large diameter. Recently, this was settled in the affirmative by Csikvári. The construction is as follows. Define trees $T'(r_1,\ldots,r_m)$ by induction: $T'()$ is the tree with a single vertex x_0. $T'(r_1,\ldots,r_m)$ is the tree obtained from $T'(r_1,\ldots,r_{m-1})$ by adding r_m pendant edges to each vertex u with $d(u,x_0) = m-1 \pmod{2}$. The diameter of this tree is $2m$ (assuming $r_1 > 1$) and it has $2m+1$ distinct eigenvalues:

Proposition 5.6.3 (CSIKVÁRI [111]) *The tree* $T'(r_1,\ldots,r_m)$ *has eigenvalues 0 and* $\pm\sqrt{s_i}\,(1 \leq i \leq m)$, *where* $s_i = r_i + \cdots + r_m$.

Now all trees $T'(n_1^2 - n_2^2,\ldots,n_{m-1}^2 - n_m^2,n_m^2)$ are integral of diameter $2m$ when $n_1 > n_2 > \ldots > n_m$.

A short proof can be given using the following observation. If A and B are trees with fixed vertices x and y, respectively, then let $A \sim mB$ be the tree constructed on the union of A and m copies of B, where x is joined to the m copies of y. Now Proposition 5.1.1(i) and induction immediately yields that $T = A \sim mB$ has characteristic polynomial $\phi_T = \phi_B^{m-1}(\phi_A\phi_B - m\phi_{A\setminus x}\phi_{B\setminus y})$, where the last factor is symmetric in A and B.

Proof Induction on m. The statement holds for $m \leq 1$. With $A = T'(r_3,\ldots)$ and $B = T'(r_2,r_3,\ldots)$, we have $T'(r_1,r_2,r_3,\ldots) = A \sim r_1 B$ and $T'(r_1 + r_2,r_3,\ldots) = B \sim r_1 A$.   □

5.7 Exercises

Exercise 5.1 Show that there are six integral trees on at most ten vertices, namely (i) K_1, (ii) K_2, (iii) $K_{1,4} = \hat{D}_4$, (iv) $\hat{D}_5$, (v) $\hat{E}_6$, (vi) $K_{1,9}$. (For notation, see §3.1.1.)

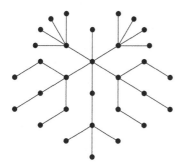

An integral tree on 31 vertices.
What is the spectrum?

Exercise 5.2 Show that the only trees that have integral Laplace spectrum are the stars $K_{1,m}$.

Exercise 5.3 ([106, 192]) The *energy* $E(\Gamma)$ of a graph Γ, as defined by Gutman, is $\sum_i |\theta_i|$, the sum of the absolute values of the eigenvalues of the adjacency matrix A. It can be expressed in terms of the characteristic polynomial $\phi(x)$ by

$$E(\Gamma) = \frac{1}{\pi} \int_{-\infty}^{+\infty} \left[n - x \frac{d}{dx} \log \phi(ix) \right] dx.$$

Show that if T is a tree on n vertices, different from the star $S = K_{1,n-1}$ and the path $P = P_n$, then

$$E(S) < E(T) < E(P).$$

Chapter 6
Groups and Graphs

6.1 $\Gamma(G,H,S)$

Let G be a finite group, H a subgroup, and S a subset of G. We can define a graph $\Gamma(G,H,S)$ by taking as vertices the cosets gH ($g \in G$) and calling g_1H and g_2H adjacent when $Hg_2^{-1}g_1H \subseteq HSH$. The group G acts as a group of automorphisms on $\Gamma(G,H,S)$ via left multiplication, and this action is transitive. The stabilizer of the vertex H is the subgroup H. A graph $\Gamma(G,H,S)$ with $H = 1$ is called a *Cayley graph*.

Conversely, let Γ be a graph with transitive group of automorphisms G. Let x be a vertex of Γ, and let $H := G_x$ be the stabilizer of x in G. Now Γ can be identified with $\Gamma(G,H,S)$, where $S = \{g \in G \mid x \sim gx\}$. If Γ is, moreover, edge-transitive, then S can be chosen to have cardinality 1.

Instead of representing each vertex as a coset, one can represent each vertex y by the subgroup G_y that fixes it. If $H = G_x$ and $y = gx$, then $G_y = gHg^{-1}$, so now G acts by conjugation.

6.2 Spectrum

Let Γ be a graph and G a group of automorphisms. Let M be a matrix with rows and columns indexed by the vertex set of Γ, and suppose that M commutes with all elements of G (so that $gM = Mg$, or, equivalently, $M_{xy} = M_{gx,gy}$). Now $\operatorname{tr} gM$ only depends on the conjugacy class of g in G, so the map $g \mapsto \operatorname{tr} gM$ defines a class function on G.

(Also the spectrum of gM only depends on the conjugacy class of g in G, but it is not clear how the spectrum should be ordered. Having the trace, however, suffices: one can retrieve the spectrum of a matrix M from the traces of the powers M^i. People also introduce the zeta function of a graph Γ by $\zeta_\Gamma(-s) = \sum \lambda^s = \operatorname{tr} L^s$, where the sum is over the eigenvalues λ of the Laplacian L, in order to have a single object that encodes the spectrum.)

If Γ has vertex set X, and $V = \mathbb{R}^X$ is the $\mathbb{R}$-vector space spanned by the vertices of Γ, then by Schur's lemma M acts as a multiple of the identity on each irreducible G-invariant subspace of V. In other words, the irreducible G-invariant subspaces are eigenspaces of M. If M acts like θI on the irreducible G-invariant subspace W with character χ, then $\operatorname{tr} gM|_W = \theta\chi(g)$.

Example Let Γ be the Petersen graph, with as vertices the unordered pairs from a 5-set, adjacent when they are disjoint, and let $M = A$, the adjacency matrix. Now $f(g) := \operatorname{tr} gA = \#\{x \mid x \sim gx\}$ defines a class function on $\operatorname{Aut}\Gamma = \operatorname{Sym}(5)$. Below we show f together with the character table of $\operatorname{Sym}(5)$ (with the top row indicating the cycle shape of the element):

	1	2	2^2	3	4	5	2.3
χ_1	1	1	1	1	1	1	1
χ_2	1	-1	1	1	-1	1	-1
χ_3	4	2	0	1	0	-1	-1
χ_4	4	-2	0	1	0	-1	1
χ_5	5	1	1	-1	-1	0	1
χ_6	5	-1	1	-1	1	0	-1
χ_7	6	0	-2	0	0	1	0
f	0	0	4	0	2	5	6

We see that $f = 3\chi_1 - 2\chi_3 + \chi_5$. It follows that Γ has spectrum $3^1 \, (-2)^4 \, 1^5$, where the eigenvalues are the coefficients of f, written as linear combination of irreducible characters, and the multiplicities are the degrees of these characters. The permutation character is $\pi = \chi_1 + \chi_3 + \chi_5$ (obtained for $M = I$). It is *multiplicity free*, that is, no coefficients larger than 1 occur. In the general case, the coefficient of an irreducible character χ in the expression for f will be the sum of the eigenvalues of M on the irreducible subspaces with character χ.

6.3 Non-Abelian Cayley graphs

Let G be a group and $S \subseteq G$. The *Cayley graph* $\operatorname{Cay}(G,S)$ is the (directed) graph Γ with vertex set G and edge set $E = \{(g,gs) \mid g \in G, \ s \in S\}$ (so that S is the set of out-neighbors of 1). Now Γ is regular with in- and outvalency $|S|$. It will be undirected if and only if S is symmetric, i.e., $S^{-1} = S$, where $S^{-1} = \{s^{-1} \mid s \in S\}$.

The graph $\operatorname{Cay}(G,S)$ is connected if and only if S generates G. If $H = \langle S \rangle$ is the subgroup of G generated by S, then $\operatorname{Cay}(G,S)$ consists of $|G/H|$ disjoint copies of $\operatorname{Cay}(H,S)$.

The spectrum of Cayley graphs in an Abelian group G was discussed in §1.4.9. More generally, one has the following.

Proposition 6.3.1 ([144, 282]) *Let G be a finite group and S a subset that is symmetric and invariant under conjugation. The graph* $\mathrm{Cay}(G,S)$ *has eigenvalues* $\theta_\chi = \frac{1}{\chi(1)} \sum_{s\in S} \chi(s)$ *with multiplicity* $\chi(1)^2$, *where* χ *ranges over the irreducible characters of G.*

Proof Since S is a union of conjugacy clases of G, the adjacency matrix A commutes with the elements of G, and the previous discussion applies. The regular representation of G decomposes into a direct sum of irreducible subspaces, where for each irreducible character χ there are $\chi(1)$ copies of V_χ. On each copy, A acts like θI, and $\dim V_\chi = \chi(1)$, so θ has multiplicity $\chi(1)^2$. We saw that $\mathrm{tr}\, Ag|_W = \theta\chi(g)$, so that in particular $\theta\chi(1) = \mathrm{tr}\, A|_W = \sum_{s\in S} \chi(s)$, where $W = V_\chi$. $\square$

For example, the graph $K_{3,3}$ can be described as the Cayley graph $\mathrm{Cay}(G,S)$, where $G = \mathrm{Sym}(3)$ and $S = \{(12),(13),(23)\}$. Its complement $2K_3$ is the Cayley graph $\mathrm{Cay}(G,S')$, where $S' = \{(123),(132)\}$. The character table of G is

	1	2	3
χ_1	1	1	1
χ_2	1	−1	1
χ_3	2	0	−1

and we read off the spectrum $3, -3, 0^4$ of $K_{3,3}$ from column 2 and the spectrum $2, 2, (-1)^4$ of $2K_3$ from column 3.

As an application, RENTELN [299] computes the smallest eigenvalue of the derangement graph (the graph on $\mathrm{Sym}(n)$ where $g_1 \sim g_2$ when $g_1^{-1}g_2$ has no fixed points) and finds $\theta_{\min} = -k/(n-1)$, providing an easy proof for the result that this graph has independence number $\alpha = (n-1)!$.

6.4 Covers

Let a *graph* $\Gamma = (X,E)$ consist of a set of vertices X and a set of edges E and an incidence relation between X and E (such that each edge is incident with one or two points). An edge incident with only one point is called a *loop*. A *homomorphism* $f : \Gamma \to \Delta$ of graphs is a map that sends vertices to vertices, edges to edges, loops to loops, and preserves incidence. For example, the chromatic number of Γ is the smallest integer m such that there is a homomorphism from Γ to K_m.

The map f is called a *covering* when it is a surjective homomorphism, and for each vertex x of Γ and each edge e of Δ that is incident with $f(x)$, there is a unique edge $\tilde{e}$ of Γ that is incident with x such that $f(\tilde{e}) = e$. Now Γ is called a *cover* of Δ.

If f is a covering, then paths in Δ starting at a vertex y of Δ lift uniquely to paths starting at a vertex x of Γ, for each $x \in f^{-1}(y)$.

The *universal cover* of a connected graph Δ is the unique tree T that is a cover. If a is a fixed vertex of Δ, then the vertices of T can be identified with the walks in Δ starting at a that never immediately retrace an edge, where two walks are adjacent when one is the extension of the other by one more edge. The tree T will be infinite when Δ contains at least one cycle. If f is the covering map (that assigns to a walk its final vertex), then T has a group of automorphisms H acting regularly on the fibers of f.

Given an arbitrary collection of cycles $\mathscr{C}$ in Δ and a positive integer n_C for each $C \in \mathscr{C}$, one may consider the most general cover satisfying the restriction that the inverse image of the walk traversing C n_C times is closed. (For example, the "universal cover modulo triangles" is obtained by requiring that the preimage of each triangle be a triangle.) There is a unique such graph, a quotient of the universal cover. Again the covering group (the group preserving the fibers) acts regularly on the fibers.

Conversely, let Γ be a graph and H a group of automorphisms. The *quotient graph* Γ/H has as vertices the H-orbits on the vertices of Γ, as edges the H-orbits on the edges of Γ, and a vertex x^H is incident with an edge e^H when some element of x^H is incident with some element of e^H.

The natural projection $\pi : \Gamma \to \Gamma/H$ is a homomorphism. It will be a covering when no vertex x of Γ is on two edges in an orbit e^H. In this case, we also say that Γ is a cover of Γ/H.

Now, let Γ be finite and $f : \Gamma \to \Delta$ a covering. Let A_Γ and A_Δ be the adjacency matrices of Γ and Δ. Then $(A_\Delta)_{f(x),z} = \sum_{y \in f^{-1}(z)} (A_\Gamma)_{xy}$. If we view A_Γ and A_Δ as linear transformations on the vector spaces V_Γ and V_Δ spanned by the vertices of Γ and Δ, and extend f to a linear map, then this equation becomes $A_\Delta \circ f = f \circ A_\Gamma$. If u is an eigenvector of Δ with eigenvalue θ, then $u \circ f$ (defined by $(u \circ f)_y = u_{f(y)}$) is an eigenvector of Γ with the same eigenvalue, and the same holds for Laplace eigenvectors and eigenvalues. (This is immediately clear, but also follows from the fact that the partition of $V\Gamma$ into fibers $f^{-1}(z)$ is an equitable partition.)

For example, let Γ be the path on six vertices with a loop added on both sides and Δ the path on two vertices with a loop added on both sides. Then the map sending vertices $1, 4, 5$ of Γ to one vertex of Δ and $2, 3, 6$ to the other, is a covering. The ordinary spectrum of Δ is $2, 0$, and hence also Γ has these eigenvalues. (It has spectrum $2, \sqrt{3}, 1, 0, -1, -\sqrt{3}$.)

Thus, the spectrum of Δ is a subset of the spectrum of Γ. We can be more precise and indicate which subset.

Let $V = \mathbb{R}^X$ be the vector space spanned by the vertices of Γ. Let G be a group of automorphisms of Γ. We can view the elements $g \in G$ as linear transformations of V (permuting the basis vectors). Let H be a subgroup of G, and let W be the subspace of V fixed by H.

Lemma 6.4.1 *Let M be a linear transformation of V that commutes with all $g \in G$. Then M preserves W and $\operatorname{tr} M|_W = (1_H, \phi_M|_H) = (1_H^G, \phi_M)$, where ϕ_M is the class function on G defined by $\phi_M(g) = \operatorname{tr} gM$.*

Proof The orthogonal projection P from V onto W is given by

$$P = \frac{1}{|H|} \sum_{h \in H} h.$$

If M commutes with all $h \in H$, then $MPu = PMu$, so M preserves the fixed space W, and its restriction $M|_W$ has trace trPM. Expanding P, we find tr$M|_W = $ tr$PM = \frac{1}{|H|} \sum_{h \in H}$ tr $hM = (1_H, \phi_M|_H)$. The second equality follows by Frobenius reciprocity. $\square$

Now assume that the map $\pi : \Gamma \to \Gamma/H$ is a covering. Then $\pi \circ A_\Gamma = A_{\Gamma/H} \circ \pi$. One can identify the vector space $V_{\Gamma/H}$ spanned by the vertices of Γ/H with the vector space W: the vertex x^H corresponds to $\frac{1}{\sqrt{|H|}} \sum_{h \in H} x^h \in W$. This identification identifies $A_{\Gamma/H}$ with $A|_W$. This means that Lemma 6.4.1 (applied with $M = A$) gives the spectrum of Γ/H. In precisely the same way, for $M = L$, it gives the Laplace spectrum of Γ/H.

We see that for a covering the spectrum of the quotient Γ/H does not depend on the choice of H but only on the permutation character 1_H^G. This is Sunada's observation and has been used to construct cospectral graphs; see §14.2.4.

6.5 Cayley sum graphs

In §1.4.9, we discussed Cayley graphs for an Abelian group G. A variation is the concept of *Cayley sum graph* with *sum set S* in an Abelian group G. It has vertex set G, and two elements $g, h \in G$ are adjacent when $g + h \in S$. (Other terms are *addition Cayley graphs* or just *sum graphs*.)

It is easy to determine the spectrum of a Cayley sum graph.

Proposition 6.5.1 ([143]) *Let Γ be the Cayley sum graph with sum set S in the finite Abelian group G. Let χ run through the $n = |G|$ characters of G. The spectrum of Γ consists of the numbers $\chi(S)$ for each real χ, and $\pm|\chi(S)|$ for each pair $\chi, \overline{\chi}$ of conjugate nonreal characters, where $\chi(S) = \sum_{s \in S} \chi(s)$.*

Proof If $\chi : G \to \mathbb{C}^*$ is a character of G, then $\sum_{y \sim x} \chi(y) = \sum_{s \in S} \chi(s - x) = (\sum_{s \in S} \chi(s)) \chi(-x) = \chi(S) \overline{\chi}(x)$. Now Γ is undirected, so the spectrum is real. If χ is a real character, then we found an eigenvector χ with eigenvalue $\chi(S)$. If χ is nonreal, then pick a constant α with $|\chi(S)| = \alpha^2 \chi(S)$. Then Re$(\alpha\chi)$ and Im$(\alpha\chi)$ are eigenvectors with eigenvalues $|\chi(S)|$ and $-|\chi(S)|$, respectively. $\square$

CHUNG [92] constructs Cayley sum graphs that are good expanders. For further material on Cayley sum graphs, see [6], [90], [183], [188].

6.5.1 (3,6)-fullerenes

An amusing application was given by DEVOS et al. [143]. A *(3,6)-fullerene* is a cubic plane graph whose faces (including the outer face) have sizes 3 or 6. Fowler conjectured (cf. [162]) that such graphs have spectrum $\Phi \cup \{3, -1, -1, -1\}$ (as multiset), where $\Phi = -\Phi$, and this was proved in [143].

For example, the graph

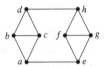

has spectrum $3, \sqrt{5}, 1, (-1)^4, -\sqrt{5}$ with eigenvalues $3, -1, -1, -1$ together with the symmetric part $\pm\sqrt{5}, \pm 1$.

The proof goes as follows. Construct the bipartite double $\Gamma \otimes K_2$ of Γ. This is a cover of Γ, and both triangles and hexagons lift to hexagons, three at each vertex, so $\Gamma \otimes K_2$ is a quotient of $\mathcal{H}$, the regular tesselation of the plane with hexagons.

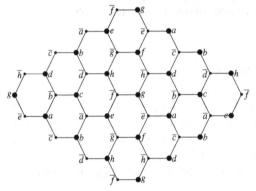

Let $\mathcal{H}$ have vertex set H, and let $\Gamma \otimes K_2$ have vertex set U, and let Γ have vertex set V. Let $\pi : H \to U$ and $\rho : U \to V$ be the quotient maps. The graph $\Gamma \otimes K_2$ is bipartite with bipartite halves U_1 and U_2, say. Fix a vertex $a_1 \in U_1$ and call it 0. Now $\pi^{-1}(U_1)$ is a lattice in $\mathbb{R}^2$, and $\pi^{-1}(a_1)$ is a sublattice (because the concatenation of two walks of even length in Γ starting and ending in a again is such a walk), so the quotient $G = \pi^{-1}(U_1)/\pi^{-1}(a_1)$ is an Abelian group, and G can be naturally identified with V. The automorphism of $\Gamma \otimes K_2$ that for each $u \in V$ interchanges the two vertices u_1, u_2 of $\rho^{-1}(u)$ lifts (for each choice of $\bar{a} \in \pi^{-1}(a_2)$) to an isometry of $\mathcal{H}$ with itself that is a point reflection $x \mapsto v - x$ (where $v = \bar{a}$). It follows that if two edges $x_1 y_2$ and $z_1 w_2$ in $\mathcal{H}$ are parallel, then $x + y = z + w$. Hence Γ is the Cayley sum graph for G where the sum set S is the set of three neighbors of a in Γ.

Now the spectrum follows. By the foregoing, the spectrum consists of the values $\pm|\chi(S)|$ for nonreal characters χ of G, and $\chi(S)$ for real characters. Since $\operatorname{tr}A = 0$ and Γ is cubic and not bipartite (it has four triangles) it suffices to show that there are precisely four real characters (then the corresponding eigenvalues must be $3, -1, -1, -1$). But this is clear since the number of real characters is the number of elements of order 2 in G, an Abelian group with (at most) two generators, hence at most four, and fewer than four would force nonzero $\operatorname{tr}A$. This proves Fowler's conjecture.

6.6 Exercises

Exercise 6.1 Show that a (3,6)-fullerene has precisely four triangles.

Chapter 7
Topology

In our discussion of the Shannon capacity (§3.7), we encountered the Haemers invariant, the minimum possible rank for certain matrices that fit a given graph. By far the most famous such invariant is the Colin de Verdière invariant of a graph, an algebraic invariant that turns out to have a topological meaning.

7.1 Embeddings

An *embedding* of a loopless graph in $\mathbb{R}^n$ consists of a representation of the vertices by distinct points in $\mathbb{R}^n$ and a representation of the edges by curve segments between the endpoints such that these curve segments only intersect in endpoints. (A curve segment between $\mathbf{x}$ and $\mathbf{y}$ is the range of an injective continuous map ϕ from $[0,1]$ to $\mathbb{R}^n$ with $\phi(0) = \mathbf{x}$ and $\phi(1) = \mathbf{y}$.)

Every finite graph can be embedded in $\mathbb{R}^m$ if $m \geq 3$. A graph is *planar* if it admits an embedding in $\mathbb{R}^2$. A graph is *outerplanar* if it admits an embedding in $\mathbb{R}^2$ such that the points are on the unit circle and the representations of the edges are contained in the unit disk. A graph Γ is *linklessly embeddable* if it admits an embedding in $\mathbb{R}^3$ such that no two disjoint circuits of Γ are linked. (Two disjoint Jordan curves in $\mathbb{R}^3$ are linked if there is no topological 2-sphere in $\mathbb{R}^3$ separating them.)

Examples of outerplanar graphs are all trees, C_n, and $\overline{P_5}$. Examples of graphs that are planar but not outerplanar are K_4, $\overline{3K_2}$, $\overline{C_6}$, and $K_{2,n-2}$ for $n \geq 5$. Examples of graphs that are not planar, but linklessly embeddable are K_5 and $K_{3,n-3}$ for $n \geq 6$. The Petersen graph and K_n for $n \geq 6$ are not linklessly embeddable.

7.2 Minors

A *graph minor* of a graph Γ is any graph that can be obtained from Γ by a sequence of edge deletions and contractions, and deletion of isolated vertices. Here the *contraction* of an edge e of the graph $(V\Gamma, E\Gamma)$ is the operation that merges the endpoints of e in $V\Gamma$ and deletes e from $E\Gamma$. A deep theorem of Robertson and Seymour [300] states that for every graph property $\mathscr{P}$ that is closed under taking graph minors, there exists a finite list of graphs such that a graph Γ has property $\mathscr{P}$ if and only if no graph from the list is a graph minor of Γ. Graph properties such as being planar, being outerplanar, being embeddable in some given surface, or being linklessly embeddable are closed under taking graph minors. For example, the Kuratowski-Wagner theorem ([248, 343]) states that a graph is planar if and only if no minor is isomorphic to K_5 or $K_{3,3}$.

The *Hadwiger conjecture* [193] says that if a graph has chromatic number m, then it has a K_m minor.

7.3 The Colin de Verdière invariant

A symmetric real matrix M is said to satisfy the *strong Arnold hypothesis* whenever there exists no symmetric nonzero matrix X with zero diagonal such that $MX = O$ and $M \circ X = O$, where $\circ$ denotes the componentwise (Hadamard, Schur) multiplication.

The Colin de Verdière parameter $\mu(\Gamma)$ of a graph Γ is defined by (see [101, 226])

$$\mu(\Gamma) = \max_{M \in \mathscr{L}_\Gamma} \operatorname{corank} M,$$

where $\mathscr{L}_\Gamma$ is the set of symmetric real matrices M indexed by $V\Gamma$ that satisfy

(a) the strong Arnold hypothesis,

(b) $M_{uv} < 0$ if $u \sim v$, and $M_{uv} = 0$ if $u \not\sim v$ (nothing is required for the diagonal entries of M), and

(c) M has exactly one negative eigenvalue, of multiplicity 1.

We agree that $\mu(\Gamma) = 0$ if Γ has no vertices.

Although $\mu(\Gamma)$ is an algebraic parameter, it is directly related to some important topological graph properties, as we shall see below. It is easily seen that $\mu(K_n) = n - 1$ (take $M = -J$), and that $\mu(\Gamma) = 1$ if $n > 1$ and Γ has no edges (M must be a diagonal matrix with exactly one negative entry, and the strong Arnold hypothesis forbids two or more diagonal entries to be 0). If Γ has at least one edge, then $\mu(\Gamma + \Delta) = \max\{\mu(\Gamma), \mu(\Delta)\}$.

Theorem 7.3.1 ([101]) *The Colin de Verdière parameter $\mu(\Gamma)$ is graph minor monotone, that is, if Δ is a graph minor of Γ, then $\mu(\Delta) \leq \mu(\Gamma)$.*

In other words, for a given integer k, the property $\mu(\Gamma) \leq k$ is closed under taking graph minors (see [226]).

Theorem 7.3.2 ([101, 263, 301]) *The Colin de Verdière parameter* $\mu(\Gamma)$ *satisfies:*

   (i) $\mu(\Gamma) \leq 1$ *if and only if* Γ *is the disjoint union of paths.*
  (ii) $\mu(\Gamma) \leq 2$ *if and only if* Γ *is outerplanar.*
 (iii) $\mu(\Gamma) \leq 3$ *if and only if* Γ *is planar.*
  (iv) $\mu(\Gamma) \leq 4$ *if and only if* Γ *is linklessly embeddable.*
   (v) *If* Γ *is embeddable in the real projective plane or in the Klein bottle, then* $\mu(\Gamma) \leq 5.$
  (vi) *If* Γ *is embeddable in the torus, then* $\mu(\Gamma) \leq 6.$
 (vii) *If* Γ *is embeddable in a surface S with negative Euler characteristic* $\chi(S)$, *then* $\mu(\Gamma) \leq 4 - 2\chi(S).$ □

COLIN DE VERDIÈRE [101] conjectures that $\chi(\Gamma) \leq \mu(\Gamma) + 1$ for all Γ, where $\chi(\Gamma)$ is the chromatic number of Γ. (This would follow immediately from the Hadwiger conjecture.) If true, this would imply the four-color theorem.

VAN DER HOLST [224] gave a self-contained proof for (iii) above:

Proposition 7.3.3 *If* Γ *is planar, then* $\mu(\Gamma) \leq 3.$

Proof Suppose Γ is a counterexample. Add edges until Γ is maximally planar— $\mu(\Gamma)$ does not decrease. Now Γ is 3-connected and contains a triangle xyz that is a face. Let $M \in \mathscr{L}_\Gamma$ satisfy $\mu(\Gamma) = \operatorname{corank} M$. Then $\operatorname{corank} M > 3$ so that $\ker M$ contains a nonzero vector u with $u_x = u_y = u_z = 0$. Choose u with minimal support.

Since Γ is 3-connected, there exist three pairwise disjoint paths P_i $(i = 1, 2, 3)$, where each P_i starts in a vertex v_i outside $\operatorname{supp} u$ adjacent to some vertex of $\operatorname{supp} u$, and ends in xyz.

Since $Mu = 0$, the vertices v_i are all adjacent to both $\operatorname{supp}^> u$ and $\operatorname{supp}^< u$. By Proposition 5.3.2, both $\operatorname{supp}^> u$ and $\operatorname{supp}^< u$ induce connected subgraphs of Γ and hence can be contracted to a point. Also contract each path P_i to a point, and add a vertex a inside the triangle xyz adjacent to each of its vertices. The resulting graph is still planar but contains $K_{3,3}$, a contradiction. □

7.4 The Van der Holst-Laurent-Schrijver invariant

VAN DER HOLST, LAURENT & SCHRIJVER [225] define the graph invariant $\lambda(\Gamma)$ of a graph $\Gamma = (V, E)$ as the largest integer d for which there exists a d-dimensional subspace X of $\mathbb{R}^V$ such that for each nonzero $x \in X$ the positive support $\operatorname{supp}^>(x)$ (cf. §5.3) induces a (nonempty) connected subgraph of Γ. (All results in this section are from [225].)

Lemma 7.4.1 *One has* $\lambda(\Gamma) = d$ *if and only if there is a map* $\phi : V \to \mathbb{R}^d$ *such that for each open halfspace H in $\mathbb{R}^d$ the set* $\phi^{-1}(H)$ *induces a (nonempty) connected subgraph of* Γ.

Proof   Given X, with basis $x_1, \ldots, x_d$, let $\phi(v) = (x_1(v), \ldots, x_d(v))$. Conversely, given ϕ, define X to be the collection of maps sending $v \in V$ to $c^\top \phi(v)$, where $c \in \mathbb{R}^d$.   □

Proposition 7.4.2 *If Δ is a minor of Γ, then $\lambda(\Delta) \leq \lambda(\Gamma)$.*

Proof   Given a suitable map $\psi : V(\Delta) \to \mathbb{R}^d$ as above, we construct a suitable map ϕ. There are three cases. (i) If Δ arises from Γ by deletion of an isolated vertex v, then let $\phi(u) = \psi(u)$ for $u \neq v$, and $\phi(v) = 0$. (ii) If Δ arises from Γ by deletion of an edge e, then let $\phi = \psi$. (iii) If Δ arises from Γ by contraction of an edge $e = uv$ to a single vertex w, then let $\phi(u) = \phi(v) = \psi(w)$, and $\phi(z) = \psi(z)$ for $z \neq u, v$.   □

One has $\lambda(K_n) = n - 1$. More generally, if Γ is the 1-skeleton of a d-dimensional convex polytope, then $\lambda(\Gamma) \geq d$. In particular, $\lambda(\Gamma) \geq 3$ if Γ is a 3-connected planar graph. If Δ is obtained from Γ by deleting a single vertex, then $\lambda(\Gamma) \leq \lambda(\Delta) + 1$. Let V_8 be the Cayley graph with vertex set $\mathbb{Z}_8$ and difference set $\{\pm 1, 4\}$. We have the analogue of Theorem 7.3.2:

Proposition 7.4.3   *(i) $\lambda(\Gamma) \leq 1$ if and only if Γ has no K_3 minor.*
  (ii) $\lambda(\Gamma) \leq 2$ if and only if Γ has no K_4 minor.
  (iii) $\lambda(\Gamma) \leq 3$ if and only if Γ has no K_5 or V_8 minor.
  (iv) $\lambda(\Gamma) \leq 4$ if Γ is linklessly embeddable.

Many further minor-monotone algebraic graph invariants have been proposed. Often these can be related to topological embeddability properties.

Chapter 8
Euclidean Representations

The main goal of this chapter is the famous result of CAMERON, GOETHALS, SEIDEL & SHULT [84] characterizing graphs with smallest eigenvalue not less than -2.

8.1 Examples

We have seen examples of graphs with smallest eigenvalue $\theta_{\min} \geq -2$. The most important example is formed by the line graphs (see §1.4.5), and people wanted to characterize line graphs by this condition and possibly some additional hypotheses.

Another series of examples are the so-called *cocktail party graphs*, the graphs $K_{m \times 2}$, i.e., $\overline{mK_2}$, with spectrum $2m - 2$, 0^m, $(-2)^{m-1}$. For $m \geq 4$, these are not line graphs.

And there are exeptional examples like the Petersen graph (with spectrum 3 1^5 $(-2)^4$), lots of them. It is easy to see that the Petersen graph is not a line graph. More generally, no line graph can have a *3-claw*, that is, an induced $K_{1,3}$ subgraph, as is immediately clear from the definition.

8.2 Euclidean representation

Suppose the graph Γ has smallest eigenvalue $\theta_{\min} \geq -2$. Then $A + 2I$ is positive semidefinite, so that $A + 2I$ is the Gram matrix of a collection of vectors in some Euclidean space $\mathbb{R}^m$ (where $m = \text{rk}\,(A + 2I)$), cf. §2.9.

In this way we obtain a map $x \mapsto \bar{x}$ from vertices of Γ to vectors in $\mathbb{R}^m$, where

$$(\bar{x}, \bar{y}) = \begin{cases} 2 \text{ if } x = y \\ 1 \text{ if } x \sim y \\ 0 \text{ if } x \nsim y. \end{cases}$$

The additive subgroup of $\mathbb{R}^m$ generated by the vectors $\bar{x}$, for x in the vertex set X of Γ, is a *root lattice*: an integral lattice generated by *roots*: vectors with squared length 2. Root lattices have been classified. That classification is the subject of the next section.

8.3 Root lattices

We start with an extremely short introduction to lattices.

Lattice

A *lattice* Λ is a discrete additive subgroup of $\mathbb{R}^n$. Equivalently, it is a finitely generated free $\mathbb{Z}$-module with positive definite symmetric bilinear form.

Basis

Assume that our lattice Λ has dimension n, i.e., spans $\mathbb{R}^n$. Let $\{a_1,\ldots,a_n\}$ be a $\mathbb{Z}$-basis of Λ. Let A be the matrix with the vectors a_i as rows. If we choose a different $\mathbb{Z}$-basis $\{b_1,\ldots,b_n\}$, so that $b_i = \sum s_{ij}a_j$, and B is the matrix with the vectors b_i as rows, then $B = SA$, with $S = (s_{ij})$. Since S is integral and invertible, it has determinant ± 1. It follows that $|\det A|$ is uniquely determined by Λ, independent of the choice of basis.

Volume

$\mathbb{R}^n/\Lambda$ is an n-dimensional torus, compact and with finite volume. Its volume is the volume of the fundamental domain, which equals $|\det A|$. If Λ' is a sublattice of Λ, then $\mathrm{vol}(\mathbb{R}^n/\Lambda') = \mathrm{vol}(\mathbb{R}^n/\Lambda).|\Lambda/\Lambda'|$.

Gram matrix

Let G be the matrix (a_i,a_j) of inner products of basis vectors for a given basis. Then $G = AA^\top$, so $\mathrm{vol}(\mathbb{R}^n/\Lambda) = \sqrt{\det G}$.

Dual lattice

The *dual* Λ^* of a lattice Λ is the lattice of vectors having integral inner products with all vectors in Λ: $\Lambda^* = \{x \in \mathbb{R}^n \mid (x,r) \in \mathbb{Z} \text{ for all } r \in \Lambda\}$. It has a basis $\{a_1^*,\ldots,a_n^*\}$ defined by $(a_i^*,a_j) = \delta_{ij}$. Now $A^*A^\top = I$, so $A^* = (A^{-1})^\top$ and Λ^* has Gram matrix $G^* = G^{-1}$. It follows that $\mathrm{vol}(\mathbb{R}^n/\Lambda^*) = \mathrm{vol}(\mathbb{R}^n/\Lambda)^{-1}$. We have $\Lambda^{**} = \Lambda$.

Integral lattice

The lattice Λ is called *integral* when every two lattice vectors have an integral inner product. For an integral lattice Λ one has $\Lambda \subseteq \Lambda^*$.

The lattice Λ is called *even* when (x,x) is an even integer for each $x \in \Lambda$. An even lattice is integral.

Discriminant

The *determinant*, or *discriminant*, disc Λ of a lattice Λ is defined by disc $\Lambda = \det G$. When Λ is integral, we have disc $\Lambda = |\Lambda^*/\Lambda|$.

A lattice is called *self-dual* or *unimodular* when $\Lambda = \Lambda^*$, i.e., when it is integral with discriminant 1. An even unimodular lattice is called *type II*, and the remaining unimodular lattices are called *type I*. It can be shown that if there is an even unimodular lattice in $\mathbb{R}^n$, then n is divisible by 8.

Direct sums

If Λ and Λ' are lattices in $\mathbb{R}^m$ and $\mathbb{R}^n$, respectively, then $\Lambda \perp \Lambda'$, the *orthogonal direct sum* of Λ and Λ', is the lattice $\{(x,y) \in \mathbb{R}^{m+n} \mid x \in \Lambda \text{ and } y \in \Lambda'\}$. A lattice is called *irreducible* when it is not the orthogonal direct sum of two nonzero lattices.

8.3.1 Examples

(i) $\mathbb{Z}^n$

The lattice $\mathbb{Z}^n$ is unimodular, type I.

(ii) A_2

The triangular lattice in the plane $\mathbb{R}^2$ has basis $\{r,s\}$. Choose the scale such that r has length $\sqrt{2}$. Then the Gram matrix is $G = \begin{pmatrix} 2 & -1 \\ -1 & 2 \end{pmatrix}$, so $\det G = 3$ and $p,q \in A_2^*$. A fundamental region for A_2 is the parallelogram on $0,r,s$. A fundamental region for A_2^* is the parallelogram on $0,p,q$. Note that the area of the former is thrice that of the latter.

The representation of this lattice in $\mathbb{R}^2$ has nonintegral coordinates. It is easier to work in $\mathbb{R}^3$, on the hyperplane $\sum x_i = 0$, and choose $r = (1,-1,0)$, $s = (0,1,-1)$. Then A_2 consists of the points (x_1,x_2,x_3) with $x_i \in \mathbb{Z}$ and $\sum x_i = 0$. The dual lattice A_2^* consists of the points (x_1,x_2,x_3) with $x_1 \equiv x_2 \equiv x_3 \pmod 1$ and $\sum x_i = 0$ (so that $3x_1 \in \mathbb{Z}$). It contains, for example, $p = \frac{1}{3}(2r+s) = (\frac{2}{3}, -\frac{1}{3}, -\frac{1}{3})$.

(iii) E_8

Let $\rho : \mathbb{Z}^n \to 2^n$ be coordinatewise reduction mod 2. Given a binary linear code C, the lattice $\rho^{-1}(C)$ is integral, since it is contained in $\mathbb{Z}^n$, but never unimodular, unless it is all of $\mathbb{Z}^n$, a boring situation.

Now suppose that C is self-orthogonal, so that any two code words have an even inner product. Then $\Lambda(C) = \frac{1}{\sqrt{2}}\rho^{-1}(C)$ is an integral lattice. If $\dim C = k$, then we have $\mathrm{vol}(\mathbb{R}^n / \rho^{-1}(C)) = 2^{n-k}$ and hence $\mathrm{vol}(\mathbb{R}^n / \Lambda(C)) = 2^{n/2-k}$. In particular, $\Lambda(C)$ will be unimodular when C is self-dual, and even when C is "doubly even", i.e., has weights divisible by 4.

Let C be the [8,4,4] extended Hamming code. Then $\Lambda(C)$ is an even unimodular 8-dimensional lattice known as E_8.

The code C has weight enumerator $1 + 14X^4 + X^8$ (that is, has one word of weight 0, 14 words of weight 4, and one word of weight 8). It follows that the *roots* (vectors r with $(r,r) = 2$) in this incarnation of E_8 are the 16 vectors $\pm\frac{1}{\sqrt{2}}(2,0,0,0,0,0,0,0)$ (with 2 in any position), and the $16 \cdot 14 = 224$ vectors $\frac{1}{\sqrt{2}}(\pm 1, \pm 1, \pm 1, \pm 1, 0, 0, 0, 0)$ with ± 1 in the nonzero positions of a weight 4 vector. Thus, there are 240 roots.

8.3.2 Root lattices

A *root lattice* is an integral lattice generated by *roots* (vectors r with $(r,r) = 2$). For example, A_2 and E_8 are root lattices.

The set of roots in a root lattice is a (reduced) *root system* Φ, i.e., satisfies

(i) If $r \in \Phi$ and $\lambda r \in \Phi$, then $\lambda = \pm 1$.

(ii) Φ is closed under the reflection w_r that sends s to $s - 2\frac{(r,s)}{(r,r)}r$ for each $r \in \Phi$.

(iii) $2\frac{(r,s)}{(r,r)} \in \mathbb{Z}$.

Since Φ generates Λ and Φ is invariant under $W = \langle w_r \mid r \in \Phi \rangle$, the same holds for Λ, so root lattices have a large group of automorphisms.

A *fundamental system* of roots Π in a root lattice Λ is a set of roots generating Λ and such that $(r,s) \leq 0$ for distinct $r,s \in \Pi$. A *reduced fundamental system* of roots is a fundamental system that is linearly independent. A nonreduced fundamental system is called *extended*.

For example, in A_2 the set $\{r,s\}$ is a reduced fundamental system, and $\{r,s,-r-s\}$ is an extended fundamental system.

The *Dynkin diagram* of a fundamental system Π such that $(r,s) \neq -2$ for $r,s \in \Pi$ is the graph with vertex set Π where r and s are joined by an edge when $(r,s) = -1$. (The case $(r,s) = -2$ happens only for a nonreduced system with A_1 component. In that case we do not define the Dynkin diagram.)

Every root lattice has a reduced fundamental system: Fix some vector u, not orthogonal to any root. Put $\Phi^+(u) = \{r \in \Phi \mid (r,u) > 0\}$ and $\Pi(u) = \{r \in \Phi^+(u) \mid r$ cannot be written as $s + t$ with $s,t \in \Phi^+(u)\}$. Then $\Pi(u)$ is a reduced fundamental system of roots, and written on this basis each root has only positive or only negative coefficients.

(Indeed, if $r, s \in \Pi(u)$ and $(r, s) = 1$, then say $r - s \in \Phi^+(u)$ and $r = (r - s) + s$, contradiction. This shows that $\Pi(u)$ is a fundamental system. If $\sum \gamma_r r = 0$, then separate the γ_r into positive and negative ones to get $\sum \alpha_r r = \sum \beta_s s = x \neq 0$, where all coefficients α_r, β_s are positive. Now $0 < (x, x) = \sum \alpha_r \beta_s (r, s) \leq 0$, a contradiction. This shows that $\Pi(u)$ is reduced. Each root in $\Phi^+(u)$ has an expression over $\Pi(u)$ with only positive coefficients.)

Proposition 8.3.1 *Let Π be a reduced fundamental system.*
(i) For all $x \in \mathbb{R}^n$ there is a $w \in W$ such that $(w(x), r) \geq 0$ for all $r \in \Pi$.
(ii) $\Pi = \Pi(u)$ for some u. (That is, W is transitive on reduced fundamental systems.)
(iii) If Λ is irreducible, then there is a unique $\tilde{r} \in \Phi$ such that $\Pi \cup \{\tilde{r}\}$ is an extended fundamental system.

Proof (i) Let G be the Gram matrix of Π, and write $A = 2I - G$. Since G is positive definite, A has largest eigenvalue less than 2. Using the Perron-Frobenius theorem, let $\gamma = (\gamma_r)_{r \in \Pi}$ be a positive eigenvector of A. If $(x, s) < 0$ for some $s \in \Pi$, then put $x' = w_s(x) = x - (x, s)s$. Now

$$(x', \sum_r \gamma_r r) = (x, \sum_r \gamma_r r) - (G\gamma)_s (x, s) > (x, \sum_r \gamma_r r).$$

But W is finite, so after finitely many steps we reach the desired conclusion.

(ii) Induction on $|\Pi|$. Fix x with $(x, r) \geq 0$ for all $r \in \Pi$. Then $\Pi_0 = \Pi \cap x^{\perp}$ is a fundamental system of a lattice in a lower-dimensional space, so of the form $\Pi_0 = \Pi_0(u_0)$. Take $u = x + \varepsilon u_0$ for small $\varepsilon > 0$. Then $\Pi = \Pi(u)$.

(iii) If $r \in \Phi^+(u)$ has maximal (r, u), then $\tilde{r} = -r$ is the unique root that can be added. It can be added, since $(\tilde{r}, s) \geq 0$ means $(r, s) < 0$, so that $r + s$ is a root, contradicting the maximality of r. And it is unique because linear dependencies of an extended system correspond to an eigenvector with eigenvalue 2 of the extended Dynkin diagram, and by the Perron-Frobenius theorem there is, up to a constant, a unique such eigenvector when the diagram is connected, that is, when Λ is irreducible.   □

8.3.3 Classification

The irreducible root lattices one finds are A_n ($n \geq 0$), D_n ($n \geq 4$), E_6, E_7, E_8. Each is defined by its Dynkin diagram.

(1) A_n: The lattice vectors are the $x \in \mathbb{Z}^{n+1}$ with $\sum x_i = 0$. There are $n(n + 1)$ roots: $e_i - e_j$ ($i \neq j$). The discriminant is $n + 1$, and $A_n^*/A_n \cong \mathbb{Z}_{n+1}$, with the quotient generated by $\frac{1}{n+1}(e_1 + \cdots + e_n - ne_{n+1}) \in A_n^*$.

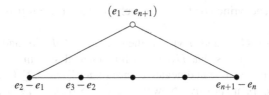

(2) D_n: The lattice vectors are the $x \in \mathbb{Z}^n$ with $\sum x_i \equiv 0 \pmod 2$. There are $2n(n-1)$ roots $\pm e_i \pm e_j$ $(i \neq j)$. The discriminant is 4, and D_n^*/D_n is isomorphic to $\mathbb{Z}_4$ when n is odd, and to $\mathbb{Z}_2 \times \mathbb{Z}_2$ when n is even. D_n^* contains e_1 and $\frac{1}{2}(e_1 + \cdots + e_n)$. Note that $D_3 \cong A_3$.

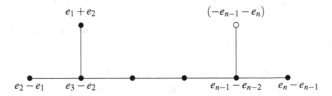

(3) E_8: (Recall that we already gave a construction of E_8 from the Hamming code.) The lattice is the span of D_8 and $c := \frac{1}{2}(e_1 + \cdots + e_8)$. There are $240 = 112 + 128$ roots, of the forms $\pm e_i \pm e_j$ $(i \neq j)$ and $\frac{1}{2}(\pm e_1 \pm \cdots \pm e_8)$ with an even number of minus signs. The discriminant is 1, and $E_8^* = E_8$.

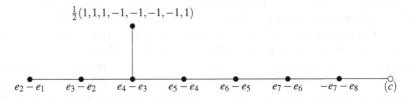

(4) E_7: Take $E_7 = E_8 \cap c^{\perp}$. There are $126 = 56 + 70$ roots. The discriminant is 2, and E_7^* contains $\frac{1}{4}(1,1,1,1,1,1,-3,-3)$.

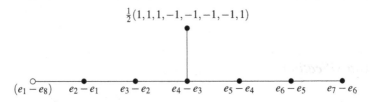

(5) E_6: For the vector $d = -e_7 - e_8$, take $E_6 = E_8 \cap \{c, d\}^{\perp}$. There are $72 = 32 + 40$ roots. The discriminant is 3, and E_6^* contains the vector $\frac{1}{3}(1,1,1,1,-2,-2,0,0)$.

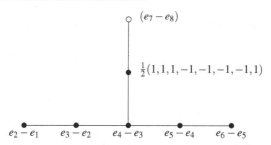

That this is all is an easy consequence of the Perron-Frobenius theorem: $A = 2I - G$ is the adjacency matrix of a graph, namely the Dynkin diagram, and this graph has largest eigenvalue at most 2. These graphs were determined in Theorem 3.1.3. The connected graphs with largest eigenvalue less than 2 are the Dynkin diagrams of reduced fundamental systems of irreducible root systems, and the connected graphs with largest eigenvalue 2 are the Dynkin diagrams of extended root systems.

In the pictures above, the reduced fundamental systems were drawn with black dots, and the additional element of the extended system with an open dot (and a name given in parentheses).

8.4 The Cameron-Goethals-Seidel-Shult theorem

Now return to the discussion of connected graphs Γ with smallest eigenvalue $\theta_{\min} \geq -2$. In §8.2 we found a map $x \mapsto \bar{x}$ from the vertex set X of Γ to some Euclidean space $\mathbb{R}^m$ such that the inner product $(\bar{x}, \bar{y})$ takes the value 2, 1, 0 when $x = y$, $x \sim y$ and $x \not\sim y$, respectively.

Let Σ be the image of X under this map. Then Σ generates a root lattice Λ. Since Γ is connected, the root lattice is irreducible.

By the classification of root lattices, it follows that the root lattice is one of A_n, D_n, E_6, E_7, or E_8. Note that the graph is determined by Σ, so the classification of graphs with $\theta_{\min} \geq -2$ is equivalent to the classification of subsets Φ of the root system with the property that all inner products are 2, 1, or 0, i.e., are nonnegative.

Now A_n and D_n can be chosen to have integral coordinates, and $E_6 \subset E_7 \subset E_8$, so we have the two cases (i) $\Sigma \subset \mathbb{Z}^{m+1}$ and (ii) $\Sigma \subset E_8$. A graph is called *exceptional* in case (ii). Since E_8 has a finite number of roots, there are only finitely many exceptional graphs.

In case (i) one quickly sees what the structure of Γ has to be. Something like a line graph with attached cocktail party graphs. This structure has been baptised *generalized line graph*. The precise definition will be clear from the proof of the theorem below.

Theorem 8.4.1 *(i) Let Γ be a connected graph with smallest eigenvalue $\theta_{\min} \geq -2$. Then Γ is either a generalized line graph or one of finitely many exceptions, represented by roots in the E_8 lattice.*

(ii) A regular generalized line graph is either a line graph or a cocktail party graph.

(iii) A graph represented by roots in the E_8 lattice has at most 36 vertices, and every vertex has valency at most 28.

Proof (i) Consider the case $\Sigma \subset \mathbb{Z}^{m+1}$. Roots in $\mathbb{Z}^{m+1}$ have shape $\pm e_i \pm e_j$. If some e_i has the same sign in all $\sigma \in \Sigma$ in which it occurs, then choose the basis such that this sign is $+$. Let I be the set of all such indices i. Then $\{x \mid \bar{x} = e_i + e_j \text{ for some } i, j \in I\}$ induced a line graph in Γ, with x corresponding to the edge ij on I. If $j \notin I$, then e_j occurs with both signs, and there are $\sigma, \tau \in \Sigma$ with $\sigma = \pm e_i + e_j$ and $\tau = \pm e_{i'} - e_j$. Since all inner products in Σ are nonnegative, $i = i'$ with $i \in I$, and $\sigma = e_i + e_j$, $\tau = e_i - e_j$. Thus, i is determined by j and we have a map $\phi : j \mapsto i$ from indices outside I to indices in I. Now for each $i \in I$ the set $\{x \mid \bar{x} = e_i \pm e_j \text{ for some } j \text{ with } \phi(j) = i\}$ induces a cocktail party graph. Altogether we see in what way Γ is a line graph with attached cocktail party graphs.

(ii) Now let Γ be regular. A vertex x with $\bar{x} = e_i - e_j$ is adjacent to all vertices with image $e_i \pm e_k$ different from $e_i + e_j$. But a vertex y with $\bar{y} = e_i + e_k$ where $i, k \in I$ is adjacent to all vertices with image $e_i \pm e_k$ without exception (and also to vertices with image $e_k \pm e_l$). Since Γ is regular, both types of vertices cannot occur together, so that Γ is either a line graph or a cocktail party graph.

(iii) Suppose $\Sigma \subset E_8$. Consider the 36-dimensional space of symmetric 8×8 matrices, equipped with the positive definite inner product $(P, Q) = \operatorname{tr} PQ$. Associated with the 240 roots r of E_8 are 120 rank 1 matrices $P_r = rr^\top$ with mutual inner products $(P_r, P_s) = \operatorname{tr} rr^\top ss^\top = (r, s)^2$. The Gram matrix of the set of P_r for $r \in \Sigma$ is $G = 4I + A$. Since G is positive definite (it has smallest eigenvalue ≥ 2), the vectors P_r are linearly independent, and hence $|\Sigma| \leq 36$.

Finally, let r be a root of E_8. The 56 roots s of E_8 that satisfy $(r, s) = 1$ fall into 28 pairs s, s' with $(s, s') = -1$. So, Σ can contain at most one member from each of these pairs, and each vertex of Γ has valency at most 28.   □

The bounds in (ii) are best possible: Take the graph $K_8 + L(K_8)$ and add edges joining $i \in K_8$ with $jk \in L(K_8)$ whenever i, j, k are distinct. This graph has 36 vertices, the vertices in K_8 have 28 neighbors, and the smallest eigenvalue is -2. A representation in E_8 is given by $i \mapsto \frac{1}{2}(e_1 + \cdots + e_8) - e_i$ and $jk \mapsto e_j + e_k$.

There is a large amount of literature on exceptional graphs.

8.5 Further applications

The basic observation of this chapter is that if M is a symmetric positive semidefinite matrix, then M is the Gram matrix of a collection of vectors in some Euclidean space. Now one can use the geometry of Euclidean space to study the situation.

If the adjacency matrix A of a graph has smallest eigenvalue $-m$, then $A + mI$ is positive semidefinite, and this technique can be used.

More generally, if f is a polynomial such that $f(\theta) \geq 0$ for all eigenvalues θ of A, then $f(A)$ is positive semidefinite, hence a Gram matrix. For example, in §9.7.4 below we sketch a uniqueness proof for a graph using the fact that $4J - (A - 2I)(A + 6I)$ is positive semidefinite.

This applies in particular to distance-regular graphs, where the idempotents provide Euclidean representations, cf. §12.10.

8.6 Exercises

Exercise 8.1 Show that the following describes a root system of type E_6. Take the following 72 vectors in $\mathbb{R}^9$: 18 vectors $\pm(u,0,0)$, $\pm(0,u,0)$, $\pm(0,0,u)$, where $u \in \{(1,-1,0),(0,1,-1),(-1,0,1)\}$, and 54 vectors $\pm(u,v,w)$, where $u,v,w \in \{(\frac{2}{3},-\frac{1}{3},-\frac{1}{3}),(-\frac{1}{3},\frac{2}{3},-\frac{1}{3}),(-\frac{1}{3},-\frac{1}{3},\frac{2}{3})\}$.

Exercise 8.2 Show that the following describes a root system of type E_7. Take the following 126 vectors in $\mathbb{R}^7$: 60 vectors $\pm e_i \pm e_j$ with $1 \leq i < j \leq 6$, and 64 vectors $\pm(x_1,\ldots,x_6,\frac{1}{\sqrt{2}})$ with $x_i = \pm\frac{1}{2}$, where an even number of x_i has $+$ sign, and 2 vectors $\pm(0,\ldots,0,\sqrt{2})$.

Chapter 9
Strongly Regular Graphs

9.1 Strongly regular graphs

A graph (simple, undirected, and loopless) of order v is called *strongly regular* with parameters v, k, λ, μ whenever it is not complete or edgeless and

- (i) each vertex is adjacent to k vertices,
- (ii) for each pair of adjacent vertices there are λ vertices adjacent to both,
- (iii) for each pair of nonadjacent vertices there are μ vertices adjacent to both.

We require that both edges and nonedges occur, so that the parameters are well-defined.

In association scheme terminology (cf. §11.1), a strongly regular graph is a symmetric association scheme with two (nonidentity) classes, in which one relation is singled out to be the adjacency relation.

9.1.1 Simple examples

Easy examples of strongly regular graphs:

- (i) A quadrangle is strongly regular with parameters $(4, 2, 0, 2)$.
- (ii) A pentagon is strongly regular with parameters $(5, 2, 0, 1)$.
- (iii) The 3×3 grid, the Cartesian product of two triangles, is strongly regular with parameters $(9, 4, 1, 2)$.
- (iv) The Petersen graph is strongly regular with parameters $(10, 3, 0, 1)$.

(Each of these graphs is uniquely determined by its parameters, so if you do not know what a pentagon is, or what the Petersen graph is, this defines it.)

Each of these examples can be generalized in numerous ways. For example,

- (v) Let $q = 4t + 1$ be a prime power. The *Paley graph* $\text{Paley}(q)$ is the graph with the finite field $\mathbb{F}_q$ as vertex set, where two vertices are adjacent when they differ by a (nonzero) square. It is strongly regular with parameters $(4t+1,2t,t-1,t)$, as we shall see below. Doing this for $q=5$ and $q=9$, we find Examples (ii) and (iii) again. For $q=13$ we find a graph that is locally a hexagon. For $q=17$ we find a graph that is locally an 8-gon + diagonals.

(vi) The $m \times m$ grid, the Cartesian product of two complete graphs on m vertices, is strongly regular with parameters $(m^2,2(m-1),m-2,2)$ (for $m>1$). For $m=2$ and $m=3$, we find Examples (i) and (iii) again.

(vii) The complete multipartite graph $K_{m \times a}$, with vertex set partitioned into m groups of size a, where two points are adjacent when they are from different groups, is strongly regular with parameters $(ma,(m-1)a,(m-2)a,(m-1)a)$ (for $m>1$ and $a>1$). For $m=a=2$, we find Example (i) again.

The *complement* of a graph Γ is the graph $\overline{\Gamma}$ with the same vertex set as Γ, where two vertices are adjacent if and only if they are nonadjacent in Γ. The complement of a strongly regular graph with parameters (v,k,λ,μ) is again strongly regular, and has parameters $(v,v-k-1,v-2k+\mu-2,v-2k+\lambda)$. (Indeed, we keep the same association scheme, but now single out the other nonidentity relation.)

(viii) The Paley graph Paley(q) is isomorphic to its complement. (Indeed, an isomorphism is given by multiplication by a nonsquare.) In particular, we see that the pentagon and the 3×3 grid are (isomorphic to) their own complements.

(ix) The disjoint union mK_a of m complete graphs of size a is strongly regular with parameters $(ma,a-1,a-2,0)$ (for $m>1$ and $a>1$). These graphs are the complements of those in Example (vii).

(x) The *triangular graph* on the pairs in an m-set, denoted by $T(m)$, or by $\binom{m}{2}$, has these pairs as vertices, where two pairs are adjacent whenever they meet in one point. These graphs are strongly regular, with parameters $(\binom{m}{2},2(m-2),m-2,4)$, if $m \geq 4$. For $m=4$ we find $K_{3 \times 2}$. For $m=5$ we find the complement of the Petersen graph.

The four parameters are not independent. Indeed, if $\mu \neq 0$ we find the relation

$$v = 1+k+\frac{k(k-1-\lambda)}{\mu}$$

by counting vertices at distance 0, 1, and 2 from a given vertex.

9.1.2 The Paley graphs

We claimed above that the Paley graphs (with vertex set $\mathbb{F}_q$, where q is a prime power congruent 1 mod 4, and where two vertices are adjacent when their difference is a nonzero square) are strongly regular. Let us verify this.

Proposition 9.1.1 *The Paley graph Paley(q) with $q=4t+1$ is strongly regular with parameters $(v,k,\lambda,\mu) = (4t+1,2t,t-1,t)$. It has eigenvalues k, $(-1 \pm \sqrt{q})/2$ with multiplicities 1, 2t, 2t, respectively.*

Proof The values for v and k are clear. Let $\chi : \mathbb{F}_q \to \{-1,0,1\}$ be the quadratic residue character defined by $\chi(0) = 0$, $\chi(x) = 1$ when x is a (nonzero) square, and $\chi(x) = -1$ otherwise. Note that $\sum_x \chi(x) = 0$, and that for nonzero a we have $\sum_z \chi(z^2 - az) = \sum_{z \neq 0} \chi(1 - \frac{a}{z}) = -1$. Now λ and μ follow from

$$4 \sum_{\substack{z \\ x \sim z \sim y}} 1 = \sum_{z \neq x,y} (\chi(z-x)+1)(\chi(z-y)+1) = -1 - 2\chi(x-y) + (q-2).$$

For the spectrum, see Theorem 9.1.3 below. $\qquad\square$

9.1.3 Adjacency matrix

For convenience we call an eigenvalue *restricted* if it has an eigenvector perpendicular to the all-1 vector **1**.

Theorem 9.1.2 *For a simple graph Γ of order v, not complete or edgeless, with adjacency matrix A, the following are equivalent:*

(i) *Γ is strongly regular with parameters (v,k,λ,μ) for certain integers k, λ, μ.*
(ii) *$A^2 = (\lambda - \mu)A + (k - \mu)I + \mu J$ for certain real numbers k, λ, μ.*
(iii) *A has precisely two distinct restricted eigenvalues.*

Proof The equation in (ii) can be rewritten as

$$A^2 = kI + \lambda A + \mu(J - I - A).$$

Now (i) $\Longleftrightarrow$ (ii) is obvious.

(ii) $\Rightarrow$ (iii): Let ρ be a restricted eigenvalue, and u a corresponding eigenvector perpendicular to **1**. Then $Ju = 0$. Multiplying the equation in (ii) on the right by u yields $\rho^2 = (\lambda - \mu)\rho + (k - \mu)$. This quadratic equation in ρ has two distinct solutions. (Indeed, $(\lambda - \mu)^2 = 4(\mu - k)$ is impossible since $\mu \leq k$ and $\lambda \leq k - 1$.)

(iii) $\Rightarrow$ (ii): Let r and s be the restricted eigenvalues. Then $(A - rI)(A - sI) = \alpha J$ for some real number α. So A^2 is a linear combination of A, I, and J. $\qquad\square$

9.1.4 Imprimitive graphs

A strongly regular graph is called *imprimitive* if it, or its complement, is disconnected, and *primitive* otherwise. Imprimitive strongly regular graphs are boring.

If a strongly regular graph is not connected, then $\mu = 0$ and $k = \lambda + 1$. And conversely, if $\mu = 0$ or $k = \lambda + 1$ then the graph is a disjoint union aK_m of some number a of complete graphs K_m. In this case $v = am$, $k = m - 1$, $\lambda = m - 2$, $\mu = 0$ and the spectrum is $(m-1)^a$, $(-1)^{a(m-1)}$.

If the complement of a strongly regular graph is not connected, then $k = \mu$. And conversely, if $k = \mu$ then the graph is the complete multipartite graph $K_{a \times m}$, the complement of aK_m, with parameters $v = am$, $k = \mu = (a-1)m$, $\lambda = (a-2)m$ and spectrum $(a-1)m^1$, $0^{a(m-1)}$, $(-m)^{a-1}$.

Let r and s $(r > s)$ be the restricted eigenvalues of A. For a primitive strongly regular graph one has $k > r > 0$ and $s < -1$.

9.1.5 Parameters

Theorem 9.1.3 *Let Γ be a strongly regular graph with adjacency matrix A and parameters (v, k, λ, μ). Let r and s $(r > s)$ be the restricted eigenvalues of A and let f, g be their respective multiplicities. Then*

(i) $k(k-1-\lambda) = \mu(v-k-1)$.
(ii) $rs = \mu - k$, $r+s = \lambda - \mu$.
(iii) $f, g = \frac{1}{2}(v-1 \mp \frac{(r+s)(v-1)+2k}{r-s})$.
(iv) *If r and s are nonintegral, then $f = g$ and $(v,k,\lambda,\mu) = (4t+1, 2t, t-1, t)$ for some integer t.*

Proof (i) Fix a vertex x of Γ. Let $\Gamma(x)$ and $\Delta(x)$ be the sets of vertices adjacent and nonadjacent to x, respectively. Counting in two ways the number of edges between $\Gamma(x)$ and $\Delta(x)$ yields (i). The equations (ii) are direct consequences of Theorem 9.1.2(ii), as we saw in the proof. Formula (iii) follows from $f + g = v - 1$ and $0 = \operatorname{trace} A = k + fr + gs = k + \frac{1}{2}(r+s)(f+g) + \frac{1}{2}(r-s)(f-g)$. Finally, if $f \neq g$, then one can solve for r and s in (iii) (using (ii)) and find that r and s are rational, and hence integral. But $f = g$ implies $(\mu - \lambda)(v-1) = 2k$, which is possible only for $\mu - \lambda = 1$, $v = 2k+1$. ☐

These relations imply restrictions for the possible values of the parameters. Clearly, the right-hand sides of (iii) must be positive integers. These are the so-called *rationality conditions*.

9.1.6 The half case and cyclic strongly regular graphs

The case of a strongly regular graph with parameters $(v, k, \lambda, \mu) = (4t+1, 2t, t-1, t)$ for some integer t is called the *half case*. Such graphs are also called *conference graphs*. If such a graph exists, then v is the sum of two squares, see Theorem 10.4.2 below. The Paley graphs (§9.1.2, §10.4, §13.6) belong to this case, but there are many further examples.

A characterization of the Paley graphs of prime order is given by

Proposition 9.1.4 (KELLY [240], BRIDGES & MENA [44]) *A strongly regular graph with a regular cyclic group of automorphisms is a Paley graph with a prime number of vertices.*

(See the discussion of translation association schemes in [54], §2.10. This result has been rediscovered several times.)

9.1.7 Strongly regular graphs without triangles

As an example of the application of the rationality conditions we classify the strongly regular graphs of girth 5.

Theorem 9.1.5 (HOFFMAN & SINGLETON [220]) *Suppose $(v,k,0,1)$ is the parameter set of a strongly regular graph. Then $(v,k) = (5,2)$, $(10,3)$, $(50,7)$ or $(3250,57)$.*

Proof The rationality conditions imply that either $f = g$, which leads to $(v,k) = (5,2)$, or $r - s$ is an integer dividing $(r+s)(v-1) + 2k$. By Theorem 9.1.3(i)–(ii) we have

$$s = -r - 1, \ k = r^2 + r + 1, \ v = r^4 + 2r^3 + 3r^2 + 2r + 2,$$

and thus we obtain $r = 1, 2$ or 7. □

The first three possibilities are uniquely realized by the pentagon, the Petersen graph and the Hoffman-Singleton graph. For the last case existence is unknown (but see §11.5.1).

More generally we can look at strongly regular graphs of girth at least 4. Seven examples are known.

(i) The *pentagon*, with parameters $(5,2,0,1)$.

(ii) The *Petersen graph*, with parameters $(10,3,0,1)$. This is the complement of the triangular graph $T(5)$.

(iii) The *folded 5-cube*, with parameters $(16,5,0,2)$. This graph is obtained from the 5-cube 2^5 on 32 vertices by identifying antipodal vertices. (The complement of this graph is known as the *Clebsch graph*.)

(iv) The *Hoffman-Singleton graph*, with parameters $(50,7,0,1)$. There are many constructions for this graph, cf., e.g., [54], §13.1. A short one, due to N. Robertson, is the following. Take 25 vertices (i,j) and 25 vertices $(i,j)'$ with $i,j \in \mathbb{Z}_5$, and join (i,j) with $(i,j+1)$, $(i,j)'$ with $(i,j+2)'$, and (i,k) with $(j,ij+k)'$ for all $i,j,k \in \mathbb{Z}_5$. Now the subsets $(i,*)$ become pentagons, the $(i,*)'$ become pentagons (drawn as pentagrams), and each of the 25 unions of $(i,*)$ with $(j,*)'$ induces a Petersen subgraph.

(v) The *Gewirtz graph*, with parameters $(56,10,0,2)$. This is the graph with as vertices the $77 - 21 = 56$ blocks of the unique Steiner system $S(3,6,22)$ not containing a given symbol, where two blocks are adjacent when they are disjoint. It is a subgraph of the following.

(vi) The M_{22} *graph*, with parameters $(77,16,0,4)$. This is the graph with as vertices the 77 blocks of the unique Steiner system $S(3,6,22)$, adjacent when they are disjoint. It is a subgraph of the following.

(vii) The *Higman-Sims graph*, with parameters $(100,22,0,6)$. This is the graph with as $1+22+77$ vertices an element ∞, the 22 symbols of $S(3,6,22)$, and the 77 blocks of $S(3,6,22)$. The element ∞ is adjacent to the 22 symbols, each symbol is adjacent to the 21 blocks containing it, and blocks are adjacent when disjoint. The (rank 3) automorphism group of this graph is $HS.2$, where HS is the sporadic simple group of Higman and Sims. This graph can be partitioned into two halves, each inducing a Hoffman-Singleton graph, cf. [54], §13.1.

Each of these seven graphs is uniquely determined by its parameters. It is unknown whether there are any further examples. There are infinitely many feasible parameter sets. For the parameters $(324,57,0,12)$ nonexistence was shown in GAVRILYUK & MAKHNEV [168] and in KASKI & ÖSTERGÅRD [239].

9.1.8 Further parameter restrictions

Except for the rationality conditions, a few other restrictions on the parameters are known. We mention two of them. The *Krein conditions*, due to SCOTT [311], can be stated as follows:

$$(r+1)(k+r+2rs) \le (k+r)(s+1)^2,$$
$$(s+1)(k+s+2rs) \le (k+s)(r+1)^2.$$

When equality holds in one of these, the subconstituents of the graph (the induced subgraphs on the neighbors and on the nonneighbors of a given point) are both strongly regular (in the wide sense) again. For example, in the Higman-Sims graph with parameters $(v,k,\lambda,\mu) = (100,22,0,6)$ and $k,r,s = 22,2,-8$, the second subconstituent of any point has parameters $(77,16,0,4)$.

Seidel's *absolute bound* for the number of vertices of a primitive strongly regular graph (see Corollary 10.6.8 below) reads

$$v \le f(f+3)/2, \ v \le g(g+3)/2.$$

For example, the parameter set $(28,9,0,4)$ (spectrum $9^1 \ 1^{21} \ (-5)^6$) is ruled out both by the second Krein condition and by the absolute bound.

A useful identity is an expression for the *Frame quotient* (cf. [54], 2.2.4 and 2.7.2). One has

$$fg(r-s)^2 = vk(v-1-k)$$

(as is easy to check directly from the expressions for f and g given in Theorem 9.1.3 (iii)). From this one immediately concludes that if v is prime, then $r-s = \sqrt{v}$ and we are in the "half case" $(v,k,\lambda,\mu) = (4t+1,2t,t-1,t)$.

The Frame quotient, Krein conditions, and absolute bound are special cases of general (in)equalities for association schemes—see also §11.4 below. In BROUWER

& VAN LINT [62] one may find a list of known restrictions and constructions. It is a sequel to HUBAUT [229]'s earlier survey of constructions.

Using the above parameter conditions, NEUMAIER [284] derives the μ-bound:

Theorem 9.1.6 *For a primitive strongly regular graph* $\mu \leq s^3(2s+3)$. *If equality holds, then* $r = -s^2(2s+3)$.

Examples of equality in the μ-bound are known for $s = -2$ (the Schläfli graph, with $(v,k,\lambda,\mu) = (27,16,10,8)$) and $s = -3$ (the complement of the McLaughlin graph, with $(v,k,\lambda,\mu) = (275,162,105,81)$).

BROUWER & NEUMAIER [64] showed that a connected partial linear space with girth at least 5 and more than one line, in which every point is collinear with m other points, contains at least $\frac{1}{2}m(m+3)$ points. It follows that a strongly regular graph with $\mu = 2$ either has $k \geq \frac{1}{2}\lambda(\lambda+3)$ or has $(\lambda+1)|k$.

BAGCHI [15] showed that any $K_{1,1,2}$-free strongly regular graph is either the collinearity graph of a generalized quadrangle (cf. §9.6 below) or satisfies $k \geq (\lambda+1)(\lambda+2)$. (It follows that in the above condition on $\mu = 2$ the $(\lambda+1)|k$ alternative only occurs for the $m \times m$ grid, where $m = \lambda + 2$.)

9.1.9 Strongly regular graphs from permutation groups

Suppose G is a permutation group, acting on a set Ω. The *rank* of the action is the number of orbits of G on $\Omega \times \Omega$. (These latter orbits are called *orbitals*.) If R is an orbital, or a union of orbits, then (Ω,R) is a directed graph that admits G as group of automorphisms.

If G is transitive of rank 3 and its orbitals are symmetric (for all $x,y \in \Omega$ the pairs (x,y) and (y,x) belong to the same orbital), say with orbitals I, R, S, where $I = \{(x,x) \mid x \in \Omega\}$, then (Ω,R) and (Ω,S) form a pair of complementary strongly regular graphs.

For example, let G be $\mathrm{Sym}(n)$ acting on a set Σ of size 5. This action induces an action on the set Ω of unordered pairs of elements in Σ, and this latter action is rank 3, and gives the pair of graphs $T(5)$ and $\overline{T(5)}$, where this latter graph is the Petersen graph.

The rank 3 groups have been classified by the combined effort of many people, including Foulser, Kallaher, Kantor, Liebler, Liebeck and Saxl, see [238, 254, 255, 73].

9.1.10 Strongly regular graphs from quasisymmetric designs

As an application of Theorem 9.1.2, we show that quasisymmetric block designs give rise to strongly regular graphs. A *quasisymmetric design* is a 2-(v,k,λ) design (see §4.9) such that any two blocks meet in either x or y points, for certain fixed distinct x, y. Given this situation, we may define a graph Γ on the set of blocks, and call two blocks adjacent when they meet in x points. Let N be the point-block matrix of the design and A the adjacency matrix of Γ. Then $N^\top N = kI + xA + y(J - I - A)$. Since each of $NN^\top$, NJ, and JN is a linear combination of I and J, we see that A^2 can be expressed in terms of A, I, J, so that Γ is strongly regular by part (ii) of Theorem 9.1.2. (For an application, see §10.3.2.)

A large class of quasisymmetric block designs is provided by the 2-(v,k,λ) designs with $\lambda = 1$ (also known as Steiner systems $S(2,k,v)$). Such designs have only two intersection numbers since no two blocks can meet in more than one point. This leads to a substantial family of strongly regular graphs, including the triangular graphs $T(m)$ (derived from the trivial design consisting of all pairs from an m-set).

9.1.11 Symmetric 2-designs from strongly regular graphs

Conversely, some families of strongly regular graphs lead to designs. Let A be the adjacency matrix of a strongly regular graph with parameters (v,k,λ,λ) (i.e., with $\lambda = \mu$; such a graph is sometimes called a (v,k,λ) graph). Then, by Theorem 9.1.2,

$$AA^\top = A^2 = (k-\lambda)I + \lambda J,$$

which reflects that A is the incidence matrix of a symmetric 2-(v,k,λ) design. (And in this way one obtains precisely all symmetric 2-designs possessing a polarity without absolute points.) For instance, the triangular graph $T(6)$ provides a symmetric 2-$(15,8,4)$ design, the complementary design of the design of points and planes in the projective space $PG(3,2)$. Similarly, if A is the adjacency matrix of a strongly regular graph with parameters $(v,k,\lambda,\lambda+2)$, then $A+I$ is the incidence matrix of a symmetric 2-$(v,k+1,\lambda+2)$ design (and in this way one obtains precisely all symmetric 2-designs possessing a polarity with all points absolute). For instance, the Gewirtz graph with parameters $(56,10,0,2)$ provides a biplane 2-$(56,11,2)$.

9.1.12 Latin square graphs

A *transversal design* of *strength* t and *index* λ is a triple $(X,\mathscr{G},\mathscr{B})$, where X is a set of points, $\mathscr{G}$ is a partition of X into *groups*, and $\mathscr{B}$ is a collection of subsets of X called *blocks* such that (i) $t \leq |\mathscr{G}|$, (ii) every block meets every group in precisely one point, and (iii) every t-subset of X that meets each group in at most one point is contained in precisely λ blocks.

Suppose X is finite and $t < |\mathscr{G}|$. Then all groups $G \in \mathscr{G}$ have the same size m, and the number of blocks is λm^t. Given a point $x_0 \in X$, the groups not on x_0 together with the blocks $B \setminus \{x_0\}$ for $x_0 \in B \in \mathscr{B}$ form a transversal design of strength $t-1$ with the same index λ.

Equivalent to the concept of transversal design is that of *orthogonal array*. An orthogonal array with strength t and index λ over an alphabet of size m is a $k \times N$ array (with $N = \lambda m^t$) such that for any choice of t rows and prescribed symbols on these rows there are precisely λ columns that satisfy the demands.

When $t = 2$ the strength is usually not mentioned, and one talks about transversal designs $TD_\lambda(k,m)$ or orthogonal arrays $OA_\lambda(m,k)$, where k is the block size and m the group size.

When $\lambda = 1$ the index is suppressed from the notation. Now a $TD(k,m)$ or $OA(m,k)$ is equivalent to a set of $k-2$ mutually orthogonal Latin squares of order m. (The k rows of the orthogonal array correspond to row index, column index, and Latin square number; the columns correspond to the m^2 positions.)

The dual of a transversal design is a *net*. An (m,k)-*net* is a set of m^2 *points* together with km *lines*, partitioned into k parallel classes, where two lines from different parallel classes meet in precisely one point.

Given a point-line incidence structure, the *point graph* or *collinearity graph* is the graph with the points as vertices, adjacent when they are collinear. Dually, the *block graph* is the graph with the lines as vertices, adjacent when they have a point in common.

The collinearity graph of an (m,t)-net, that is, the block graph of a transversal design $TD(t,m)$ (note the new use of t here!), is strongly regular with parameters $v = m^2$, $k = t(m-1)$, $\lambda = m-2+(t-1)(t-2)$, $\mu = t(t-1)$ and eigenvalues $r = m-t$, $s = -t$. One says that a strongly regular graph "is a pseudo Latin square graph", or "has Latin square parameters" when there are t and m such that (v,k,λ,μ) have the above values. One also says that it has "$OA(m,t)$ parameters".

There is extensive literature on nets and transversal designs.

Proposition 9.1.7 *Suppose Γ is a strongly regular graph with $OA(m,t)$ parameters with a partition into cocliques of size m. Then the graph Δ obtained from Γ by adding edges so that these cocliques become cliques is again strongly regular and has $OA(m,t+1)$ parameters.*

Proof More generally, let Γ be a strongly regular graph with a partition into cocliques that meet the Hoffman bound. Then the graph Δ obtained from Γ by adding edges so that these cocliques become cliques has spectrum $k + m - 1$, $(r-1)^f$, $(s+m-1)^h$, $(s-1)^{g-h}$, where m is the size of the cocliques, and $h = v/m - 1$. The proposition is the special case $m = r - s$. □

For example, from the Hall-Janko graph with $OA(10,4)$ parameters $(100, 36, 12, 14)$ and a partition into ten 10-cocliques (which exists) one obtains a strongly regular graph with $OA(10,5)$ parameters $(100,45,20,20)$, and hence also a symmetric design 2-$(100,45,20)$. But an $OA(10,5)$ (three mutually orthogonal Latin squares of order 10) is unknown.

9.1.13 Partial geometries

A *partial geometry* with parameters (s,t,α) is a point-line geometry (any two points
are on at most one line) such that all lines have size $s+1$, there are $t+1$ lines on each
point, and given a line and a point outside, the point is collinear with α points on the
given line. One calls this structure a $pg(s,t,\alpha)$. Partial geometries were introduced
by BOSE [36].

Note that the dual of a $pg(s,t,\alpha)$ is a $pg(t,s,\alpha)$ (where "dual" means that the
names "point" and "line" are swapped).

One immediately computes the number of points $v = (s+1)(st+\alpha)/\alpha$ and lines
$b = (t+1)(st+\alpha)/\alpha$. The collinearity graph of a $pg(s,t,\alpha)$ is complete if $\alpha = s+1$,
and otherwise strongly regular with parameters $v = (s+1)(st+\alpha)/\alpha$, $k = s(t+1)$,
$\lambda = s-1+t(\alpha-1)$, $\mu = \alpha(t+1)$, and eigenvalues $\theta_1 = s-\alpha$, $\theta_2 = -t-1$. (Note:
earlier we used s for the smallest eigenvalue, but here s has a different meaning!)

The extreme examples of partial geometries are *generalized quadrangles* (partial
geometries with $\alpha = 1$) and Steiner systems $S(2,K,V)$ (partial geometries with $\alpha =
s+1$). Many examples are also provided by nets (with $t = \alpha$) or their duals, the
transversal designs (with $s = \alpha$).

A strongly regular graph is called *geometric* when it is the collinearity graph of
a partial geometry. It is called *pseudogeometric* when there are integers s,t,α such
that the parameters (v,k,λ,μ) have the values given above.

BOSE [36] showed that a pseudogeometric graph with given t and sufficiently
large s must be geometric. NEUMAIER [284] showed that the same conclusion works
in all cases, and hence derives a contradiction in the nonpseudogeometric case.

Theorem 9.1.8 (Bose-Neumaier) *A strongly regular graph with $s < -1$ and $r >
\frac{1}{2}s(s+1)(\mu+1) - 1$ is the block graph of an $S(2,K,V)$ or a transversal design.*

It follows immediately (from this and the μ-bound) that

Theorem 9.1.9 *For any fixed $s = -m$, there are only finitely many primitive strongly
regular graphs with smallest eigenvalue s, that are not the block graph of an
$S(2,K,V)$ or a transversal design.*   □

9.2 Strongly regular graphs with eigenvalue -2

For later use, we give SEIDEL [315]'s classification of the strongly regular graphs
with $s = -2$.

Theorem 9.2.1 *Let Γ be a strongly regular graph with smallest eigenvalue -2.
Then Γ is one of*

  (i) *the complete n-partite graph $K_{n\times 2}$, with parameters $(v,k,\lambda,\mu) = (2n, 2n-
2, 2n-4, 2n-2)$, $n \geq 2$,*

  (ii) *the lattice graph $L_2(n) = K_n \square K_n$, with parameters* $(v,k,\lambda,\mu) = (n^2, 2(n-1), n-2, 2)$, $n \geq 3$,

  (iii) *the Shrikhande graph, with parameters* $(v,k,\lambda,\mu) = (16,6,2,2)$,

  (iv) *the triangular graph $T(n)$ with parameters* $(v,k,\lambda,\mu) = (\binom{n}{2}, 2(n-2), n-2, 4)$, $n \geq 5$,

  (v) *one of the three Chang graphs, with parameters* $(v,k,\lambda,\mu) = (28,12,6,4)$,

  (vi) *the Petersen graph, with parameters* $(v,k,\lambda,\mu) = (10,3,0,1)$,

  (vii) *the Clebsch graph, with parameters* $(v,k,\lambda,\mu) = (16,10,6,6)$, *or*

  (viii) *the Schläfli graph, with parameters* $(v,k,\lambda,\mu) = (27,16,10,8)$.

Proof If Γ is imprimitive, then we have case (i). Otherwise, the μ-bound gives $\mu \leq 8$, and the rationality conditions give $(r+2)|(\mu-2)(\mu-4)$, and integrality of v gives $\mu | 2r(r+1)$. For $\mu = 2$ we find the parameters of $L_2(n)$, for $\mu = 4$ those of $T(n)$, and for the remaining values for μ only the parameter sets $(v,k,\lambda,\mu) = (10,3,0,1)$, $(16,10,6,6)$, and $(27,16,10,8)$ survive the parameter conditions and the absolute bound. It remains to show that the graph is uniquely determined by its parameters in each case. Now SHRIKHANDE [325] proved uniqueness of the graph with $L_2(n)$ parameters, with the single exception of $n = 4$, where there is one more graph, now known as the Shrikhande graph, and CHANG [86, 87] proved uniqueness of the graph with $T(n)$ parameters, with the single exception of $n = 8$, where there are three more graphs, now known as the Chang graphs. In the remaining three cases uniqueness is easy to see.   □

Let us give definitions for the graphs involved.

The *Shrikhande graph* is the result of Seidel switching the lattice graph $L_2(4)$ with respect to an induced circuit of length 8. It is the complement of the Latin square graph for the cyclic Latin square of order 4. It is locally a hexagon. Drawn

on a torus: 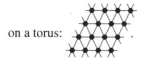 .

The three *Chang graphs* are the result of switching $T(8)$ (the line graph of K_8) with respect to (a) a 4-coclique $\overline{K_4}$, that is, 4 pairwise disjoint edges in K_8; (b) $K_3 + K_5$, that is, 8 edges forming a triangle and a (disjoint) pentagon in K_8; (c) the line graph of the cubic graph formed by an 8-circuit plus edges between opposite vertices.

The *Clebsch graph* is the complement of the folded 5-cube.

The *Schläfli graph* is the complement of the collinearity graph of $GQ(2,4)$ (cf. §9.6).

9.3 Connectivity

For a graph Γ, let $\Gamma_i(x)$ denote the set of vertices at distance i from x in Γ. Instead of $\Gamma_1(x)$ we write $\Gamma(x)$.

Proposition 9.3.1 *If Γ is a primitive strongly regular graph, then for each vertex x the subgraph $\Gamma_2(x)$ is connected.*

Proof Note that $\Gamma_2(x)$ is regular of valency $k - \mu$. If it is not connected, then its eigenvalue $k - \mu$ would have multiplicity at least 2, and hence would be not larger than the second largest eigenvalue r of Γ. Then $x^2 + (\mu - \lambda)x + \mu - k \leq 0$ for $x = k - \mu$, i.e., $(k - \mu)(k - \lambda - 1) \leq 0$, a contradiction.   □

The *vertex connectivity* $\kappa(\Gamma)$ of a connected noncomplete graph Γ is the smallest integer m such that Γ can be disconnected by removing m vertices.

Theorem 9.3.2 (BROUWER & MESNER [63]) *Let Γ be a connected strongly regular graph of valency k. Then $\kappa(\Gamma) = k$, and the only disconnecting sets of size k are the sets of all neighbors of some vertex x.*

Proof Clearly, $\kappa(\Gamma) \leq k$. Let S be a disconnecting set of vertices not containing all neighbors of some vertex. Let $\Gamma \setminus S = A + B$ be a separation of $\Gamma \setminus S$. Since the eigenvalues of $A \cup B$ interlace those of Γ, it follows that at least one of A and B, say B, has largest eigenvalue at most r. It follows that the average valency of B is at most r. Since B has an edge, $r > 0$.

Now let $|S| \leq k$. Since B has average valency at most r, we can find two points x, y in B such that $|S \cap \Gamma(x)| + |S \cap \Gamma(y)| \geq 2(k - r)$, so that these points have at least $k - 2r$ common neighbors in S.

If Γ has nonintegral eigenvalues, then we have $(v, k, \lambda, \mu) = (4t + 1, 2t, t - 1, t)$ for some t, and $r = (-1 + \sqrt{v})/2$. The inequality $\max(\lambda, \mu) \geq k - 2r$ gives $t \leq 2$, but for $t = 2$ the eigenvalues are integral, so we have $t = 1$ and Γ is the pentagon. But the claim is true in that case.

Now let r, s be integral. If $s \leq -3$, then $\mu = k + rs \leq k - 3r$ and $\lambda = \mu + r + s \leq k - 2r - 3$, so no two points can have $k - 2r$ common neighbors.

Therefore $s = -2$, and we have one of the eight cases in Seidel's classification. But not case (i), since $r > 0$.

Since both A and B contain an edge, both B and A have size at most $\bar{\mu} = v - 2k + \lambda$, so that both A and B have size at least $k - \lambda$, and $v \geq 3k - 2\lambda$. This eliminates cases (vii) and (viii).

If B is a clique, then $|B| \leq r + 1 = k - \lambda - 1$, a contradiction. So, B contains two nonadjacent vertices, and their neighbors must be in $B \cup S$, so $2k - \mu \leq |B| + |S| - 2$ and $k - \mu + 2 \leq |B| \leq \bar{\mu}$.

In cases (iii), (v), and (vi) we have $\bar{\mu} = k - \mu + 2$, so equality holds and $|B| = \bar{\mu}$ and $|S| = k$. Since $v < 2\bar{\mu} + k$, we have $|A| < \bar{\mu}$ and A must be a clique (of size $v - k - \bar{\mu} = k - \lambda$). But the Petersen graph does not contain a 3-clique, and the Shrikhande graph does not contain 4-cliques; also, if A is a 6-clique in a Chang graph, and $a, b, c \in A$, then $\Gamma(a) \cap S$, $\Gamma(b) \cap S$, and $\Gamma(c) \cap S$ are three 7-sets in the 12-set S that pairwise meet in precisely two points, which is impossible. This eliminates cases (iii), (v), and (vi).

We are left with the two infinite families of lattice graphs and triangular graphs. In both cases it is easy to see that if x, y are nonadjacent, then there exist k paths joining x and y, vertex disjoint apart from x, y, and entirely contained in $\{x, y\} \cup \Gamma(x) \cup \Gamma(y)$. Hence $|S| = k$, and if S separates x and y, then $S \subseteq \Gamma(x) \cup \Gamma(y)$.

The subgraph $\Delta := \Gamma \setminus (\{x,y\} \cup \Gamma(x) \cup \Gamma(y))$ is connected (if one removes a point and its neighbors from a lattice graph, the result is a smaller lattice graph, and the same holds for a triangular graph), except in the case of the triangular graph $\binom{5}{2}$, where Δ is empty.

Each vertex of $(\Gamma(x) \cup \Gamma(y)) \setminus S$ has a neighbor in Δ and we find a path of length 4 disjoint from S joining x and y, except in the case of the triangular graph $\binom{5}{2}$, where each vertex of $\Gamma(x) \setminus S$ is adjacent to each vertex of $\Gamma(y) \setminus S$, and we find a path of length 3 disjoint from S joining x and y. $\qquad\qquad\qquad\square$

We remark that it is not true that for every strongly regular graph Γ with vertex x the vertex connectivity of the subgraph $\Gamma_2(x)$ equals its valency $k - \mu$. A counterexample is given by the graph Γ that is the complement of the strongly regular graph Δ with parameters $(96, 19, 2, 4)$ constructed by Haemers for $q = 4$, see [197], p. 76, or [62], §8A. Indeed, we have $\Delta(x) \cong K_3 + 4C_4$, so that $\Gamma_2(x)$ has degree 16 and vertex connectivity 15.

One might guess that the cheapest way to disconnect a strongly regular graph such that all components have at least two vertices would be by removing the $2k - \lambda - 2$ neighbours of an edge. Cioabă, Kim and Koolen recently observed that this is false (the simplest counterexample is probably $T(6)$, where edges have 10 neighbours and certain triangles only 9), but proved it for several infinite classes of strongly regular graphs.

CHVÁTAL & ERDŐS [95] showed that if a graph Γ on at least three vertices has vertex connectivity κ and largest independent set of size α, and $\alpha \leq \kappa$, then Γ has a Hamiltonian circuit. BIGALKE & JUNG [29] showed that if Γ is 1-tough, with $\alpha \leq \kappa + 1$ and $\kappa \geq 3$, and Γ is not the Petersen graph, then Γ is Hamiltonian. These results imply that if Γ is strongly regular with smallest eigenvalue s, and s is not integral, or $-s \leq \mu + 1$, then Γ is Hamiltonian. This, together with explicit inspection of the Hoffman-Singleton graph, the Gewirtz graph, and the M_{22} graph, shows that all connected strongly regular graphs on fewer than 99 vertices are Hamiltonian, except for the Petersen graph.

9.4 Cocliques and colorings

In §2.5 we derived some bounds for the size of a coclique in terms of eigenvalues. These bounds are especially useful for strongly regular graphs. Moreover, strongly regular graphs for which the bounds of Hoffman and Cvetković are tight have a very special structure:

Theorem 9.4.1 *Let Γ be a strongly regular graph with eigenvalues k (degree), r and s $(r > s)$ and multiplicities 1, f and g, respectively. Suppose that Γ is not complete multipartite (i.e., $r \neq 0$), and let C be a coclique in Γ. Then*

(i) $|C| \leq g$,
(ii) $|C| \leq ns/(s-k)$,

(iii) if $|C| = g = ns/(s-k)$, then the subgraph Γ' of Γ induced by the vertices not in C is strongly regular with eigenvalues $k' = k+s$ (degree), $r' = r$ and $s' = r+s$ and respective multiplicities 1, $f-g+1$ and $g-1$.

Proof Parts (i) and (ii) follow from Theorems 3.5.1 and 3.5.2. Assume $|C| = g = ns/(s-k)$. By Theorem 2.5.4, Γ' is regular of degree $k+s$. Apply Lemma 2.11.1 to $P = A - \frac{k-r}{n}J$, where A is the adjacency matrix of Γ. Since Γ is regular, A and J commute and therefore P has eigenvalues r and s with multiplicities $f+1$ and g, respectively. We take $Q = -\frac{k-r}{n}J$ of size $|C| = g$ and $R = A' - \frac{k-r}{n}J$, where A' is the adjacency matrix of Γ'. Lemma 2.11.1 gives the eigenvalues of R: r ($f+1-g$ times), s (0 times), $r+s$ ($g-1$ times), and $r+s+g(k-r)/n$ (1 time). Since Γ' is regular of degree $k+s$ and A' commutes with J, we obtain the required eigenvalues for A'. By Theorem 9.1.2, Γ' is strongly regular. □

For instance, an $(m-1)$-coclique in $\overline{T(m)}$ is tight for both bounds and the graph on the remaining vertices is $\overline{T(m-1)}$.

Also for the chromatic number we can say more in the case of a strongly regular graph.

Theorem 9.4.2 *If Γ is a primitive strongly regular graph, not the pentagon, then*

$$\chi(\Gamma) \geq 1 - \frac{s}{r}.$$

Proof Since Γ is primitive, $r > 0$ and by Corollary 3.6.4 it suffices to show that the multiplicity g of s satisfies $g \geq -s/r$ for all primitive strongly regular graphs but the pentagon. First we check this claim for all feasible parameter sets with at most 23 vertices. Next we consider strongly regular graphs with $v \geq 24$ and $r < 2$. The complements of these graphs have $s > -3$, and by Theorem 9.1.3 (iv), $s = -2$. By use of Theorem 9.2 we easily find that all these graphs satisfy the claim.

Assume that Γ is primitive, that $r \geq 2$, and that the claim does not hold (that is, $g < -s/r$). Now $(v-1-g)r+gs+k = 0$ gives

$$g^2 < -sg/r = v-1-g+k/r \leq v-1-g+k/2 < 3v/2-g.$$

This implies $g(g+3) \leq 3v/2 = 2\sqrt{3v/2}$. By use of the absolute bound $v \leq g(g+3)/2$, we get $v/2 < 2\sqrt{3v/2}$, so $v < 24$, a contradiction. □

For example, if Γ is the complement of the triangular graph $T(m)$, then Γ is strongly regular with eigenvalues $k = \frac{1}{2}(m-2)(m-3)$, $r = 1$, and $s = 3-m$ (for $m \geq 4$). The above bound gives $\chi(\Gamma) \geq m-2$, which is tight, while Hoffman's lower bound (Theorem 3.6.2) equals $\frac{1}{2}m$. On the other hand, if m is even, Hoffman's bound is tight for the complement of Γ while the above bound is much smaller. We saw (see §3.6) that a Hoffman coloring (i.e., a coloring with $1-k/s$ classes) corresponds to an equitable partition of the adjacency matrix. For the complement this gives an equitable partition into maximal cliques, which is called a spread of the strongly regular graph. For more applications of eigenvalues to the chromatic number we refer to [155] and [177]. See also §9.7.

9.5 Automorphisms

Let A be the adjacency matrix of a graph Γ, and P the permutation matrix that describes an automorphism ϕ of Γ. Then $AP = PA$. If ϕ has order m, then $P^m = I$, so that the eigenvalues of AP are m-th roots of unity times eigenvalues of A.

Apply this in the special case of strongly regular graphs. Suppose ϕ has f fixed points, and moves g points to a neighbor. Then $f = \operatorname{tr} P$ and $g = \operatorname{tr} AP$. Now consider $M = A - sI$. It has eigenvalues $k - s$, $r - s$, and 0. Hence MP has eigenvalues $k - s$, $(r - s)\zeta$ for certain m-th roots of unity ζ, and 0. It follows that $g - sf = \operatorname{tr} MP \equiv k - s \pmod{r - s}$.

For example, for the Petersen graph every automorphism satisfies $f \equiv g + 1 \pmod 3$. Or, for a hypothetical Moore graph on 3250 vertices (cf. §11.5.1), every automorphism satisfies $8f + g \equiv 5 \pmod{15}$.

In some cases, where a structure is given locally, it must either be a universal object, or a quotient, where the quotient map preserves local structure, that is, only identifies points that are far apart. In the finite case arguments like those in this section can be used to show that $f = g = 0$ is impossible, so that nontrivial quotients do not exist. For an example, see [53].

9.6 Generalized quadrangles

A *generalized n-gon* is a connected bipartite graph of diameter n and girth $2n$. (The girth of a graph is the length of a shortest circuit.)

It is common to call the vertices in one color class of the unique 2-coloring *points*, and the other vertices *lines*. For example, a generalized 3-gon is the same thing as a projective plane: any two points have an even distance at most 3, hence are joined by a line, and similarly any two lines meet in a point; finally, two lines cannot meet in two points since that would yield a quadrangle, but the girth is 6.

A *generalized quadrangle* is a generalized 4-gon. In terms of points and lines, the definition becomes: a *generalized quadrangle* is an incidence structure (P, L) with set of points P and set of lines L, such that two lines meet in at most one point, and if p is a point not on the line m, then there is a unique point q on m and a unique line n on p such that q is on n.

9.6.1 Parameters

A generalized n-gon is called *firm* (*thick*) when each vertex has at least two (resp. three) neighbors, that is, when each point is on at least two (three) lines, and each line is on at least two (three) points.

An example of a nonfirm generalized quadrangle is a pencil of lines on one common point x_0. Each point different from x_0 is on a unique line, and $\Gamma_3(x_0) = \emptyset$.

Proposition 9.6.1 *(i) If a generalized n-gon Γ has a pair of opposite vertices x,y where x has degree at least two, then every vertex has an opposite, and Γ is firm.*

(ii) A thick generalized n-gon has parameters: each line has the same number of points, and each point is on the same number of lines. When moreover n is odd, then the number of points on each line equals the number of lines through each point.

Proof For a vertex x of a generalized n-gon, let $k(x)$ be its degree. Call two vertices of a generalized n-gon *opposite* when they have distance n. If x and y are opposite, then each neighbor of one is on a unique shortest path to the other, and we find $k(x) = k(y)$.

(i) Being nonopposite gives a bijection between $\Gamma(x)$ and $\Gamma(y)$, and hence if $k(x) > 1$ then also each neighbor z of x has an opposite and satisfies $k(z) > 1$. Since Γ is connected, it is firm.

(ii) Let x,z be two points joined by the line y. Let w be opposite to y. Since $k(w) > 2$, there is a neighbor u of w opposite to both x and z. Now $k(x) = k(u) = k(z)$. Since Γ is connected and bipartite, this shows that $k(p)$ is independent of the point p. If n is odd, then a vertex opposite a point is a line. $\square$

A firm, nonthick generalized quadrangle is the vertex-edge incidence graph of a complete bipartite graph.

The *halved graph* of a bipartite graph Γ is the graph on the same vertex set, where two vertices are adjacent when they have distance 2 in Γ. The *point graph* and *line graph* of a generalized n-gon are the two components of its halved graph containing the points and lines, respectively.

The point graph and line graph of a finite thick generalized n-gon are distance-regular of diameter $\lfloor n/2 \rfloor$ (see Chapter 12). In particular, the point graph and line graph of a thick generalized quadrangle are strongly regular (see Theorem 9.6.2).

It is customary to let $GQ(s,t)$ denote a finite generalized quadrangle with $s+1$ points on each line and $t+1$ lines on each point. Note that it is also customary to use s to denote the smallest eigenvalue of a strongly regular graph, so in this context one has to be careful to avoid confusion.

It is a famous open problem whether a thick generalized n-gon can have finite s and infinite t. In the special case of generalized quadrangles a little is known: Cameron, Kantor, Brouwer, and Cherlin [81, 49, 88] show that this cannot happen for $s+1 \leq 5$.

9.6.2 Constructions of generalized quadrangles

Suppose V is a vector space provided with a nondegenerate quadratic form f of Witt index 2 (that is, such that the maximal totally singular subspaces have vector space dimension 2). Consider in the projective space PV the singular projective points and the totally singular projective lines. These will form a generalized quadrangle.

Indeed, f defines a bilinear form B on V via $B(x,y) = f(x+y) - f(x) - f(y)$. Call x and y *orthogonal* when $B(x,y) = 0$. When two singular vectors are orthogonal, the subspace spanned by them is totally singular. And conversely, in a totally singular subspace any two vectors are orthogonal. The collection of all vectors orthogonal to a given vector is a hyperplane. We have to check that if $P = \langle x \rangle$ is a singular projective point, and L is a totally singular projective line not containing P, then P has a unique neighbor on L. But the hyperplane of vectors orthogonal to x meets L, and cannot contain L otherwise f would have larger Witt index.

This construction produces generalized quadrangles over arbitrary fields. If V is a vector space over a finite field $\mathbb{F}_q$, then a nondegenerate quadratic form can have Witt index 2 in dimensions 4, 5, and 6. A hyperbolic quadric in 4 dimensions yields a generalized quadrangle with parameters $GQ(q,1)$, a parabolic quadric in 5 dimensions yields a generalized quadrangle with parameters $GQ(q,q)$, and an elliptic quadric in 6 dimensions yields a generalized quadrangle with parameters $GQ(q,q^2)$.

Other constructions, and other parameters occur.

In the below we'll meet $GQ(2,t)$ for $t = 1,2,4$ and $GQ(3,9)$. Let us give simple direct descriptions for $GQ(2,1)$ and $GQ(2,2)$. The unique $GQ(2,1)$ is the 3-by-3 grid: 9 points, 6 lines. Its point graph is $K_3 \square K_3$. The unique $GQ(2,2)$ is obtained by taking as points the 15 pairs from a 6-set, and as lines the 15 partitions of that 6-set into three pairs. Now collinearity is being disjoint. Given a point ac, and a line $\{ab,cd,ef\}$, the two points ab and cd on this line are not disjoint from ac, so that ef is the unique point on this line collinear with ac, and the line joining ac and ef is $\{ac,bd,ef\}$.

9.6.3 Strongly regular graphs from generalized quadrangles

As mentioned before, the point graph (collinearity graph) of a finite thick generalized quadrangle is strongly regular. The parameters and eigenvalues can be obtained in a straightforward way (Exercise 9.4).

Theorem 9.6.2 *The collinearity graph of a finite generalized quadrangle with parameters $GQ(s,t)$ is strongly regular with parameters*

$$v = (s+1)(st+1), \quad k = s(t+1), \quad \lambda = s-1, \quad \mu = t+1$$

and spectrum

$$s(t+1) \text{ with multiplicity } 1,$$
$$s-1 \quad \text{with multiplicity } st(s+1)(t+1)/(s+t),$$
$$-t-1 \quad \text{with multiplicity } s^2(st+1)/(s+t).$$

In particular, if a $GQ(s,t)$ exists, then $(s+t)|s^2(st+1)$.

9.6.4 Generalized quadrangles with lines of size 3

Let a *weak generalized quadrangle* be a point-line geometry with the properties that two lines meet in at most one point, and given a line m and a point p outside there is a unique pair (q,n) such that $p \sim n \sim q \sim m$, where $\sim$ denotes incidence. The difference with the definition of a generalized quadrangle is that connectedness is not required. (But, of course, as soon as there are a point and a line, the geometry is connected.)

Theorem 9.6.3 *A weak generalized quadrangle where all lines have size* 3 *is one of the following:*

(i) *a coclique (no lines),*
(ii) *a pencil (all lines passing through a fixed point),*
(iii) *the unique GQ(2,1),*
(iv) *the unique GQ(2,2),*
(v) *the unique GQ(2,4).*

Proof After reducing to the case of $GQ(2,t)$ one finds $(t+2)|(8t+4)$, i.e., $(t+2)|12$, i.e., $t \in \{1,2,4,10\}$, and $t = 10$ is ruled out by the Krein conditions. Alternatively, or afterwards, notice that the point graphs have eigenvalue 1, so that their complements have smallest eigenvalue -2, and apply Seidel's classification. Cases (iii), (iv), and (v) here have point graphs that are the complements of the lattice graph $K_3 \square K_3$, the triangular graph $T(6)$, and the Schläfli graph, respectively. $\square$

This theorem can be used in the classification of root lattices, where the five cases correspond to A_n, D_n, E_6, E_7, and E_8 (cf. [54], p. 102). And the classification of root lattices can be used in the classification of graphs with smallest eigenvalue -2. See Chapter 8, especially Theorem 8.4.1.

9.7 The $(81, 20, 1, 6)$ strongly regular graph

Large parts of this section are taken from [57]. Sometimes the graph of this section is called the *Brouwer-Haemers graph*.

Let $\Gamma = (X,E)$ be a strongly regular graph with parameters $(v,k,\lambda,\mu) = (81,20,1,6)$. Then Γ has spectrum $\{20^1, 2^{60}, -7^{20}\}$, where the exponents denote multiplicities. We will show that up to isomorphism there is a unique such graph Γ. More generally we give a short proof for the fact (due to IVANOV & SHPECTOROV [233]) that a strongly regular graph with parameters $(v,k,\lambda,\mu) = (q^4,(q^2+1)(q-1),q-2,q(q-1))$ that is the collinearity graph of a partial quadrangle (that is, in which all maximal cliques have size q) is the second subconstituent of the collinearity graph of a generalized quadrangle $GQ(q,q^2)$. In the special case $q = 3$, this will imply our previous claim, since $\lambda = 1$ implies that all maximal cliques have size 3 and it is known (see CAMERON, GOETHALS & SEIDEL [83]) that there is a unique generalized quadrangle $GQ(3,9)$ (and this generalized quadrangle has an automorphism group transitive on the points).

9.7.1 Descriptions

Let us first give a few descriptions of our graph on 81 vertices. Note that the uniqueness shows that all constructions below give isomorphic graphs, something which is not immediately obvious from the description in all cases.

A. Let X be the point set of $AG(4,3)$, the 4-dimensional affine space over $\mathbb{F}_3$, and join two points when the line connecting them hits the hyperplane at infinity (a $PG(3,3)$) in a fixed elliptic quadric Q. This description shows immediately that $v = 81$ and $k = 20$ (since $|Q| = 10$). Also $\lambda = 1$ since no line meets Q in more than two points, so the affine lines are the only triangles. Finally $\mu = 6$, since a point outside Q in $PG(3,3)$ lies on 4 tangents, 3 secants, and 6 exterior lines with respect to Q, and each secant contributes 2 to μ. We find that the group of automorphisms contains $G = 3^4 \cdot PGO_4^- \cdot 2$, where the last factor 2 accounts for the linear transformations that do not preserve the quadratic form Q, but multiply it by a constant. In fact, this is the full group, as will be clear from the uniqueness proof.

B. A more symmetric form of this construction is found by starting with $X = \mathbf{1}^\perp / \langle \mathbf{1} \rangle$ in $\mathbb{F}_3^6$ provided with the standard bilinear form. The corresponding quadratic form ($Q(x) = \mathrm{wt}(x)$, the number of nonzero coordinates of x) is elliptic, and if we join two vertices $x + \langle \mathbf{1} \rangle$, $y + \langle \mathbf{1} \rangle$ of X when $Q(x - y) = 0$, i.e., when their difference has weight 3, we find the same graph as under **A**. This construction shows that the automorphism group contains $G = 3^4 \cdot (2 \times \mathrm{Sym}\,(6)) \cdot 2$, and again this is the full group.

C. There is a unique strongly regular graph with parameters $(112, 30, 2, 10)$, the collinearity graph of the unique generalized quadrangle with parameters $GQ(3,9)$. Its second subconstituent is an $(81, 20, 1, 6)$ strongly regular graph, and hence isomorphic to our graph Γ. (See CAMERON, GOETHALS & SEIDEL [83].) We find that Aut Γ contains (and in fact equals) the point stabilizer in $U_4(3) \cdot D_8$ acting on $GQ(3,9)$.

D. The graph Γ is the coset graph of the truncated ternary Golay code C: take the 3^4 cosets of C and join two cosets when they contain vectors differing in only one place.

E. The graph Γ is the Hermitean forms graph on $\mathbb{F}_9^2$; more generally, take the q^4 matrices M over $\mathbb{F}_{q^2}$ satisfying $M^\top = \overline{M}$, where $^-$ denotes the field automorphism $x \to x^q$ (applied entrywise), and join two matrices when their difference has rank 1. This will give us a strongly regular graph with parameters $(v, k, \lambda, \mu) = (q^4, (q^2 + 1)(q - 1), q - 2, q(q - 1))$.

F. The graph Γ is the graph with vertex set $\mathbb{F}_{81}$, where two vertices are joined when their difference is a fourth power. (This construction was given by VAN LINT & SCHRIJVER [257].)

There is a unique strongly regular graph with parameters $(275,112,30,56)$ known as the McLaughlin graph. Its first subconstituent is the 112-point graph mentioned under **C**. Its second subconstituent is the unique strongly regular graph with parameters $(162,56,10,24)$. In §3.13.7 we discussed how to find all splits of this latter graph into two copies of Γ.

9.7.2 Uniqueness

Now let us embark upon the uniqueness proof. Let $\Gamma = (X,E)$ be a strongly regular graph with parameters $(v,k,\lambda,\mu) = (q^4,(q^2+1)(q-1),q-2,q(q-1))$ and assume that all maximal cliques (we shall just call them lines) of Γ have size q. Let Γ have adjacency matrix A. Using the spectrum of A, which is $\{k^1,(q-1)^f,(q-1-q^2)^g\}$, where $f = q(q-1)(q^2+1)$ and $g = (q-1)(q^2+1)$, we can obtain some structure information. Let $\mathbf{T}$ be the collection of subsets of X of cardinality q^3 inducing a subgraph that is regular of degree $q-1$.

Claim 1. *If $T \in \mathbf{T}$, then each point of $X \setminus T$ is adjacent to q^2 points of T.*

Look at the matrix B of average row sums of A, with sets of rows and columns partitioned according to $\{T, X \setminus T\}$. We have

$$B = \begin{bmatrix} q-1 & q^2(q-1) \\ q^2 & k-q^2 \end{bmatrix}$$

with eigenvalues k and $q-1-q^2$, so interlacing is tight, and by Corollary 2.5.4(ii) it follows that the row sums are constant in each block of A.

Claim 2. *Given a line L, there is a unique $T_L \in \mathbf{T}$ containing L.*

Let Z be the set of vertices in $X \setminus L$ without a neighbor in L. Then $|Z| = q^4 - q - q(k-q+1) = q^3 - q$. Let $T = L \cup Z$. Each vertex of Z is adjacent to $q\mu = q^2(q-1)$ vertices with a neighbor in L, so T induces a subgraph that is regular of degree $q-1$.

Claim 3. *If $T \in \mathbf{T}$ and $x \in X \setminus T$, then x is on at least one line L disjoint from T, and T_L is disjoint from T for any such line L.*

The point x is on q^2+1 lines, but has only q^2 neighbors in T. Each point of L has q^2 neighbors in T, so each point of T has a neighbor on L and hence is not in T_L.

Claim 4. *Any $T \in \mathbf{T}$ induces a subgraph Δ isomorphic to $q^2 K_q$.*

It suffices to show that the multiplicity m of the eigenvalue $q-1$ of Δ is (at least) q^2 (it cannot be more). By interlacing we find $m \geq q^2 - q$, so we need some additional work. Let $M := A - (q-1/q^2)J$. Then M has spectrum $\{(q-1)^{f+1},(q-1-q^2)^g\}$, and we want that M_T, the submatrix of M with rows and columns indexed by T, has eigenvalue $q-1$ with multiplicity (at least) $q^2 - 1$, or, equivalently (by Lemma 2.11.1), that $M_{X \setminus T}$ has eigenvalue $q-1-q^2$ with multiplicity (at least) $q-2$. But for each $U \in \mathbf{T}$ with $U \cap T = \emptyset$ we find an eigenvector $x_U = (2-q)\chi_U + \chi_{X \setminus (T \cup U)}$ of $M_{X \setminus T}$ with eigenvalue $q-1-q^2$. A collection $\{x_U \mid U \in \mathbf{U}\}$ of such eigenvectors cannot be linearly dependent when $\mathbf{U} = \{U_1, U_2, \ldots\}$ can be ordered such that $U_i \not\subset \bigcup_{j<i} U_j$ and $\bigcup U \neq X \setminus T$, so we can find (using Claim 3) at least $q-2$ such linearly independent eigenvectors, and we are done.

Claim 5. *Any $T \in \mathbf{T}$ determines a unique partition of X into members of* $\mathbf{T}$.
Indeed, we saw this in the proof of the previous step.

Let Π be the collection of partitions of X into members of $\mathbf{T}$. We have $|\mathbf{T}| = q(q^2+1)$ and $|\Pi| = q^2+1$. Construct a generalized quadrangle $GQ(q,q^2)$ with point set $\{\infty\} \cup \mathbf{T} \cup X$ as follows: The q^2+1 lines on ∞ are $\{\infty\} \cup \pi$ for $\pi \in \Pi$. The q^2 remaining lines on each $T \in \mathbf{T}$ are $\{T\} \cup L$ for $L \subset T$. It is completely straightforward to check that we really have a generalized quadrangle $GQ(q,q^2)$.

9.7.3 Independence and chromatic numbers

Let Γ again be the strongly regular graph with parameters $(v,k,\lambda,\mu) = (81,20,1,6)$. We have $\alpha(\Gamma) = 15$ and $\chi(\Gamma) = 7$.

Clearly, the independence number of our graph is one less than the independence number of the unique $GQ(3,9)$ of which it is the second subconstituent. So it suffices to show that $\alpha(\Delta) = 16$, where Δ is the collinearity graph of $GQ(3,9)$.

It is easy to indicate a 16-coclique: define $GQ(3,9)$ in $PG(5,3)$ provided with the nondegenerate elliptic quadratic form $\sum_{i=1}^{6} x_i^2$. There are 112 isotropic points, 80 of weight 3 and 32 of weight 6. Among the 32 of weight 6, 16 have coordinate product 1, and 16 have coordinate product -1, and these two 16-sets are cocliques.

That there is no larger coclique can be seen by cubic counting. Let C be a 16-coclique in Δ. Let there be n_i vertices outside that have i neighbors inside. Then

$$\sum n_i = 96, \quad \sum i n_i = 480, \quad \sum \binom{i}{2} n_i = 1200, \quad \sum \binom{i}{3} n_i = 2240,$$

so

$$\sum (i-4)^2 (i-10) n_i = 0.$$

(Here the quadratic counting is always possible in a strongly regular graph, and the last equation can be written because the second subconstituent is itself strongly regular.) Now each point is on 10 lines, and hence cannot have more than 10 neighbors in C. It follows that each point has either 4 or 10 neighbors in C. In particular, C is maximal.

As an aside: Solving these equations gives $n_4 = 80$, $n_{10} = 16$. Let D be the set of 16 vertices with 10 neighbors in C. If two vertices $d_1, d_2 \in D$ are adjacent, then they can have only 2 common neighbors in C, but each has 10 neighbors in C, a contradiction. So, also D is a 16-coclique, which means that 16-cocliques in Δ come in pairs.

Since $81/15 > 5$, we have $\chi(\Gamma) \geq 6$. Since Δ has a split into two Gewirtz graphs, and the Gewirtz graph has chromatic number 4, it follows that $\chi(\Delta) \leq 8$. (And in fact equality holds.) This shows that for our graph $6 \leq \chi(\Gamma) \leq 8$. In fact, $\chi(\Gamma) = 7$ can be seen by computer (Edwin van Dam, pers. comm.).

Since $\lambda = 1$, the maximum clique size equals 3. And from the uniqueness proof it is clear that Γ admits a partition into 27 triangles. So the complement of Γ has chromatic number 27.

9.7.4 Second subconstituent

The second subconstituent of Γ has spectrum $14^1 \, 2^{40} \, (-4)^{10} \, (-6)^9$ (as can be seen using Lemma 2.11.1) and is uniquely determined by its spectrum ([33]). The proof is an example where "partially tight" interlacing (Theorem 2.5.1 (ii)) is used. We give a very brief indication of the uniqueness proof.

Let Σ be a graph with spectrum $14^1 \, 2^{40} \, (-4)^{10} \, (-6)^9$. Then Σ is regular of valency 14 and is connected. Since $(A - 2I)(A + 4I)(A + 6I) = 72J$, we find that each vertex is in 4 triangles. Partition the vertex set into $\{x\}$, a set T of 8 neighbors of x such that $\{x\} \cup T$ contains the 4 triangles on x, the 6 remaining neighbors of x, and the rest. The quotient matrix for this partition has second-largest eigenvalue equal to 2, and by "partially tight" interlacing the vector that is constant 15, 3, 1, -1 on the four parts is a 2-eigenvector of A. It follows that each vertex of T has precisely one neighbor in T, that is, two triangles on x have only x in common. It also follows that a nonneighbor y of x has $6 - \mu(x, y)$ neighbors on the triangles on x. The rank 10 matrix $B = 4J - (A - 2I)(A + 6I)$ is positive semidefinite and hence can be written as $B = N^\top N$ for a 10×60 matrix N. The map ϕ sending a vertex x to column x of N maps the vertices of Σ to vectors with squared norm 2 and integral inner products, so that these images span a root lattice in $\mathbb{R}^{10}$. After some work, one finds that this root lattice must be $A_5 + A_5$, and the graph is uniquely determined.

9.8 Strongly regular graphs and two-weight codes

9.8.1 Codes, graphs, and projective sets

In this section we show the equivalence of three kinds of objects:

(i) projective two-weight codes,
(ii) subsets X of a projective space such that $|X \cap H|$ takes two values when H ranges through the hyperplanes of the projective space,
(iii) strongly regular graphs defined by a difference set that is a cone in a vector space.

This equivalence is due to DELSARTE [138]. An extensive survey of this material was given by CALDERBANK & KANTOR [80].

A *linear code* is a linear subspace of some finite vector space with fixed basis. For basic terminology and results on codes, see MACWILLIAMS & SLOANE [267] and VAN LINT [256]. A linear code C is called *projective* when its dual $C^\perp$ has minimum weight at least 3, that is, when no two coordinate positions of C are linearly dependent. The *weight* of a vector is its number of nonzero coordinates. A *two-weight* code is a linear code in which precisely two nonzero weights occur.

Let us first discuss the correspondence between linear codes and subsets of projective spaces.

9.8.2 The correspondence between linear codes and subsets of a projective space

A linear code C of word length n over the alphabet $\mathbb{F}_q$ is a linear subspace of the vector space $\mathbb{F}_q^n$. The *weight* of a vector is its number of nonzero coordinates. We call C an $[n, m, w]$-code if C has dimension m and minimum nonzero weight w. We say that C has *effective length* (or *support*) $n - z$ when there are precisely z coordinate positions j such that $c_j = 0$ for all $c \in C$. The *dual* $C^\perp$ of a code C is the linear code $\{d \in \mathbb{F}_q^n \mid \langle c, d \rangle = 0 \text{ for all } u \in C\}$, where $\langle c, d \rangle = \sum c_i d_i$ is the standard inner product (bilinear form).

Let us call two linear codes of length n over $\mathbb{F}_q$ *equivalent* when one arises from the other by permutation of coordinates or multiplication of coordinates by a nonzero constant. For example, the $\mathbb{F}_3$-codes generated by $\begin{pmatrix} 1111 \\ 0012 \end{pmatrix}$ and $\begin{pmatrix} 1212 \\ 1100 \end{pmatrix}$ are equivalent. If we study codes up to equivalence, and assume that n is chosen minimal, i.e., that the generator matrix has no zero columns, we may identify the set of columns in an $m \times n$ generator matrix with points of a projective space $PG(m-1, q)$. In this way, we find a subset X of $PG(m-1, q)$, possibly with repeated points, or, if you prefer, a weight function $w : PG(m-1, q) \to \mathbb{N}$.

Choosing one code in an equivalence class means choosing a representative in $\mathbb{F}_q^m$ for each $x \in X$, and fixing an order on X. Now the code words can be identified with the linear functionals f, and the x-coordinate position is $f(x)$.

Clearly, the code has word length $n = |X|$. Note that the code will have dimension m if and only if X spans $PG(m-1, q)$, i.e., if and only if X is not contained in a hyperplane.

The weight of the code word f equals the number of x such that $f(x) \neq 0$. But a nonzero f vanishes on a hyperplane of $PG(m-1, q)$. Consequently, the number of words of nonzero weight w in the code equals $q - 1$ times the number of hyperplanes H that meet X in $n - w$ points. In particular, the minimum distance of the code is n minus the maximum size of $H \cap X$ for a hyperplane H.

The minimum weight of the dual code equals the minimum number of points of X that are dependent. So, it is 2 if and only if X has repeated points, and 3 when X has no repeated points but has three collinear points.

Example Take for X the entire projective space $PG(m-1, q)$, so that $n = |X| = (q^m - 1)/(q - 1)$. We find the so-called simplex code: all words have weight q^{m-1}, and we have an $[n, m, q^{m-1}]$-code over $\mathbb{F}_q$. Its dual is the $[n, n-m, 3]$ Hamming code. It is perfect!

9.8.3 The correspondence between projective two-weight codes, subsets of a projective space with two intersection numbers, and affine strongly regular graphs

Given a subset X of size n of $PG(m-1,q)$, let us define a graph Γ with vertex set $\mathbb{F}_q^m$, with $x \sim y$ if and only if $\langle y - x \rangle \in X$. Then clearly Γ is regular of valency $k = (q-1)n$. We show below that this graph has eigenvalues $k - qw_i$ when the linear code has weights w_i. Hence if a linear code has only two nonzero weights, and its dual has minimum weight at least 3, then we have a strongly regular graph.

Let us look at the details.

Let $F = \mathbb{F}_q$ and $K = \mathbb{F}_{q^k}$ and let tr $: K \to F$ be the trace map defined by $\text{tr}(x) = x + x^q + \cdots + x^{q^{k-1}}$. Then the F-linear maps $f : F \to K$ are precisely the maps f_a defined by $f_a(x) = \text{tr}(ax)$, for $a \in K$. If $\text{tr}(ax) = 0$ for all x, then $a = 0$.

(Indeed, first of all, these maps are indeed F-linear. If $a \neq 0$, then $\text{tr}(ax)$ is a polynomial of degree q^{k-1} and cannot have q^k zeros. It follows that we find q^k distinct maps f_a. But this is the total number of F-linear maps from K to F (since K is a vector space of dimension k over F, and such a map is determined by its values on a basis).)

Let G be a finite Abelian group. If $a : G \to \mathbb{C}$ is any function, and we define the matrix A by $A_{xy} = a(y-x)$, then the eigenspaces of A have a basis consisting of characters of G.

(Indeed, if $\chi : G \to \mathbb{C}^*$ is a character (a homomorphism from the additively written group G into the multiplicative group of nonzero complex numbers), then

$$(A\chi)_x = \sum_{y \in G} a(y-x)\chi(y) = \left(\sum_{z \in G} a(z)\chi(z) \right) \chi(x)$$

so that χ (regarded as column vector) is an eigenvector of A with eigenvalue $\sum_{z \in G} a(z)\chi(z)$. But G has $|G|$ characters, and these are linearly independent, so this gives us the full spectrum of A.)

Example. The matrix $A = \begin{pmatrix} a & b & c \\ c & a & b \\ b & c & a \end{pmatrix}$ has eigenvectors $\begin{pmatrix} 1 \\ \omega \\ \omega^2 \end{pmatrix}$ with eigenvalues $a + b\omega + c\omega^2$, where ω runs through the cube roots of unity.

Now apply this to the adjacency matrix A of the graph Γ. Let $D := \{d \in \mathbb{F}_q^m \mid \langle d \rangle \in X\}$, so that $|D| = (q-1).|X|$. Then the neighbors of the vertex x of Γ are the points $x + d$ for $d \in D$, and we see that Γ has valency $k = |D| = (q-1)n$. The eigenvalues of A are the sums $\sum_{d \in D} \chi(d)$, where χ is a character of the additive group of $\mathbb{F}_q^m$. Let $\zeta = e^{2\pi i/p}$ be a primitive p-th root of unity, and let tr $: \mathbb{F}_q \to \mathbb{F}_p$ be the trace function. Then the characters χ are of the form

$$\chi_a(x) = \zeta^{\mathrm{tr}(\langle a,x \rangle)}.$$

Now

$$\sum_{\lambda \in \mathbb{F}_q} \chi_a(\lambda x) = \begin{cases} q & \text{if } \langle a,x \rangle = 0 \\ 0 & \text{otherwise.} \end{cases}$$

(Indeed, if S denotes this sum, then $\chi_a(\mu x)S = S$ for all μ, so if $S \neq 0$, then $\mathrm{tr}(\langle a, \mu x \rangle) = \mathrm{tr}(\mu \langle a,x \rangle) = 0$ for all μ, and by the above $\langle a,x \rangle = 0$.)

Thus, we find, if D_0 is a set of representatives for X,

$$\sum_{d \in D} \chi_a(d) = \sum_{d \in D_0} \sum_{\lambda \in \mathbb{F}_q \setminus \{0\}} \chi_a(\lambda d) = q.|H_a \cap X| - |X|$$

where H_a is the hyperplane $\{\langle x \rangle \mid \langle a,x \rangle = 0\}$ in $PG(m-1,q)$. This shows that if H_a meets X in m_a points, so that the corresponding $q-1$ code words have weight $w_a = n - m_a$, then the corresponding eigenvalue is $qm_a - n = (q-1)n - qw_a = k - qw_a$.

We have proved:

Theorem 9.8.1 *There is a 1-1-1 correspondence between*

(i) *linear codes C of effective word length n and dimension m and $(q-1)f_i$ words of weight w_i, and*

(ii) *weighted subsets X of total size n of the projective space $PG(m-1,q)$ such that for f_i hyperplanes H we have $|X \setminus H| = w_i$, and*

(iii) *graphs Γ, without loops but possibly with multiple edges, with vertex set $\mathbb{F}_q^m$, invariant under translation and dilatation, and with eigenvalues $k - qw_i$ of multiplicity $(q-1)f_i$, where $k = n(q-1)$.*

If the code C is *projective*, that is, if no two coordinate positions are dependent (so that the dual code has minimum weight at least 3), then X has no repeated points, and we find an ordinary subset under (ii), and a simple graph under (iii) (that is, without multiple edges).

Corollary 9.8.2 *There is a 1-1-1 correspondence between*

(i) *projective linear codes C of effective word length n and dimension m with precisely two nonzero weights w_1 and w_2, and*

(ii) *subsets X of size n of the projective space $PG(m-1,q)$ such that for each hyperplane H we have $|X \setminus H| = w_i$, $i \in \{1,2\}$, and*

(iii) *strongly regular graphs Γ, with vertex set $\mathbb{F}_q^m$, invariant under translation and dilatation, and with eigenvalues $k - qw_i$, where $k = n(q-1)$.*

For example, if we take a hyperoval in $PG(2,q)$, q even, we find a two-weight $[q+2,3,q]$-code over $\mathbb{F}_q$. If we take the curve $\{(1,t,t^2,\ldots,t^{m-1}) \mid t \in \mathbb{F}_q\} \cup \{(0,0,\ldots,0,1)\}$ in $PG(m-1,q)$, q arbitrary, we find a $[q+1,m,q-m+2]$-code over $\mathbb{F}_q$. (These codes are optimal: they reach the Singleton bound.)

A 1-1 correspondence between projective codes and two-weight codes was shown in BROUWER & VAN EUPEN [56].

9.8.4 Duality for affine strongly regular graphs

Let X be a subset of $PG(m-1,q)$ such that all hyperplanes meet it in either m_1 or m_2 points. In the dual projective space (where the roles of points and hyperplanes have been interchanged), the collection Y of hyperplanes that meet X in m_1 points is a set with the same property: there are numbers n_1 and n_2 such that each point is in either n_1 or n_2 hyperplanes from Y.

Indeed, let $x \in X$ be in n_1 hyperplanes from Y. We can find n_1 (independent of the choice of x) by counting hyperplanes on pairs x, y of distinct points in X:

$$n_1 \cdot (m_1 - 1) + \left(\frac{q^{m-1}-1}{q-1} - n_1 \right) \cdot (m_2 - 1) = (|X| - 1) \cdot \frac{q^{m-2}-1}{q-1}.$$

In a similar way we find n_2, the number of hyperplanes from Y on a point outside X. Computation yields $(m_1 - m_2)(n_1 - n_2) = q^{k-2}$. This proves:

Proposition 9.8.3 *The difference of the weights in a projective 2-weight code, and the difference of the nontrivial eigenvalues of an affine strongly regular graph, are a power of p, where p is the characteristic of the field involved.*

Let Γ and Δ be the strongly regular graphs corresponding to X and Y, respectively. We see that Γ and Δ both have q^k vertices; Γ has valency $k = (q-1)|X|$ and multiplicity $f = (q-1)|Y|$, and for Δ these values have interchanged roles. We call Δ the *dual* of Γ. (More generally, it is possible to define the dual of an association scheme with a regular Abelian group of automorphisms, cf. [54], p. 68.)

Example. The ternary Golay code is a perfect $[11,6,5]$ code over $\mathbb{F}_3$, and its dual C is a $[11,5,6]$ code with weights 6 and 9. The corresponding strongly regular graph Γ has parameters $(v,k,v-k-1,\lambda,\mu,r,s,f,g) = (243, 22, 220, 1, 2, 4, -5, 132, 110)$ (it is the *Berlekamp-van Lint-Seidel graph*) and its dual has parameters $(243, 110, 132, 37, 60, 2, -25, 220, 22)$, and we see that $k, v-k-1$ interchange place with g, f. The code corresponding to Δ is a $[55,5,36]$ ternary code.

Example. The quaternary Hill code ([217]) is a $[78,6,56]$ code over $\mathbb{F}_4$ with weights 56 and 64. The corresponding strongly regular graph has parameters $(4096, 234, 3861, 2, 14, 10, -22, 2808, 1287)$. Its dual has parameters $(4096, 1287, 2808, 326, 440, 7, -121, 3861, 234)$, corresponding to a quaternary $[429,6,320]$ code with weights 320 and 352. This code lies outside the range of the tables, but its residue is a world record $[109,5,80]$ code. The binary $[234,12,112]$ code derived from the Hill code has a $[122,11,56]$ code as residue—also this is a world record.

9.8.5 Cyclotomy

In this section, we take D to be a union of cosets of a subgroup of the multiplicative group of a field $\mathbb{F}_q$. (Thus, the q here corresponds to the q^k of the previous sections.)

Let $q = p^\kappa$, p prime and $e|(q-1)$, say $q = em+1$. Let $K \subseteq \mathbb{F}_q^*$ be the subgroup of the e-th powers (so that $|K| = m$). Let α be a primitive element of $\mathbb{F}_q$. For $J \subseteq \{0, 1, \ldots, e-1\}$, put $u := |J|$ and $D := D_J := \bigcup\{\alpha^j K \mid j \in J\} = \{\alpha^{ie+j} \mid j \in J, 0 \le i < m\}$. Define a (directed) graph $\Gamma = \Gamma_J$ with vertex set $\mathbb{F}_q$ and edges (x, y) whenever $y - x \in D$. Note that Γ will be undirected iff either -1 is an e-th power (i.e., q is even or $e|(q-1)/2$) or $J + (q-1)/2 = J$ (arithmetic in $\mathbb{Z}_e$).

Let $A = A_J$ be the adjacency matrix of Γ defined by $A(x, y) = 1$ if (x, y) is an edge of Γ and $A(x, y) = 0$ otherwise. Let us compute the eigenvalues of A. For each (additive) character χ of $\mathbb{F}_q$ we have

$$(A\chi)(x) = \sum_{y \sim x} \chi(y) = \left(\sum_{u \in D} \chi(u)\right) \chi(x).$$

So each character gives us an eigenvector, and since these are all independent we know all eigenvalues. Their explicit determination requires some theory of Gauss sums. Let us write $A\chi = \theta(\chi)\chi$. Clearly, $\theta(1) = mu$, the valency of Γ. Now assume $\chi \ne 1$. Then $\chi = \chi_g$ for some g, where

$$\chi_g(\alpha^j) = \exp\left(\frac{2\pi i}{p} \mathrm{tr}(\alpha^{j+g})\right)$$

and $\mathrm{tr} : \mathbb{F}_q \to \mathbb{F}_p$ is the trace function.

If μ is any multiplicative character of order e (say, $\mu(\alpha^j) = \zeta^j$, where $\zeta = \exp(\frac{2\pi i}{e})$), then

$$\sum_{i=0}^{e-1} \mu^i(x) = \begin{cases} e & \text{if } \mu(x) = 1 \\ 0 & \text{otherwise.} \end{cases}$$

Hence,

$$\theta(\chi_g) = \sum_{u \in D} \chi_g(u) = \sum_{j \in J} \sum_{u \in K} \chi_{j+g}(u) = \frac{1}{e} \sum_{j \in J} \sum_{x \in \mathbb{F}_q^*} \chi_{j+g}(x) \sum_{i=0}^{e-1} \mu^i(x) =$$

$$= \frac{1}{e} \sum_{j \in J} \left(-1 + \sum_{i=1}^{e-1} \sum_{x \ne 0} \chi_{j+g}(x) \mu^i(x) \right) = \frac{1}{e} \sum_{j \in J} \left(-1 + \sum_{i=1}^{e-1} \mu^{-i}(\alpha^{j+g}) G_i \right)$$

where G_i is the Gauss sum $\sum_{x \ne 0} \chi_0(x) \mu^i(x)$.

In general, determination of Gauss sums seems to be complicated, but there are a few explicit results. For our purposes the most interesting is the following:

Proposition 9.8.4 (Stickelberger, and Davenport and Hasse; see MCELIECE & RUMSEY [275]) *Suppose $e > 2$ and p is semiprimitive mod e (i.e., there exists an l such that $p^l \equiv -1 \pmod{e}$). Choose l minimal and write $\kappa = 2lt$. Then*

$$G_i = (-1)^{t+1} \varepsilon^{it} \sqrt{q},$$

where

$$\varepsilon = \begin{cases} -1 & \text{if } e \text{ is even and } (p^l + 1)/e \text{ is odd} \\ +1 & \text{otherwise.} \end{cases}$$

Under the hypotheses of this proposition, we have

$$\sum_{i=1}^{e-1} \mu^{-i}(\alpha^{j+g}) G_i = \sum_{i=1}^{e-1} \zeta^{-i(j+g)}(-1)^{t+1} \varepsilon^{it} \sqrt{q} = \begin{cases} (-1)^t \sqrt{q} & \text{if } r \neq 1 \\ (-1)^{t+1} \sqrt{q}(e-1) & \text{if } r = 1, \end{cases}$$

where $\zeta = \exp(2\pi i/e)$ and $r = r_{g,j} = \zeta^{-j-g} \varepsilon^t$ (so that $r^e = \varepsilon^{et} = 1$), and hence

$$\theta(\chi_g) = \frac{u}{e}(-1 + (-1)^t \sqrt{q}) + (-1)^{t+1} \sqrt{q} \cdot \#\{j \in J \mid r_{g,j} = 1\}.$$

If we abbreviate the cardinality in this formula with #, then: If $\varepsilon^t = 1$ then $\# = 1$ if $g \in -J \pmod{e}$, and $\# = 0$ otherwise. If $\varepsilon^t = -1$ (then e is even and p is odd) then $\# = 1$ if $g \in \frac{1}{2}e - J \pmod{e}$, and $\# = 0$ otherwise. We proved:

Theorem 9.8.5 *Let $q = p^\kappa$, with p prime, and $e | (q-1)$, where p is semiprimitive mod e (i.e., there is an $l > 0$ such that $p^l \equiv -1 \bmod e$). Choose l minimal with this property and write $\kappa = 2lt$. Choose u, $1 \leq u \leq e-1$, and assume that q is even or u is even or $e | (q-1)/2$. Then the graphs Γ_J (where J is arbitrary for q even or $e | (q-1)/2$ and satisfies $J + (q-1)/2 = J \bmod e$ otherwise) are strongly regular, with eigenvalues*

$$\begin{aligned} k = \tfrac{q-1}{e}u \qquad & \text{with multiplicity } 1, \\ \theta_1 = \tfrac{u}{e}(-1 + (-1)^t \sqrt{q}) \qquad & \text{with multiplicity } q - 1 - k, \\ \theta_2 = \tfrac{u}{e}(-1 + (-1)^t \sqrt{q}) + (-1)^{t+1} \sqrt{q} \; & \text{with multiplicity } k. \end{aligned}$$

(Obviously, when t is even we have $r = \theta_1$, $s = \theta_2$, and otherwise $r = \theta_2$, $s = \theta_1$.)

Clearly, if $e | e' | (q-1)$, then the set of e-th powers is a union of cosets of the set of e'-th powers, so when applying the above theorem we may assume that e has been chosen as large as possible, i.e., $e = p^l + 1$. Then the restriction "q is even or u is even or $e | (q-1)/2$" is empty, and J can always be chosen arbitrarily.

The above construction can be generalized. Pick several values e_i ($i \in I$) with $e_i | (q-1)$. Let K_i be the subgroup of $\mathbb{F}_q^*$ of the e_i-th powers. Let J_i be a subset of $\{0,1,\ldots,e_i - 1\}$. Let $D_i := D_{J_i} := \bigcup \{\alpha^j K_i \mid j \in J_i\}$. Put $D := \bigcup D_i$. If the D_i are mutually disjoint, then D defines a graph of which we can compute the spectrum.

For example, let p be odd, and take $e_i = p^{l_i} + 1$ ($i = 1,2$) and $q = p^\kappa$, where $\kappa = 4l_i s_i$ ($i = 1,2$). Pick J_1 to consist of even numbers only, and J_2 to consist of odd numbers only. Then $D_1 \cap D_2 = \emptyset$ and $g \in -J_i \pmod{e_i}$ cannot happen for $i = 1,2$ simultaneously. This means that the resulting graph will be strongly regular with eigenvalues

$$\theta(\chi_g) = \left(\frac{|J_1|}{e_1} + \frac{|J_2|}{e_2}\right)(-1+\sqrt{q}) - \sqrt{q}.\delta(g \in -J_i(\text{mod } e_i) \text{ for } i=1 \text{ or } i=2)$$

(where $\delta(P) = 1$ if P holds, and $\delta(P) = 0$ otherwise). See also [68]. In the special case $p = 3$, $l_1 = 1$, $l_2 = 2$, $e_1 = 4$, $e_2 = 10$, $J_1 = \{0\}$, $J_2 = \{1\}$, the difference set consists of the powers α^i with $i \equiv 0$ (mod 4) or $i \equiv 1$ (mod 10) (i.e., is the set $\{1, \alpha, \alpha^4, \alpha^8, \alpha^{11}, \alpha^{12}, \alpha^{16}\}\langle\alpha^{20}\rangle$), and we found the first graph from DE LANGE [249] again. (It has parameters $(v, k, \lambda, \mu) = (6561, 2296, 787, 812)$ and spectrum $2296^1\ 28^{4264}\ (-53)^{2296}$.)

9.9 Table of parameters for strongly regular graphs

Below a table with the feasible parameters for strongly regular graphs on at most 100 vertices. Here *feasible* means that the parameters v, k, λ, μ and multiplicities f, g are integers, with $0 \le \lambda < k - 1$ and $0 < \mu < k < v$. In some cases a feasible parameter set is ruled out by the absolute bound or the Krein conditions, or the restriction that the order of a conference graph must be the sum of two squares. For some explanation of the comments, see §9.9.1.

∃	v	k	λ	μ	r^f	s^g	Comments
!	5	2	0	1	0.618^2	-1.618^2	pentagon; Paley(5); Seidel 2-graph−∗
!	9	4	1	2	1^4	-2^4	Paley(9); 3^2; 2-graph−∗
!	10	3	0	1	1^5	-2^4	PETERSEN [294]; $NO_4^-(2)$; $NO_3^{-\perp}(5)$; 2-graph
	6	3	4		1^4	-2^5	$T(5)$; 2-graph
!	13	6	2	3	1.303^6	-2.303^6	Paley(13); 2-graph−∗
!	15	6	1	3	1^9	-3^5	$O_5(2)$ polar graph; $Sp_4(2)$ polar graph; $NO_4^-(3)$; 2-graph−∗
	8	4	4		2^5	-2^9	$T(6)$; 2-graph−∗
!	16	5	0	2	1^{10}	-3^5	$q_{22}^2 = 0$; vanLint-Schrijver(1); $VO_4^-(2)$ affine polar graph; projective binary [5,4] code with weights 2, 4; $RSHCD^-$; 2-graph
		10	6	6	2^5	-2^{10}	$q_{11}^1 = 0$; Clebsch graph [100, 108, 99, 315]; vanLint-Schrijver(2); 2-graph
2!	16	6	2	2	2^6	-2^9	SHRIKHANDE [325]; 4^2; from a partial spread: projective binary [6,4] code with weights 2, 4; $RSHCD^+$; 2-graph
	9	4	6		1^9	-3^6	OA(4,3); $Bilin_{2×2}(2)$; Goethals-Seidel(2,3); $VO_4^+(2)$ affine polar graph; 2-graph
!	17	8	3	4	1.562^8	-2.562^8	Paley(17); 2-graph−∗
!	21	10	3	6	1^{14}	-4^6	
	10	5	4		3^6	-2^{14}	$T(7)$
−	21	10	4	5	1.791^{10}	-2.791^{10}	Conf
!	25	8	3	2	3^8	-2^{16}	5^2
	16	9	12		1^{16}	-4^8	OA(5,4)
15!	25	12	5	6	2^{12}	-3^{12}	complete enumeration by PAULUS [290]; Paley(25); OA(5,3); 2-graph−∗
10!	26	10	3	4	2^{13}	-3^{12}	complete enumeration by PAULUS [290]; 2-graph
	15	8	9		2^{12}	-3^{13}	S(2,3,13); 2-graph
!	27	10	1	5	1^{20}	-5^6	$q_{22}^2 = 0$; $O_6^-(2)$ polar graph; $GQ(2,4)$; 2-graph−∗
	16	10	8		4^6	-2^{20}	$q_{11}^1 = 0$; Schläfli graph; unique by SEIDEL [315]; 2-graph−∗
−	28	9	0	4	1^{21}	-5^6	$Krein_2$; Absolute bound
	18	12	10		4^6	-2^{21}	$Krein_1$; Absolute bound
4!	28	12	6	4	4^7	-2^{20}	$T(8)$; Chang graphs, CHANG [87]; 2-graph
	15	6	10		1^{20}	-5^7	$NO_6^+(2)$; Goethals-Seidel(3,3); Taylor 2-graph for $U_3(3)$
41!	29	14	6	7	2.193^{14}	-3.193^{14}	complete enumeration by Bussemaker & Spence [pers.comm.]; Paley(29); 2-graph−∗

∃	v	k	λ	μ	r^f	s^g	Comments
−	33	16	7	8	2.372^{16}	-3.372^{16}	Conf
3854!	35	16	6	8	2^{20}	-4^{14}	complete enumeration by McKay & Spence [277]; 2-graph−∗
		18	9	9	3^{14}	-3^{20}	S(2,3,15); lines in $PG(3,2)$; $O_6^+(2)$ polar graph; 2-graph−∗
!	36	10	4	2	4^{10}	-2^{25}	6^2
		25	16	20	1^{25}	-5^{10}	
180!	36	14	4	6	2^{21}	-4^{14}	$U_3(3).2/L_2(7).2$—subconstituent of the Hall-Janko graph; complete enumeration by McKay & Spence [277]; $RSHCD^-$; 2-graph
		21	12	12	3^{14}	-3^{21}	2-graph
!	36	14	7	4	5^8	-2^{27}	$T(9)$
		21	10	15	1^{27}	-6^8	
32548!	36	15	6	6	3^{15}	-3^{20}	complete enumeration by McKay & Spence [277]; OA(6,3); $NO_6^-(2)$; $RSHCD^+$; 2-graph
		20	10	12	2^{20}	-4^{15}	$NO_5^-(3)$; 2-graph
+	37	18	8	9	2.541^{18}	-3.541^{18}	Paley(37); 2-graph−∗
28!	40	12	2	4	2^{24}	-4^{15}	complete enumeration by Spence [328]; $O_5(3)$ polar graph; $Sp_4(3)$ polar graph
		27	18	18	3^{15}	-3^{24}	$NU(4,2)$
+	41	20	9	10	2.702^{20}	-3.702^{20}	Paley(41); 2-graph−∗
78!	45	12	3	3	3^{20}	-3^{24}	complete enumeration by Coolsaet, Degraer & Spence [104]; $U_4(2)$ polar graph
		32	22	24	2^{24}	-4^{20}	$NO_5^+(3)$
!	45	16	8	4	6^9	-2^{35}	$T(10)$
		28	15	21	1^{35}	-7^9	
+	45	22	10	11	2.854^{22}	-3.854^{22}	Mathon [272]; 2-graph−∗
!	49	12	5	2	5^{12}	-2^{36}	7^2
		36	25	30	1^{36}	-6^{12}	OA(7,6)
−	49	16	3	6	2^{32}	-5^{16}	Bussemaker, Haemers, Mathon, & Wilbrink [77]
		32	21	20	4^{16}	-3^{32}	
+	49	18	7	6	4^{18}	-3^{30}	OA(7,3)
		30	17	20	2^{30}	-5^{18}	OA(7,5)
+	49	24	11	12	3^{24}	-4^{24}	Paley(49); OA(7,4); 2-graph−∗
!	50	7	0	1	2^{28}	-3^{21}	Hoffman & Singleton [220]; $U_3(5^2).2/Sym(7)$
		42	35	36	2^{21}	-3^{28}	
−	50	21	4	12	1^{42}	-9^7	Absolute bound
		28	18	12	8^7	-2^{42}	Absolute bound
+	50	21	8	9	3^{25}	-4^{24}	2-graph
		28	15	16	3^{24}	-4^{25}	S(2,4,25); 2-graph
+	53	26	12	13	3.140^{26}	-4.140^{26}	Paley(53); 2-graph−∗
!	55	18	9	4	7^{10}	-2^{44}	$T(11)$
		36	21	28	1^{44}	-8^{10}	
!	56	10	0	2	2^{35}	-4^{20}	Sims-Gewirtz graph [170, 171, 58]; $L_3(4).2^2/Alt(6).2^2$
		45	36	36	3^{20}	-3^{35}	intersection-2 graph of a 2-(21,6,4) design with block intersections 0, 2
−	56	22	3	12	1^{48}	-10^7	Krein$_2$; Absolute bound
		33	22	15	9^7	-2^{48}	Krein$_1$; Absolute bound
−	57	14	1	4	2^{38}	-5^{18}	Wilbrink & Brouwer [349]
		42	31	30	4^{18}	-3^{38}	
+	57	24	11	9	5^{18}	-3^{38}	S(2,3,19)
		32	16	20	2^{38}	-6^{18}	
−	57	28	13	14	3.275^{28}	-4.275^{28}	Conf
+	61	30	14	15	3.405^{30}	-4.405^{30}	Paley(61); 2-graph−∗
−	63	22	1	11	1^{55}	-11^7	Krein$_2$; Absolute bound
		40	28	20	10^7	-2^{55}	Krein$_1$; Absolute bound
+	63	30	13	15	3^{35}	-5^{27}	intersection-8 graph of a 2-(36,16,12) design with block intersections 6, 8; $O_7(2)$ polar graph; $Sp_6(2)$ polar graph; 2-graph−∗
		32	16	16	4^{27}	-4^{35}	S(2,4,28); intersection-6 graph of a 2-(28,12,11) design with block intersections 4, 6; $NU(3,3)$; 2-graph−∗
!	64	14	6	2	6^{14}	-2^{49}	8^2; from a partial spread of 3-spaces: projective binary [14,6] code with weights 4, 8

∃	v	k	λ	μ	r^f	s^g	Comments
		49	36	42	1^{49}	-7^{14}	OA(8,7)
167!	64	18	2	6	2^{45}	-6^{18}	complete enumeration by HAEMERS & SPENCE [207]; $GQ(3,5)$; from a hyperoval: projective 4-ary [6,3] code with weights 4, 6
		45	32	30	5^{18}	-3^{45}	
−	64	21	0	10	1^{56}	-11^{7}	Krein$_2$; Absolute bound
		42	30	22	10^{7}	-2^{56}	Krein$_1$; Absolute bound
+	64	21	8	6	5^{21}	-3^{42}	OA(8,3); Bilin$_{2\times3}$(2); from a Baer subplane: projective 4-ary [7,3] code with weights 4, 6; from a partial spread of 3-spaces: projective binary [21,6] code with weights 8, 12
		42	26	30	2^{42}	-6^{21}	OA(8,6)
+	64	27	10	12	3^{36}	-5^{27}	from a unital: projective 4-ary [9,3] code with weights 6, 8; $VO_6^-(2)$ affine polar graph; $RSHCD^-$; 2-graph
		36	20	20	4^{27}	-4^{36}	2-graph
+	64	28	12	12	4^{28}	-4^{35}	OA(8,4); from a partial spread of 3-spaces: projective binary [28,6] code with weights 12, 16; $RSHCD^+$; 2-graph
		35	18	20	3^{35}	-5^{28}	OA(8,5); Goethals-Seidel(2,7); $VO_6^+(2)$ affine polar graph; 2-graph
−	64	30	18	10	10^{8}	-2^{55}	Absolute bound
		33	12	22	1^{55}	-11^{8}	Absolute bound
?	65	32	15	16	3.531^{32}	-4.531^{32}	2-graph−∗?
!	66	20	10	4	8^{11}	-2^{54}	$T(12)$
		45	28	36	1^{54}	-9^{11}	
?	69	20	7	5	5^{23}	-3^{45}	
		48	32	36	2^{45}	-6^{23}	S(2,6,46) does not exist
−	69	34	16	17	3.653^{34}	-4.653^{34}	Conf
+	70	27	12	9	6^{20}	-3^{49}	S(2,3,21)
		42	23	28	2^{49}	-7^{20}	
+	73	36	17	18	3.772^{36}	-4.772^{36}	Paley(73); 2-graph−∗
?	75	32	10	16	2^{56}	-8^{18}	2-graph−∗?
		42	25	21	7^{18}	-3^{56}	2-graph−∗?
−	76	21	2	7	2^{56}	-7^{19}	HAEMERS [198]
		54	39	36	6^{19}	-3^{56}	
?	76	30	8	14	2^{57}	-8^{18}	2-graph?
		45	28	24	7^{18}	-3^{57}	2-graph?
?	76	35	18	14	7^{19}	-3^{56}	2-graph?
		40	18	24	2^{56}	-8^{19}	2-graph?
!	77	16	0	4	2^{55}	-6^{21}	S(3,6,22); $M_{22}.2/2^4$: Sym(6); unique by BROUWER [48]; subconstituent of Higman-Sims graph
		60	47	45	5^{21}	-3^{55}	intersection-2 graph of a 2-(22,6,5) design with block intersections 0, 2
−	77	38	18	19	3.887^{38}	-4.887^{38}	Conf
!	78	22	11	4	9^{12}	-2^{65}	$T(13)$
		55	36	45	1^{65}	-10^{12}	
!	81	16	7	2	7^{16}	-2^{64}	9^2; from a partial spread: projective ternary [8,4] code with weights 3, 6
		64	49	56	1^{64}	-8^{16}	OA(9,8)
!	81	20	1	6	2^{60}	-7^{20}	unique by BROUWER & HAEMERS [57]; $VO_4^-(3)$ affine polar graph; projective ternary [10,4] code with weights 6, 9
		60	45	42	6^{20}	-3^{60}	
+	81	24	9	6	6^{24}	-3^{56}	OA(9,3); $VNO_4^-(3)$ affine polar graph; from a partial spread: projective ternary [12,4] code with weights 6, 9
		56	37	42	2^{56}	-7^{24}	OA(9,7)
+	81	30	9	12	3^{50}	-6^{30}	$VNO_4^-(3)$ affine polar graph; HAMADA & HELLESETH [211]: projective ternary [15,4] code with weights 9, 12
		50	31	30	5^{30}	-4^{50}	
+	81	32	13	12	5^{32}	-4^{48}	OA(9,4); Bilin$_{2\times2}$(3); $VO_4^+(3)$ affine polar graph; from a partial spread: projective ternary [16,4] code with weights 9, 12
		48	27	30	3^{48}	-6^{32}	OA(9,6)
−	81	40	13	26	1^{72}	-14^{8}	Absolute bound
		40	25	14	13^{8}	-2^{72}	Absolute bound
+	81	40	19	20	4^{40}	-5^{40}	Paley(81); OA(9,5); projective ternary [20,4] code with weights 12, 15; 2-graph−∗

∃	v	k	λ	μ	r^f	s^g	Comments
+	82	36	15	16	4^{41}	-5^{40}	2-graph
		45	24	25	4^{40}	-5^{41}	S(2,5,41); 2-graph
?	85	14	3	2	4^{34}	-3^{50}	
		70	57	60	2^{50}	-5^{34}	
+	85	20	3	5	3^{50}	-5^{34}	$O_5(4)$ polar graph; $Sp_4(4)$ polar graph
		64	48	48	4^{34}	-4^{50}	
?	85	30	11	10	5^{34}	-4^{50}	
		54	33	36	3^{50}	-6^{34}	S(2,6,51)?
?	85	42	20	21	4.110^{42}	-5.110^{42}	2-graph−∗?
?	88	27	6	9	3^{55}	-6^{32}	
		60	41	40	5^{32}	-4^{55}	
+	89	44	21	22	4.217^{44}	-5.217^{44}	Paley(89); 2-graph−∗
!	91	24	12	4	10^{13}	-2^{77}	$T(14)$
		66	45	55	1^{77}	-11^{13}	
−	93	46	22	23	4.322^{46}	-5.322^{46}	Conf
?	95	40	12	20	2^{75}	-10^{19}	2-graph−∗?
		54	33	27	9^{19}	-3^{75}	2-graph−∗?
+	96	19	2	4	3^{57}	-5^{38}	HAEMERS [197]
		76	60	60	4^{38}	-4^{57}	
+	96	20	4	4	4^{45}	-4^{50}	$GQ(5,3)$
		75	58	60	3^{50}	-5^{45}	
?	96	35	10	14	3^{63}	-7^{32}	
		60	38	36	6^{32}	-4^{63}	
−	96	38	10	18	2^{76}	-10^{19}	DEGRAER [136]
		57	36	30	9^{19}	-3^{76}	
?	96	45	24	18	9^{20}	-3^{75}	2-graph?
		50	22	30	2^{75}	-10^{20}	2-graph?
+	97	48	23	24	4.424^{48}	-5.424^{48}	Paley(97); 2-graph−∗
?	99	14	1	2	3^{54}	-4^{44}	
		84	71	72	3^{44}	-4^{54}	
?	99	42	21	15	9^{21}	-3^{77}	
		56	28	36	2^{77}	-10^{21}	
+	99	48	22	24	4^{54}	-6^{44}	2-graph−∗
		50	25	25	5^{44}	-5^{54}	S(2,5,45); 2-graph−∗
!	100	18	8	2	8^{18}	-2^{81}	10^2
		81	64	72	1^{81}	-9^{18}	
!	100	22	0	6	2^{77}	-8^{22}	$q^2_{22}=0$; HIGMAN & SIMS [216]; $HS.2/M_{22}.2$; unique by GEWIRTZ [170]
		77	60	56	7^{22}	-3^{77}	$q^1_{11}=0$
+	100	27	10	6	7^{27}	-3^{72}	OA(10,3)
		72	50	56	2^{72}	-8^{27}	OA(10,8)?
?	100	33	8	12	3^{66}	-7^{33}	
		66	44	42	6^{33}	-4^{66}	
+	100	33	14	9	8^{24}	-3^{75}	S(2,3,25)
		66	41	48	2^{75}	-9^{24}	
−	100	33	18	7	13^{11}	-2^{88}	Absolute bound
		66	39	52	1^{88}	-14^{11}	Absolute bound
+	100	36	14	12	6^{36}	-4^{63}	Hall-Janko graph; $J_2.2/U_3(3).2$; subconstituent of $G_2(4)$ graph; OA(10,4)
		63	38	42	3^{63}	-7^{36}	OA(10,7)?
+	100	44	18	20	4^{55}	-6^{44}	JØRGENSEN & KLIN [236]; $RSHCD^-$; 2-graph
		55	30	30	5^{44}	-5^{55}	2-graph
+	100	45	20	20	5^{45}	-5^{54}	OA(10,5)?; $RSHCD^+$; 2-graph
		54	28	30	4^{54}	-6^{45}	OA(10,6)?; 2-graph

9.9.1 Comments

Comment	Explanation
$q^1_{11}=0$, $q^2_{22}=0$	Zero Krein parameter, see §11.4.
m^2	Hamming graph $H(2,m)$, a.k.a. lattice graph $L_2(m)$, or grid graph $m \times m$, or $K_m \square K_m$, see §12.4.1, §1.4.5.

continued...

Comment	Explanation
$T(m)$	Johnson graph $J(m,2)$, a.k.a. triangular graph $T(m)$, see §12.4.2, §1.4.5.
$OA(n,t)$ $(t \geq 3)$	Block graph of an orthogonal array $OA(n,t)$ (that is, $t-2$ mutually orthogonal Latin squares of order n).
$S(2,k,v)$	Block graph of a Steiner system $S(2,k,v)$ (that is, a 2-$(v,k,1)$ design).
Goethals-Seidel(k,r)	Graph constructed from a Steiner system $S(2,k,v)$ (with $r = (v-1)/(k-1)$) and a Hadamard matrix of order $r+1$ as in [179].
2-graph	Graph in the switching class of a regular 2-graph, see §10.2.
2-graph$-*$	Descendant of a regular 2-graph, see §10.2.
$RSHCD^{\pm}$	Graph derived from a regular symmetric Hadamard matrix with constant diagonal (cf. §10.5.1, [62], [179]).
Taylor 2-graph for $U_3(q)$	Graph derived from Taylor's regular 2-graph (cf. [62], [336], [337]).
Paley(q)	Paley graph on $\mathbb{F}_q$, see §10.4, §13.6.
vanLint-Schrijver(u)	Graph constructed by the cyclotomic construction of [257], taking the union of u classes.
Bilin$_{2 \times d}(q)$	Graph on the $2 \times d$ matrices over $\mathbb{F}_q$, adjacent when their difference has rank 1.
$GQ(s,t)$	Collinearity graph of a generalized quadrangle with parameters $GQ(s,t)$, see §9.6.3.
$O_{2d}^{\varepsilon}(q), O_{2d+1}(q)$	Isotropic points on a nondegenerate quadric in the projective space $PG(2d-1,q)$ or $PG(2d,q)$, joined when the connecting line is totally singular.
$Sp_{2d}(q)$	Points of $PG(2d-1,q)$ provided with a nondegenerate symplectic form, joined when the connecting line is totally isotropic.
$U_d(q)$	Isotropic points of $PG(d-1,q^2)$ provided with a nondegenerate Hermitean form, joined when the connecting line is totally isotropic.
$NO_{2d}^{\varepsilon}(2)$	Nonisotropic points of $PG(2d-1,2)$ provided with a nondegenerate quadratic form, joined when they are orthogonal, i.e., when the connecting line is a tangent.
$NO_{2d}^{\varepsilon}(3)$	One class of nonisotropic points of $PG(2d-1,3)$ provided with a nondegenerate quadratic form, joined when they are orthogonal, i.e., when the connecting line is elliptic.

continued...

Comment	Explanation
$NO_{2d+1}^{\varepsilon}(q)$	One class of nondegenerate hyperplanes of $PG(2d,q)$ provided with a nondegenerate quadratic form, joined when their intersection is degenerate.
$NO_{2d+1}^{\varepsilon\perp}(5)$	One class of nonisotropic points of $PG(2d,5)$ provided with a nondegenerate quadratic form, joined when they are orthogonal.
$NU_n(q)$	Nonisotropic points of $PG(n-1,q)$ provided with a nondegenerate Hermitean form, joined when the connecting line is a tangent.
$VO_{2d}^{\varepsilon}(q)$	Vectors of a $2d$-dimensional vector space over $\mathbb{F}_q$ provided with a nondegenerate quadratic form Q, where two vectors u and v are joined when $Q(v-u)=0$.
$VNO_{2d}^{\varepsilon}(q)$ (q odd)	Vectors of a $2d$-dimensional vector space over $\mathbb{F}_q$ provided with a nondegenerate quadratic form Q, where two vectors u and v are joined when $Q(v-u)$ is a nonzero square.

9.10 Exercises

Exercise 9.1 ([179]) Consider the graph on the set of flags (incident point-line pairs) of the projective plane $PG(2,4)$, where (p,L) and (q,M) are adjacent when $p \neq q$ and $L \neq M$ and either $p \in M$ or $q \in L$. Show that this graph is strongly regular with parameters $(v,k,\lambda,\mu) = (105,32,4,12)$.

Exercise 9.2 ([25]) Consider the graph on the cosets of the perfect ternary Golay code (an [11,6,5] code over $\mathbb{F}_3$), where two cosets are adjacent when they differ by a vector of weight 1. Show that this graph is strongly regular with parameters $(v,k,\lambda,\mu) = (243,22,1,2)$. It is known as the *Berlekamp-van Lint-Seidel graph*.

Exercise 9.3 Fix a Steiner system $S(3,6,22)$ on a 22-set Ω. Consider the graph Γ that has as vertices the pairs of symbols from Ω, where two pairs are adjacent when they are disjoint and their union is contained in a block of the Steiner system. Show that this graph is strongly regular with parameters $(v,k,\lambda,\mu) = (231,30,9,3)$. It is known as the *Cameron graph*.

Exercise 9.4 Prove Theorem 9.6.2.

Exercise 9.5 For a strongly regular graph Γ and a vertex x of Γ, let Δ be the subgraph of Γ induced on the set of vertices different from x and nonadjacent to x. If Γ has no triangles and spectrum k^1, r^f, s^g, then show that Δ has spectrum $(k-\mu)^1$, r^{f-k}, s^{g-k}, $(-\mu)^{k-1}$. Conclude if Γ is primitive that $f \geq k$ and $g \geq k$, and that if $f = k$ or $g = k$ then Δ is itself complete or strongly regular. Determine all strongly regular graphs with $\lambda = 0$ and $f = k$.

Exercise 9.6 Let Γ be a strongly regular graph with parameters (v,k,λ,μ) and spectrum k^1, r^f, s^g. Let x be a vertex of Γ, and suppose that the graph $\Gamma(x)$ induced on the set of neighbors of x is isomorphic to $(t+1)K_a$ (so that $k = a(t+1)$ and $\lambda = a-1$). Then the graph $\Gamma_2(x)$ induced on the set of nonneighbors of x has spectrum $(k-\mu)^1$, r^{f-k}, s^{g-k}, $(-\mu)^t$, $(r+s+1)^{k-t-1}$. (Hint: Use Lemma 2.11.1.)

Exercise 9.7 ([302]) Prove that if the complete graph K_v can be decomposed into an edge-disjoint union of three copies of a strongly regular graph with parameters (v,k,λ,μ), then there is an $m \in \mathbb{Z}$ such that $v = (3m-1)^2$, $k = 3m^2 - 2m$, $\lambda = m^2 - 1$, $\mu = m^2 - m$. (Hint: Apply the argument from §1.5.1.)

Exercise 9.8 ([37]) Show that having a constant k almost follows from having constant λ, μ. More precisely: Consider a graph Γ with the property that any two adjacent (nonadjacent) vertices have λ (resp. μ) common neighbors. Show that if Γ is not regular, then either $\mu = 0$ and Γ is a disjoint union of $(\lambda + 2)$-cliques, or $\mu = 1$, and Γ is obtained from a disjoint union of $(\lambda + 1)$-cliques by adding a new vertex, adjacent to all old vertices.

Exercise 9.9 A spread in a generalized quadrangle is a subset S of the lines such that every point is on exactly one line of S. Prove that a $GQ(q^2,q)$ has no spread. (Hint: A spread is a coclique in the line graph.)

Exercise 9.10 Show that the Schläfli graph is obtained from $L(K_8)$ (that is, $T(8)$) by switching one point isolated, and removing it.

Exercise 9.11 ([236]) Show that the strongly regular graph with parameters $(v,k,\lambda,\mu) = (100,45,20,20)$ obtained from the Hall-Janko graph in §9.1.12 can be switched into a strongly regular graph with parameters $(100,55,30,30)$.

Exercise 9.12 There exist strongly regular graphs in $\mathbb{F}_3^4$, invariant for translation and dilatation, with parameters $(v,k,\lambda,\mu) = (81,20,1,6)$ and $(81,30,9,12)$. Determine the corresponding ternary codes and their weight enumerators.

Exercise 9.13 With C and D as in §9.7, show that $C \cup D$ induces a distance-regular graph of diameter 3 with intersection array $\{10,9,4;1,6,10\}$.

Exercise 9.14 With Γ as in §9.7, show that $\chi(\Gamma) \geq 6$ also follows from Corollary 3.6.4 applied to the induced subgraph of Γ, obtained by deleting all vertices of one color class.

Exercise 9.15 Under what conditions is the Hamming code cyclic? Negacyclic? Constacyclic?

Exercise 9.16 A *cap* in a projective space is a collection of points, no three on a line. Show that a $[n, n-m, 4]$ code over $\mathbb{F}_q$ exists if and only if there is a cap of size n in $PG(m-1,q)$. Construct for $m > 0$ a $[2^{m-1}, 2^{m-1} - m, 4]$ binary code.

Exercise 9.17 Given a two-weight code over $\mathbb{F}_q$ of word length n, dimension m and weights w_1 and w_2. Express the parameters v, k, λ, μ, r, s, f, g of the corresponding strongly regular graph in terms of q, n, k, w_1 and w_2.

Chapter 10
Regular Two-graphs

10.1 Strong graphs

Let us call a graph *(possibly improper) strongly regular* when it is strongly regular or complete or edgeless. Above (Theorem 9.1.2) we saw that a graph Γ is (possibly improper) strongly regular if and only if its adjacency matrix A satisfies $A^2 \in \langle A, I, J \rangle$, where $\langle \ldots \rangle$ denotes the $\mathbb{R}$-span. In particular, this condition implies that Γ is regular, so that $AJ = JA$.

Consider the Seidel matrix $S = J - I - 2A$ (see §1.8.2). We have $\langle A, I, J \rangle = \langle S, I, J \rangle$. If $A^2 \in \langle A, I, J \rangle$ then also $S^2 \in \langle S, I, J \rangle$, but the converse does not hold. For example, consider the path P_3 of length 2. We have $S^2 = S + 2I$, but A only satisfies the cubic equation $A^3 = 2A$.

We call a graph *strong* whenever its Seidel matrix S satisfies $S^2 \in \langle S, I, J \rangle$. Thus a (possibly improper) strongly regular graph is strong, and conversely a regular strong graph is (possibly improper) strongly regular. As we saw, a strong graph need not be regular. Another example is given by $C_5 + K_1$, where the Seidel matrix satisfies $S^2 = 5I$. But the following properties are satisfied (recall that an eigenvalue is called *restricted* if it has an eigenvector orthogonal to the all-1 vector **1**):

Proposition 10.1.1 *For a graph Γ with v vertices and Seidel matrix S, the following holds:*

(i) *Γ is strong if and only if S has at most two restricted eigenvalues. In this case $(S - \rho_1 I)(S - \rho_2 I) = (v - 1 + \rho_1 \rho_2)J$, where ρ_1 and ρ_2 are restricted eigenvalues of S.*

(ii) *Γ is strong and regular if and only if Γ is (possibly improper) strongly regular. In this case the eigenvalue ρ_0 of S for **1** satisfies $(\rho_0 - \rho_1)(\rho_0 - \rho_2) = v(v - 1 + \rho_1 \rho_2)$.*

(iii) *If Γ is strong with restricted eigenvalues ρ_1 and ρ_2, and $v - 1 + \rho_1 \rho_2 \neq 0$, then Γ is regular, and hence (possibly improper) strongly regular.*

(iv) *S has a single restricted eigenvalue if and only if $S = \pm(J - I)$, that is, if and only if Γ is complete or edgeless.*

Proof. (i) If Γ is strong, then $S^2 + \alpha S + \beta I = \gamma J$ for some constants α, β, and γ. If ρ is a restricted eigenvalue of S with eigenvector $\mathbf{v}$ orthogonal to $\mathbf{1}$, then $(\rho^2 + \alpha\rho + \beta)\mathbf{v} = \gamma J\mathbf{v} = \mathbf{0}$, so $\rho^2 + \alpha\rho + \beta = 0$. Therefore S has at most two restricted eigenvalues. Conversely, if S has just two restricted eigenvalues ρ_1 and ρ_2, then $(S - \rho_1 I)(S - \rho_2 I) \in \langle J \rangle$, so Γ is strong. And if $(S - \rho_1 I)(S - \rho_2 I) = \gamma J$, then the diagonal entries show that $\gamma = v - 1 + \rho_1\rho_2$.

(ii) We know that (possibly improper) strongly regular implies strong and regular. Suppose Γ is strong and regular. Then $S^2 \in \langle S, I, J \rangle$ and $SJ \in \langle J \rangle$, which implies that the adjacency matrix $A = (J - S - I)/2$ of Γ satisfies $A^2 \in \langle A, I, J \rangle$, so Γ is (possibly improper) strongly regular by Theorem 9.1.2.

(iii) If Γ is not regular, then J is not a polynomial in S, so $v - 1 + \rho_1\rho_2 = 0$ follows from part (i). □

We see that $v - 1 + \rho_1\rho_2 = 0$ if and only if S has exactly two distinct eigenvalues ρ_1 and ρ_2. Recall that two graphs Γ and $\tilde{\Gamma}$ are switching equivalent (see §1.8.2) if their Seidel matrices S and $\tilde{S}$ are similar by some diagonal matrix $D = \mathrm{diag}(\pm 1, \ldots, \pm 1)$ (i.e. $\tilde{S} = DSD$). So switching-equivalent graphs have the same Seidel spectrum, and therefore the property of being strong with two Seidel eigenvalues is invariant under Seidel switching.

Suppose Γ is a strong graph on v vertices with two Seidel eigenvalues ρ_1 and ρ_2 (so $v - 1 + \rho_1\rho_2 = 0$). Clearly, Γ is regular of degree k if and only if its Seidel matrix has constant row sum $v - 1 - 2k$. Therefore $v - 1 - 2k = \rho_0$ is an eigenvalue of S, so either $\rho_0 = \rho_1$, or $\rho_0 = \rho_2$. Switching in Γ produces another strong graph, which may or may not be regular. If it is regular, then it is regular of degree either $(v - 1 - \rho_1)/2$ or $(v - 1 - \rho_2)/2$.

Examples (i) If Γ is P_3, then the Seidel eigenvalues are -1 and 2, so a regular graph that is switching equivalent must have degree either $3/2$ or 0. The former is impossible, but the latter happens: P_3 is switching equivalent to $3K_1$.

(ii) If Γ is $C_5 + K_1$, then the eigenvalues are $\pm\sqrt{5}$, and so can never be equal to the row sum. So this graph cannot be switched into a regular one.

(iii) If Γ is the 4×4 grid (the lattice graph $L_2(4)$), then $v = 16$ and $\rho_0 = \rho_1 = 3$, $\rho_2 = -5$. So Γ is strong with two eigenvalues. Switching in Γ with respect to a coclique of size 4 again gives a regular graph with the same parameters as Γ, but which is not isomorphic to Γ. This is the Shrikhande graph (see §9.2). Switching with respect to the union of two parallel lines in the grid (that is, two disjoint 4-cliques in Γ) gives a regular graph of degree 10, the Clebsch graph (see §9.2).

Strong graphs were introduced by SEIDEL [315].

10.2 Two-graphs

A *two-graph* $\Omega = (V, \Delta)$ consists of a finite set V, together with a collection Δ of unordered triples from V such that every 4-subset of V contains an even number of triples from Δ. The triples from Δ are called *coherent*.

From a graph $\Gamma = (V,E)$, one can construct a two-graph $\Omega = (V,\Delta)$ by defining a triple from V to be coherent if the three vertices induce a subgraph in Γ with an odd number of edges. It is easily checked that out of the four triples in any graph on four vertices, 0, 2, or 4 are coherent. So Ω is a two-graph. We call Ω the two-graph *associated* to Γ.

Observe that Seidel switching does not change the parity of the number of edges in any three-vertex subgraph of Γ. Therefore switching-equivalent graphs have the same associated two-graph. Conversely, from any two-graph $\Omega = (V,\Delta)$ one can construct a graph Γ as follows. Take $\omega \in V$. Define two vertices $x,y \in V \setminus \{\omega\}$ to be adjacent in Γ if $\{\omega,x,y\} \in \Delta$, and define ω to be an isolated vertex of Γ. We claim that every triple $\{x,y,z\} \in \Delta$ has an odd number of edges in Γ, which makes Ω the two-graph associated to Γ. If $\omega \in \{x,y,z\}$ this is clear. If $\omega \notin \{x,y,z\}$, the 4-subgraph condition implies that $\{x,y,z\} \in \Delta$ whenever from the triples $\{\omega,y,z\}$, $\{\omega,x,y\}$, $\{\omega,x,z\}$ just one or all three are coherent. Hence $\{x,y,x\}$ has one or three edges in Γ. Thus we have established a one-to-one correspondence between two-graphs and switching classes of graphs.

Small two-graphs were enumerated in [78]. The number of nonisomorphic two-graphs on n vertices for small n is

n	0	1	2	3	4	5	6	7	8	9	10
#	1	1	1	2	3	7	16	54	243	2038	33120

There is an explicit formula for arbitrary n. See, e.g., [268].

For the graph Γ with an isolated vertex ω, obtained from Ω as indicated above, the graph $\Gamma \setminus \omega$ plays an important role. It is called the *descendant* of Ω with respect to ω, and will be denoted by Γ_ω.

Since switching-equivalent graphs have the same Seidel spectrum, we can define the eigenvalues of a two-graph to be the Seidel eigenvalues of any graph in the corresponding switching class.

SEIDEL & TSARANOV [319] classified the two-graphs with smallest Seidel eigenvalue not less than -3:

Theorem 10.2.1 *(i) A graph Γ with smallest Seidel eigenvalue larger than -3 is switching equivalent to the void graph on n vertices, to the one-edge graph on n vertices, or to one of the following $2 + 3 + 5$ graphs on $5, 6, 7$ vertices, respectively:*

(ii) A graph Γ with smallest Seidel eigenvalue not less than -3 is switching equivalent to a subgraph of mK_2 or $\overline{T(8)}$. □

10.3 Regular two-graphs

A two-graph (V, Δ) is called *regular* (of degree a) if every unordered pair from V is contained in exactly a triples from Δ. Suppose $\Omega = (V, \Delta)$ is a two-graph, and let ∇ be the set of noncoherent triples. Then it easily follows that $\overline{\Omega} = (V, \nabla)$ is also a two-graph, called the *complement* of Ω. Moreover, Ω is regular of degree a if and only if the complement $\overline{\Omega}$ is regular of degree $\bar{a} = v - 2 - a$. The following result relates regular two-graphs with strong graphs and strongly regular graphs.

Theorem 10.3.1 *For a graph Γ with v vertices, its associated two-graph Ω, and any descendant Γ_ω of Ω the following are equivalent.*

  (i) Γ *is strong with two Seidel eigenvalues ρ_1 and ρ_2.*
  (ii) Ω *is regular of degree a.*
  (iii) Γ_ω *is (possibly improper) strongly regular with parameters $(v - 1, k, \lambda, \mu)$ with $\mu = k/2$.*

The parameters are related by $v = 1 - \rho_1 \rho_2$, $a = k = 2\mu = -(\rho_1 + 1)(\rho_2 + 1)/2$, and $\lambda = (3k - v)/2 = 1 - (\rho_1 + 3)(\rho_2 + 3)/4$. The restricted Seidel eigenvalues of Γ_ω are ρ_1 and ρ_2, and $\rho_1 + \rho_2 = v - 2a - 2 = \bar{a} - a$.

Proof. $(ii) \Rightarrow (iii)$: Let x be a vertex of Γ_ω. The number of coherent triples containing ω and x equals the number of edges in Γ_ω containing x, so Γ_ω is regular of degree a. For two vertices x and y in Γ_ω, let $p(x, y)$ denote the number of vertices z $(z \neq x, y)$ adjacent to x but not to y. If x and y are distinct and nonadjacent, then $p(x, y) + p(y, x) = a$, and the number μ of common neighbors of x and y equals $k - p(x, y) = k - p(y, x)$. Therefore $\mu = k/2 = a/2$ is independent of x and y. Similarly, if x and y are adjacent, then $p(x, y) + p(y, x) = \bar{a}$ (the degree of the complement), and the number λ of common neighbors of x and y equals $k - 1 - p(x, y) = k - 1 - p(y, x)$, which implies $\lambda = (3k - v)/2$, which is independent of x and y.

$(iii) \Rightarrow (ii)$: If Γ_ω is strongly regular and $k = 2\mu$, then Theorem 9.1.3 gives $\lambda = (3k - v)/2$. With the relations above this shows that Ω is regular of degree k.

$(i) \Rightarrow (iii)$: Switch in Γ with respect to the neigbors of ω, then ω becomes isolated, and $\Gamma \setminus \omega = \Gamma_\omega$. If S_ω is the Seidel matrix of Γ_ω, then

$$S = \begin{bmatrix} 0 & \mathbf{1}^\top \\ \mathbf{1} & S_\omega \end{bmatrix}$$

is the Seidel matrix of Γ. We know $(S - \rho_1 I)(S - \rho_2 I) = 0$. This gives $(S_\omega - \rho_1 I)(S_\omega - \rho_2 I) = -J$. Therefore Γ_ω is strongly regular with restricted Seidel eigenvalues ρ_1 and ρ_2 and $v - 1 = -\rho_1 \rho_2$ vertices. From $S = J - 2A - I$ we get the adjacency eigenvalues $r = -(\rho_1 + 1)/2$ and $s = -(\rho_2 + 1)/2$ of Γ_ω. Now the parameters of Γ_ω follow from Theorem 9.1.3.

$(iii) \Rightarrow (i)$: Suppose Γ_ω is strongly regular with $k = 2\mu$ and Seidel matrix S_ω. Then it follows readily that $S_\omega \mathbf{1} = (\rho_1 + \rho_2)\mathbf{1}$ and $(S_\omega - \rho_1 I)(S_\omega - \rho_2 I) = -J$. This implies that S satisfies $(S - \rho_1 I)(S - \rho_2 I) = 0$. $\square$

Small regular two-graphs have been classified. The table below gives the numbers of nonisomorphic nontrivial regular two-graphs with $\rho_1 = -3$ or $\rho_1 = -5$ or $v \leq 50$.

v	6	10	14	16	18	26	28	30	36
ρ_1, ρ_2	$\pm\sqrt{5}$	± 3	$\pm\sqrt{13}$	$-3,5$	$\pm\sqrt{17}$	± 5	$-3,9$	$\pm\sqrt{29}$	$-5,7$
#	1	1	1	1	1	4	1	6	227

v	38	42	46	50	76	96	126	176	276
ρ_1, ρ_2	$\pm\sqrt{37}$	$\pm\sqrt{41}$	$\pm\sqrt{45}$	± 7	$-5,15$	$-5,19$	$-5,25$	$-5,35$	$-5,55$
#	≥ 191	≥ 18	≥ 97	≥ 54	?	?	1	1	1

10.3.1 Related strongly regular graphs

Given the parameters of a regular two-graph Ω, we find three parameter sets for strongly regular graphs that may be related, namely the parameter set of the descendants and the two possible parameter sets for regular graphs in the switching class of Ω. The parameters are given by:

Proposition 10.3.2 *(i) Let Γ be strongly regular with parameters (v, k, λ, μ). The associated two-graph Ω is regular if and only if $v = 2(2k - \lambda - \mu)$. If this is the case, then it has degree $a = 2(k - \mu)$, and Γ_ω is strongly regular with parameters $(v - 1, 2(k - \mu), k + \lambda - 2\mu, k - \mu)$.*
(ii) Conversely, if Γ is regular of valency k, and the associated two-graph Ω is regular of degree a, then Γ is strongly regular with parameters $\lambda = k - (v - a)/2$ and $\mu = k - a/2$, and k satisfies the quadratic $2k^2 - (v + 2a)k + (v - 1)a = 0$.

Proof (i) By definition, Ω is regular of degree a if and only if $a = \lambda + (v - 2k + \lambda) = 2(k - \mu)$. The parameters follow immediately.
(ii) The quadratic expresses that $k - \frac{1}{2}v \in \{r, s\}$.   □

In the case of the regular two-graph on 6 vertices, the descendants are pentagons, and there are no regular graphs in the switching class.

In the case of the regular two-graph on 10 vertices, the descendants are grid graphs 3×3. The switching class contains both the Petersen graph and its complement. Therefore Ω is isomorphic to its complement (and so are the descendants).

In the case of the regular two-graph on 16 vertices, the descendants are isomorphic to the triangular graph $T(6)$ (with parameters $(15, 8, 4, 4)$ and spectrum $8^1\, 2^5$ $(-2)^9$). The switching class contains the grid graph 4×4 and the Shrikhande graph (both with parameters $(16, 6, 2, 2)$ and spectrum $6^1\, 2^6\, (-2)^9$), and the Clebsch graph (with parameters $(16, 10, 6, 6)$ and spectrum $10^1\, 2^5\, (-2)^{10}$).

It remains to specify what switching sets are needed to switch between two strongly regular graphs associated to the same regular two-graph.

Proposition 10.3.3 *Let Γ be strongly regular with parameters (v, k, λ, μ), associated with a regular two-graph.*

(i) *The graph Γ is switched into a strongly regular graph with the same parameters if and only if every vertex outside the switching set S is adjacent to half of the vertices of S.*

(ii) *The graph Γ is switched into a strongly regular graph with parameters $(v, k + c, \lambda + c, \mu + c)$ where $c = \frac{1}{2}v - 2\mu$ if and only if the switching set S has size $\frac{1}{2}v$ and is regular of valency $k - \mu$.* $\qquad\square$

For example, in order to switch the 4×4 grid graph into the Shrikhande graph, we can switch with respect to a 4-coclique. And in order to switch the 4×4 grid graph into the Clebsch graph, we need a split into two halves that are regular with valency 4, and the union of two disjoint K_4's works.

Regular two-graphs were introduced by Graham Higman and further investigated by TAYLOR [335].

10.3.2 The regular two-graph on 276 points

If N is the point-block incidence matrix of the unique Steiner system $S(4,7,23)$, then $NN^\top = 56I + 21J$, $NJ = 77J$, $JN = 7J$. Since any two blocks in this Steiner system meet in one or three points, we have $N^\top N = 7I + A + 3(J - I - A)$ where A describes the relation of meeting in one point. As we already saw in §9.1.10, A is the adjacency matrix of a strongly regular graph—in this case one with parameters $(v, k, \lambda, \mu) = (253, 112, 36, 60)$ and spectrum $112^1\ 2^{230}\ (-26)^{22}$. The Seidel matrix $S = J - I - 2A$ has spectrum $28^1\ (-5)^{230}\ 51^{22}$ and satisfies $(S - 51I)(S + 5I) = -3J$. Now $S' = \begin{pmatrix} J-I & J-2N \\ J-2N^\top & S \end{pmatrix}$ satisfies $(S' - 55I)(S' + 5I) = 0$ and hence is the Seidel matrix of a regular two-graph on 276 vertices. This two-graph is unique (GOETHALS & SEIDEL [180]). Its group of automorphisms is Co_3, acting 2-transitively.

10.3.3 Coherent subsets

A *clique*, or *coherent* subset, in a two-graph $\Omega = (V, \Delta)$ is a subset C of V such that all triples in C are coherent. If $x \notin C$, then x determines a partition $\{C_x, C'_x\}$ of C into two possibly empty parts such that a triple xyz with $y, z \in C$ is coherent precisely when y and z belong to the same part of the partition.

Proposition 10.3.4 (TAYLOR [337]) *Let C be a nonempty coherent subset of the regular two-graph Ω with eigenvalues ρ_1, ρ_2, where $\rho_2 < 0$. Then*

    (i) $|C| \le 1 - \rho_2$, *with equality iff for each $x \notin C$ we have $|C_x| = |C'_x|$,*
and

    (ii) $|C| \le m(\rho_2)$.

Proof   (i) Let $c = |C|$. Counting incoherent triples that meet C in two points, we find $\frac{1}{2}c(c-1)\bar{a} = \sum_{x \notin C} |C_x| \cdot |C'_x| \le \sum_{x \notin C} (c/2)^2 = \frac{1}{4}c^2(v-c)$. It follows that $c^2 - (v - 2\bar{a})c - 2\bar{a} \le 0$. But the two roots of $x^2 - (v - 2\bar{a})x - 2\bar{a} = 0$ are $1 - \rho_1$ and $1 - \rho_2$, and hence $1 - \rho_1 \le c \le 1 - \rho_2$.

(ii) This follows by making a system of equiangular lines in $\mathbb{R}^m$ as in §10.6.1 corresponding to the complement of Ω. We can choose unit vectors for the points in C such that their images form a simplex (any two have the same inner product), and hence $|C|$ is bounded by the dimension $m = v - m(\rho_1) = m(\rho_2)$. $\square$

10.3.4 Completely regular two-graphs

In a regular two-graph, each pair is in $a_2 = a$ coherent triples, that is, in a_2 3-cliques, and each coherent triple is in a_3 4-cliques, where a_3 is the number of common neighbors of two adjacent vertices in any strongly regular graph Γ_ω, so that $a_3 = -\frac{1}{4}(\rho_1 + 3)(\rho_2 + 3) + 1$ by Theorem 10.3.1.

Let a *t-regular two-graph* be a regular two-graph in which every i-clique is contained in a nonzero constant number a_i of $(i+1)$-cliques, for $2 \le i \le t$. By Proposition 10.3.4 we must have $t \le -\rho_2$. A *completely regular two-graph* is a t-regular two-graph with $t = -\rho_2$. For example, the regular two-graph on 276 points (§10.3.2) is completely regular. NEUMAIER [287] introduced this concept and gave parameter restrictions strong enough to leave only a finite list of feasible parameters. There are five examples, and two open cases.

#	ρ_1	ρ_2	v	a_2	a_3	a_4	a_5	a_6	a_7	Existence
1	3	−3	10	4	1					unique [314]
2	5	−3	16	6	1					unique [314]
3	9	−3	28	10	1					unique [314]
4	7	−5	36	16	6	2	1			unique (BH)
5	19	−5	96	40	12	2	1			none (NP)
6	25	−5	126	52	15	2	1			none [287]
7	55	−5	276	112	30	2	1			unique [180]
8	21	−7	148	66	25	8	3	2	1	none [287]
9	41	−7	288	126	45	12	3	2	1	none [32]
10	161	−7	1128	486	165	36	3	2	1	?
11	71	−9	640	288	112	36	10	4	3	none (BH)
12	351	−9	3160	1408	532	156	30	4	3	?
13	253	−11	2784	1270	513	176	49	12	5	none [287]

Table 10.1 Parameters of completely regular two-graphs

Here (BH) refers to an unpublished manuscript by Blokhuis and Haemers, while (NP) is the combination of NEUMAIER [287] who showed that a derived graph on 95 vertices must be locally $GQ(3,3)$, and PASECHNIK [289] who classified such graphs and found none on 95 vertices.

10.4 Conference matrices

The Seidel matrix of $C_5 + K_1$ is an example of a so-called *conference matrix*. An $n \times n$ matrix S is a *conference matrix* if all diagonal entries are 0, the off-diagonal entries are ± 1, and $SS^{\top} = (n-1)I$.

Multiplying a row or column by -1 (switching) does not affect the conference matrix property. It was shown in [140] that any conference matrix can be switched into a form where it is either symmetric or skew-symmetric:

Lemma 10.4.1 *Let S be a conference matrix of order n with $n > 2$. Then n is even and one can find diagonal matrices D and E with diagonal entries ± 1 such that $(DSE)^{\top} = DSE$ if and only if $n \equiv 2$ (mod 4). One can find such D and E with $(DSE)^{\top} = -DSE$ if and only if $n \equiv 0$ (mod 4).*

Proof Switch rows and columns so as to make all nondiagonal entries of the first row and column equal to 1. The second row now has $n/2$ entries 1 and equally many entries -1 (since it has inner product zero with the first row). So, n is even, say $n = 2m + 2$. Let there be a, b, c, d entries $1, -1, 1, -1$ in the third row below the entries $1, 1, -1, -1$ of the second row, respectively. We may assume (by switching the first column and all rows except the first if required) that $S_{23} = 1$. If $S_{32} = 1$, then $a + b = m - 1$, $c + d = m$, $a + c + 1 = m$, $a - b - c + d + 1 = 0$ imply $a + 1 = b = c = d = \frac{1}{2}m$ so that m is even. If $S_{32} = -1$, then $a + b = m - 1$, $c + d = m$, $a + c = m$, $a - b - c + d + 1 = 0$ imply $a = b = c - 1 = d = \frac{1}{2}(m - 1)$ so that m is odd. This proves that after switching the first row and column to 1, the matrix S has become symmetric in case $n \equiv 2$ (mod 4), while after switching the first row to 1 and the first column to -1, the matrix S has become skew-symmetric in case $n \equiv 0$ (mod 4). $\square$

Thus, if $n \equiv 2$ (mod 4), S gives rise to a strong graph with two eigenvalues, and its associated two-graph is regular of degree $(n-2)/2$. The descendants are strongly regular with parameters $(n-1, (n-2)/2, (n-6)/4, (n-2)/4)$. We call these graphs *conference graphs*. Conference graphs are characterized among the strongly regular graphs by $f = g$ (f and g are the multiplicities of the restricted eigenvalues), and are the only cases in which nonintegral eigenvalues can occur.

The following condition is due to BELEVITCH [24].

Theorem 10.4.2 *If n is the order of a symmetric conference matrix, then $n - 1$ is the sum of two integral squares.*

Proof. $CC^{\top} = (n-1)I$ implies that I and $(n-1)I$ are rationally congruent (two matrices A and B are rationally congruent if there exists a rational matrix R such that $RAR^{\top} = B$). A well-known property (essentially Lagrange's four squares theorem) states that for every positive rational number α, the 4×4 matrix αI_4 is rationally congruent to I_4. This implies that the $n \times n$ matrix αI_n is rationally congruent to $\mathrm{diag}(1, \ldots, 1, \alpha, \ldots, \alpha)$ where the number of ones is divisible by 4. Since $n \equiv 2$ (mod 4), I must be rationally congruent to $\mathrm{diag}(1, \ldots, 1, n-1, n-1)$. This implies that $n - 1$ is the sum of two squares. $\square$

Note that this theorem also gives a necessary condition for the existence of conference graphs. For example, 21 is not the sum of two squares, therefore there exists no conference matrix of order 22, and no strongly regular graph with parameters $(21, 10, 4, 5)$.

For many values of n, conference matrices are known to exist, see for example [178]. The following construction, where $n - 1$ is an odd prime power, is due to PALEY [288]. Let S_ω be a matrix whose rows and columns are indexed by the elements of a finite field $\mathbb{F}_q$ of order q, q odd. by $(S_\omega)_{i,j} = \chi(i - j)$, where χ is the quadratic residue character (that is, $\chi(0) = 0$ and $\chi(x) = 1$ if x is a square, and -1 if x is not a square). It follows that S is symmetric if $q \equiv 1 \pmod 4$, and S is skew symmetric if $q \equiv 3 \pmod 4$. In both cases

$$S = \begin{bmatrix} 0 & \mathbf{1}^\top \\ \mathbf{1} & S_\omega \end{bmatrix}$$

is a conference matrix. If $n \equiv 2 \pmod 4$, S represents a regular two-graph and all its descendants are isomorphic. They are the *Paley graphs*, which we already encountered in §9.1.2.

10.5 Hadamard matrices

Closely related to conference matrices are Hadamard matrices. A matrix H of order n is called a *Hadamard matrix* if every entry is 1 or -1, and $HH^\top = nI$. If H is a Hadamard matrix, then so is $H^\top$. If a row or a column of a Hadamard matrix is multiplied by -1, the matrix remains a Hadamard matrix. The *core* of a Hadamard matrix H (with respect to the first row and column) is the matrix C of order $n - 1$ obtained by first multiplying rows and columns of H by ± 1 so as to obtain a Hadamard matrix of which the first row and column consist of ones only, and then deleting the first row and column. Now all entries of C are ± 1, and we have $CC^\top = C^\top C = nI - J$, and $C\mathbf{1} = C^\top \mathbf{1} = -\mathbf{1}$. This implies that the $(0, 1)$ matrix $N = \frac{1}{2}(C + J)$ satisfies $N^\top \mathbf{1} = (\frac{1}{2}n - 1)\mathbf{1}$ and $NN^\top = \frac{1}{4}nI + (\frac{1}{4}n - 1)J$, so that, for $n > 2$, N is the incidence matrix of a symmetric $2\text{-}(n - 1, \frac{1}{2}n - 1, \frac{1}{4}n - 1)$ design. Conversely, if N is the incidence matrix of a 2-design with these parameters, then $2N - J$ is the core of a Hadamard matrix. Note that the design parameters imply that n is divisible by 4 if $n > 2$. The famous Hadamard conjecture states that this condition is sufficient for existence of a Hadamard matrix of order n. Many constructions are known (see below), but the conjecture is still far from being solved.

A Hadamard matrix H is *regular* if H has constant row and column sum (ℓ say). Now $-H$ is a regular Hadamard matrix with row sum $-\ell$. From $HH^\top = nI$ we get that $\ell^2 = n$, so $\ell = \pm\sqrt{n}$, and n is a square. If H is a regular Hadamard matrix with row sum ℓ, then $N = \frac{1}{2}(H + J)$ is the incidence matrix of a symmetric $2\text{-}(n, (n + \ell)/2, (n + 2\ell)/4)$ design. Conversely, if N is the incidence matrix of a 2-design with these parameters (a *Menon design*), then $2N - J$ is a regular Hadamard matrix.

A Hadamard matrix H is *graphical* if it is symmetric with constant diagonal. Without loss of generality we assume that the diagonal elements are 1 (otherwise we replace H by $-H$). If H is a graphical Hadamard matrix of order n, then $S = H - I$ is the Seidel matrix of a strong graph Γ with two Seidel eigenvalues: $-1 \pm \sqrt{n}$. In other words, Γ is in the switching class of a regular two-graph. The descendant of Γ with respect to some vertex has Seidel matrix $C - I$, where C is the corresponding core of H. It is a strongly regular graph with parameters $(v, k, \lambda, \mu) = (n - 1, \frac{1}{2}n - 1, \frac{1}{4}n - 1, \frac{1}{4}n - 1)$. From $\operatorname{tr} S = 0$ it follows that also for a graphical Hadamard matrix n is a square. If, in addition, H is regular with row sum $\ell = \pm\sqrt{n}$, then Γ is a strongly regular graph with parameters $(n, (n - \ell)/2, (n - 2\ell)/4, (n - 2\ell)/4)$. And conversely, a strongly regular graph with one of the above parameter sets gives rise to a Hadamard matrix of order n.

There is an extensive literature on Hadamard matrices. See, e.g., [312, 313, 109].

10.5.1 Constructions

There is a straightforward construction of Hadamard matrices from conference matrices. If S is a skew-symmetric conference matrix, then $H = S + I$ is a Hadamard matrix, and if S is a symmetric conference matrix, then

$$H = \begin{bmatrix} S+I & S-I \\ S-I & -S-I \end{bmatrix}$$

is a Hadamard matrix. Thus the conference matrices constructed in the previous section give Hadamard matrices of order $n = 4m$ if $4m - 1$ is a prime power, and if m is odd and $2m - 1$ is a prime power. Some small Hadamard matrices are:

$$\begin{bmatrix} 1 & 1 \\ 1 & -1 \end{bmatrix}, \quad \begin{bmatrix} 1 & 1 & 1 & -1 \\ 1 & 1 & -1 & 1 \\ 1 & -1 & 1 & 1 \\ -1 & 1 & 1 & 1 \end{bmatrix}, \quad \text{and} \quad \begin{bmatrix} 1 & -1 & -1 & -1 \\ -1 & 1 & -1 & -1 \\ -1 & -1 & 1 & -1 \\ -1 & -1 & -1 & 1 \end{bmatrix}.$$

Observe that the two Hadamard matrices of order 4 are regular and graphical. One easily verifies that if H_1 and H_2 are Hadamard matrices, then so is the Kronecker product $H_1 \otimes H_2$. Moreover, if H_1 and H_2 are regular with row sums ℓ_1 and ℓ_2, respectively, then $H_1 \otimes H_2$ is regular with row sum $\ell_1 \ell_2$. Similarly, the Kronecker product of two graphical Hadamard matrices is graphical again. With the small Hadamard matrices given above, we can make Hadamard matrices of order $n = 2^t$ and regular graphical Hadamard matrices of order $n = 4^t$ with row sum $\ell = \pm 2^t$.

Let *RSHCD* be the set of pairs (n, ε) such that there exists a regular symmetric Hadamard matrix H with row sums $\ell = \varepsilon\sqrt{n}$ and constant diagonal, with diagonal entries 1. If $(m, \delta), (n, \varepsilon) \in RSHCD$, then $(mn, \delta\varepsilon) \in RSHCD$.

We mention some direct constructions:

(i) $(4, \pm 1), (36, \pm 1), (100, \pm 1), (196, \pm 1) \in RSHCD$.

(ii) If there exists a Hadamard matrix of order m, then $(m^2, \pm 1) \in RSHCD$.

(iii) If both $a - 1$ and $a + 1$ are odd prime powers, then $(a^2, 1) \in RSHCD$.

(iv) If $a + 1$ is a prime power and there exists a symmetric conference matrix of order a, then $(a^2, 1) \in RSHCD$.

(v) If there is a set of $t - 2$ mutually orthogonal Latin squares of order $2t$, then $(4t^2, 1) \in RSHCD$.

(vi) $(4t^4, \pm 1) \in RSHCD$.

See [179], [62] and [312], §5.3. For the third part of (i), see [236]. For the fourth part of (i), cf. [179], Theorem 4.5 (for $k = 7$) and [231]. For (ii), cf. [179], Theorem 4.4, and [201]. For (iii), cf. [312], Corollary 5.12. For (iv), cf. [312], Corollary 5.16. For (v), consider the corresponding Latin square graph. For (vi), see [209].

10.6 Equiangular lines

10.6.1 Equiangular lines in $\mathbb{R}^d$ and two-graphs

Seidel (cf. [250, 258, 141]) studied systems of lines in Euclidean space $\mathbb{R}^d$, all passing through the origin 0, with the property that any two make the same angle φ. The cases $\varphi = 0$ (only one line) and $\varphi = \frac{\pi}{2}$ (at most d lines, mutually orthogonal) being trivial, we assume $0 < \varphi < \frac{\pi}{2}$. Let $\alpha = \cos\varphi$, so that $0 < \alpha < 1$. Choose for each line ℓ_i a unit vector x_i on ℓ_i (determined up to sign). Then $x_i^\top x_i = 1$ for each i, and $x_i^\top x_j = \pm\cos\varphi = \pm\alpha$ for $i \neq j$.

For the Gram matrix G of the vectors x_i this means that $G = I + \alpha S$, where S is the *Seidel adjacency matrix* of a graph Γ. (That is, S is symmetric with zero diagonal, and has entries -1 and 1 for adjacent and nonadjacent vertices, respectively.) Note that changing the signs of some of the x_i corresponds to *Seidel switching* of Γ.

Conversely, let S be the Seidel adjacency matrix of a graph on at least two vertices, and let θ be the smallest eigenvalue of S. (Then $\theta < 0$ since $S \neq 0$ and $\operatorname{tr} S = 0$.) Now $S - \theta I$ is positive semidefinite, and $G = I - \frac{1}{\theta} S$ is the Gram matrix of a set of vectors in $\mathbb{R}^d$, where $d = \operatorname{rk}(S - \theta I) = n - m(\theta)$ where n is the number of vertices of the graph, and $m(\theta)$ the multiplicity of θ as eigenvalue of S.

We see that there is a 1-1 correspondence between dependent equiangular systems of n lines and two-graphs on n vertices, and more precisely between equiangular systems of n lines spanning $\mathbb{R}^d$ (with $d < n$) and two-graphs on n vertices such that the smallest eigenvalue has multiplicity $n - d$.

Thus, in order to find large sets of equiangular lines, one has to find large graphs where the smallest Seidel eigenvalue has large multiplicity (or, rather, small comultiplicity).

10.6.2 Bounds on equiangular sets of lines in $\mathbb{R}^d$ or $\mathbb{C}^d$

An upper bound for the size of an equiangular system of lines (and hence an upper bound for the multiplicity of the smallest Seidel eigenvalue of a graph) is given by the so-called *absolute bound*, due to M. Gerzon (cf. [250]):

Theorem 10.6.1 ("Absolute bound") *The cardinality n of a system of equiangular lines in Euclidean space $\mathbb{R}^d$ is bounded by $\frac{1}{2}d(d+1)$.*

Proof Let $X_i = x_i x_i^\top$ be the rank 1 matrix that is the projection onto the line ℓ_i. Then $X_i^2 = X_i$ and

$$\operatorname{tr} X_i X_j = (x_i^\top x_j)^2 = \begin{cases} 1 & \text{if } i = j \\ \alpha^2 & \text{otherwise.} \end{cases}$$

We prove that the matrices X_i are linearly independent. Since they are symmetric, that will show that there are at most $\frac{1}{2}d(d+1)$. So, suppose that $\sum c_i X_i = 0$. Then $\sum_i c_i X_i X_j = 0$ for each j, so that $c_j(1 - \alpha^2) + \alpha^2 \sum c_i = 0$ for each j. This means that all c_j are equal, and since $\sum c_i = \operatorname{tr} \sum c_i X_i = 0$, they are all zero. □

In $\mathbb{C}^d$ one can study lines (1-spaces) in the same way, choosing a spanning unit vector in each and agreeing that $\langle x \rangle$ and $\langle y \rangle$ make angle $\phi = \arccos \alpha$ where $\alpha = |x^*y|$. (Here x^* stands for $\bar{x}^\top$.) The same argument now proves

Proposition 10.6.2 *The cardinality n of a system of equiangular lines in $\mathbb{C}^d$ is bounded by d^2.* □

There are very few systems of lines in $\mathbb{R}^d$ that meet the absolute bound, but it is conjectured that systems of d^2 equiangular lines in $\mathbb{C}^d$ exist for all d. Such systems are known for $d = 1, 2, 3, 4, 5, 6, 7, 8, 19$ ([355, 222, 223, 182, 11]). In quantum information theory, they are known as SICPOVMs.

The special bound gives an upper bound for n in terms of the angle ϕ, or an upper bound for ϕ (equivalently, a lower bound for $\alpha = \cos \phi$) in terms of n.

Proposition 10.6.3 ("Special bound") *If there is a system of $n > 1$ lines in $\mathbb{R}^d$ or $\mathbb{C}^d$ such that the cosine of the angle between any two lines is at most α, then $\alpha^2 \geq (n-d)/(n-1)d$, or, equivalently, $n \leq d(1 - \alpha^2)/(1 - \alpha^2 d)$ if $1 - \alpha^2 d > 0$.*

Proof Let x_i $(1 \leq i \leq n)$ be unit vectors in $\mathbb{R}^d$ or $\mathbb{C}^d$ with $|x_i^* x_j| \leq \alpha$ for $i \neq j$. Put $X_i = x_i x_i^*$ and $Y = \sum_i X_i - \frac{n}{d}I$. Then $\operatorname{tr} X_i X_j = |x_i^* x_j|^2 \leq \alpha^2$ for $i \neq j$, and $\operatorname{tr} X_i = \operatorname{tr} X_i^2 = 1$. Now $0 \leq \operatorname{tr} YY^* \leq n(n-1)\alpha^2 + n - \frac{n^2}{d}$. □

Complex systems of lines with equality in the special bound are known as *equiangular tight frames*. There is a lot of recent literature.

If equality holds in the absolute bound, then the X_i span the vector space of all symmetric matrices, and in particular I is a linear combination of the X_i. If equality holds in the special bound, the same conclusion follows. In both cases, the following proposition shows (in the real case) that the graph Γ belongs to a regular two-graph.

Proposition 10.6.4 *Suppose x_i $(1 \leq i \leq n)$ are unit vectors in $\mathbb{R}^d$ or $\mathbb{C}^d$ with $|x_i^* x_j| = \alpha$ for $i \neq j$, where $0 < \alpha < 1$. Put $X_i = x_i x_i^*$ and suppose that there are constants c_i such that $I = \sum c_i X_i$. Then $c_i = d/n$ for all i and $n = d(1 - \alpha^2)/(1 - \alpha^2 d)$.*

If the x_i are vectors in $\mathbb{R}^d$, and G is the Gram matrix of the x_i, and $G = I + \alpha S$, then S has eigenvalues $(n - d)/(\alpha d)$ and $-1/\alpha$ with multiplicities d and $n - d$, respectively. If $n > d + 1$ and $n \neq 2d$, then these eigenvalues are odd integers.

Proof If $I = \sum c_i X_i$ then $X_j = \sum_i c_i X_i X_j$ for each j, so $c_j(1 - \alpha^2) + \alpha^2 \sum c_i = 1$ for each j. This means that all c_j are equal, and since $\sum c_i = \operatorname{tr} \sum c_i X_i = \operatorname{tr} I = d$, they all equal d/n. Our equation now becomes $(d/n)(1 - \alpha^2) + \alpha^2 d = 1$, so $n = d(1 - \alpha^2)/(1 - \alpha^2 d)$.

If F is the $d \times n$ matrix whose columns are the vectors x_i, then $G = F^\top F$, and $FF^\top = \sum x_i x_i^\top = \sum X_i = (n/d)I$. It follows that $FF^\top$ has eigenvalue n/d with multiplicity d, and $G = F^\top F$ has the same eigenvalues, and in addition 0 with multiplicity $n - d$. The spectrum of S follows. If the two eigenvalues of the integral matrix S are not integers, they are conjugate algebraic integers, and then have the same multiplicity, so $n = 2d$. Since $S = J - I - 2A$, the eigenvalues of S, when integral, are odd. $\square$

Graphs for which the Seidel adjacency matrix S has only two eigenvalues are strong (cf. §10.1, Proposition 10.1.1) and belong to the switching class of a regular two-graph (Theorem 10.3.1).

The known lower and upper bounds for the maximum number of equiangular lines in $\mathbb{R}^d$ are given in the table below. For these bounds, see VAN LINT & SEIDEL [258], LEMMENS & SEIDEL [250], and SEIDEL [318] (p. 884).

d	1	2	3	4	5	6	7–14	15	16	17–18
$N_{\max}$	1	3	6	6	10	16	28	36	40	48

d	19	20	21	22	23–42	43
$N_{\max}$	72–76	90–96	126	176	276	344

Bounds for the size of systems of lines in $\mathbb{R}^d$ or $\mathbb{C}^d$ with only a few distinct, specified, angles, or just with a given total number of distinct angles, were given by DELSARTE, GOETHALS & SEIDEL [141].

10.6.3 Bounds on sets of lines with few angles and sets of vectors with few distances

In the case of equiangular lines the absolute value of the inner product took only one value. Generalizing that, one has

Theorem 10.6.5 ([141]) *For a set of n unit vectors in $\mathbb{R}^d$ such that the absolute value of the inner product between distinct vectors takes s distinct values different from 1, one has $n \leq \binom{d+2s-1}{d-1}$. If one of the inner products is 0, then $n \leq \binom{d+2s-2}{d-1}$.*

There are several examples of equality. For example, from the root system of E_8 one gets 120 lines in $\mathbb{R}^8$ with $|\alpha| \in \{0, \frac{1}{2}\}$.

Theorem 10.6.6 ([141]) *For a set of n unit vectors in $\mathbb{C}^d$ such that the absolute value of the inner product between distinct vectors takes s distinct values different from 1, one has $n \leq \binom{d+s-1}{d-1}^2$. If one of the inner products is 0, then $n \leq \binom{d+s-1}{d-1}\binom{d+s-2}{d-1}$.*

For example, there are systems of 40 vectors in $\mathbb{C}^4$ with $|\alpha| \in \{0, \frac{1}{3}\sqrt{3}\}$ and 126 vectors in $\mathbb{C}^6$ with $|\alpha| \in \{0, \frac{1}{2}\}$.

For sets of unit vectors instead of sets of lines it may be more natural to look at the inner product itself, instead of using the absolute value.

Theorem 10.6.7 ([142]) *For a set of n unit vectors in $\mathbb{R}^d$ such that the inner product between distinct vectors takes s distinct values, one has $n \leq \binom{d+s-1}{d-1} + \binom{d+s-2}{d-1}$. If the set is antipodal, then $n \leq 2\binom{d+s-2}{d-1}$.*

For example, in the antipodal case the upper bound is met with equality for $s = 1$ by a pair of vectors $\pm x$ (with $n = 2$), for $s = 2$ by the vectors $\pm e_i$ of a coordinate frame (with $n = 2d$), and for $s = 6$ by the set of shortest nonzero vectors in the Leech lattice in $\mathbb{R}^{24}$ (with inner products $-1, 0, \pm\frac{1}{4}, \pm\frac{1}{2}$ and size $n = 2\binom{28}{5}$).

In the general case the upper bound is met with equality for $s = 1$ by a simplex (with $n = d+1$). For $s = 2$ one has

d	2	5	6	22	23	3, 4, 7–21, 24–39
$N_{\max}$	5	16	27	275	276–277	$\frac{1}{2}d(d+1)$

with examples of equality in the bound $n \leq \frac{1}{2}d(d+3)$ for $d = 2, 6, 22$. The upper bounds for $d > 6$, $d \neq 22$, are due to Musin [283].

Corollary 10.6.8 ([142]) *Let Γ be a regular graph on n vertices, with smallest eigenvalue $\theta_{\min} < -1$ of multiplicity $n - d$. Then $n \leq \frac{1}{2}d(d+1) - 1$.*

(Earlier we saw for strongly regular graphs that $n \leq \frac{1}{2}f(f+3)$. Here $d = f+1$, so this gives the same bound, but applies to a larger class of graphs.)

Theorem 10.6.9 ([31]) *A set of vectors in $\mathbb{R}^d$ such that the distance between distinct vectors takes s values has size at most $\binom{d+s}{d}$.*

For $d \leq 8$, the maximal size of a 2-distance set in $\mathbb{R}^d$ was determined by Lisoněk [259]. The results are

d	1	2	3	4	5	6	7	8
$N_{\max}$	3	5	6	10	16	27	29	45

so that equality holds in the Blokhuis bound $\binom{d+2}{2}$ for $d = 1$ and $d = 8$.

The above gave generalizations of the absolute bound. There are also analogues of the special bound, see [141, 142].

Chapter 11
Association Schemes

11.1 Definition

An *association scheme with d classes* is a finite set X together with $d+1$ relations R_i on X such that

(i) $\{R_0, R_1, \ldots, R_d\}$ is a partition of $X \times X$;

(ii) $R_0 = \{(x,x) \mid x \in X\}$;

(iii) if $(x,y) \in R_i$, then also $(y,x) \in R_i$, for all $x, y \in X$ and $i \in \{0, \ldots, d\}$;

(iv) for any $(x,y) \in R_k$ the number p_{ij}^k of $z \in X$ with $(x,z) \in R_i$ and $(z,y) \in R_j$ depends only on i, j and k.

The numbers p_{ij}^k are called the *intersection numbers* of the association scheme. The above definition is the original definition of BOSE & SHIMAMOTO [39]; it is what DELSARTE [139] calls a symmetric association scheme. In Delsarte's more general definition, (iii) is replaced by:

(iii') for each $i \in \{0, \ldots, d\}$ there exists a $j \in \{0, \ldots, d\}$ such that $(x,y) \in R_i$ implies $(y,x) \in R_j$,

(iii'') $p_{ij}^k = p_{ji}^k$, for all $i, j, k \in \{0, \ldots, d\}$.

It is also very common to require just (i), (ii), (iii'), and (iv), and to call the scheme "commutative" when it also satisfies (iii''). Define $n = |X|$, and $n_i = p_{ii}^0$. Clearly, for each $i \in \{1, \ldots, d\}$, (X, R_i) is a simple graph which is regular of degree n_i.

Theorem 11.1.1 *The intersection numbers of an association scheme satisfy*

(i) $p_{0j}^k = \delta_{jk}$, $p_{ij}^0 = \delta_{ij} n_j$, $p_{ij}^k = p_{ji}^k$,

(ii) $\sum_i p_{ij}^k = n_j$, $\sum_j n_j = n$,

(iii) $p_{ij}^k n_k = p_{ik}^j n_j$,

(iv) $\sum_l p_{ij}^l p_{kl}^m = \sum_l p_{kj}^l p_{il}^m$.

Proof. Equations (i), (ii), and (iii) are straightforward. The expressions on both sides of (iv) count quadruples (w,x,y,z) with $(w,x) \in R_i$, $(x,y) \in R_j$, $(y,z) \in R_k$, for a fixed pair $(w,z) \in R_m$. $\qquad\square$

It is convenient to write the intersection numbers as entries of the so-called *inter-section matrices* $L_0, \ldots, L_d$:

$$(L_i)_{kj} = p_{ij}^k.$$

Note that $L_0 = I$ and $L_i L_j = \sum p_{ij}^k L_k$. From the definition it is clear that an association scheme with two classes is the same as a pair of complementary strongly regular graphs. If (X, R_1) is strongly regular with parameters (v, k, λ, μ), then the intersection matrices of the scheme are

$$L_1 = \begin{bmatrix} 0 & k & 0 \\ 1 & \lambda & k - \lambda - 1 \\ 0 & \mu & k - \mu \end{bmatrix}, \quad L_2 = \begin{bmatrix} 0 & 0 & v - k - 1 \\ 0 & k - \lambda - 1 & v - 2k + \lambda \\ 1 & k - \mu & v - 2k + \mu - 2 \end{bmatrix}.$$

11.2 The Bose-Mesner algebra

The relations R_i of an association scheme are described by their adjacency matrices A_i of order n defined by

$$(A_i)_{xy} = \begin{cases} 1 & \text{whenever } (x, y) \in R_i \\ 0 & \text{otherwise.} \end{cases}$$

In other words, A_i is the adjacency matrix of the graph (X, R_i). In terms of the adjacency matrices, the axioms (i)–(iv) become

(i) $\sum_{i=0}^d A_i = J$,
(ii) $A_0 = I$,
(iii) $A_i = A_i^\top$, for all $i \in \{0, \ldots, d\}$,
(iv) $A_i A_j = \sum_k p_{ij}^k A_k$, for all $i, j, k \in \{0, \ldots, d\}$.

From (i) we see that the $(0, 1)$ matrices A_i are linearly independent, and by use of (ii)–(iv) we see that they generate a commutative $(d + 1)$-dimensional algebra $\mathscr{A}$ of symmetric matrices with constant diagonal. This algebra was first studied by BOSE & MESNER [38] and is called the *Bose-Mesner algebra* of the association scheme.

Since the matrices A_i commute, they can be diagonalized simultaneously (see MARCUS & MINC [270]), that is, there exists a matrix S such that for each $A \in \mathscr{A}$, $S^{-1}AS$ is a diagonal matrix. Therefore $\mathscr{A}$ is semisimple and has a unique basis of minimal idempotents $E_0, \ldots, E_d$ (see BURROW [74]). These are matrices satisfying

$$E_i E_j = \delta_{ij} E_i, \quad \sum_{i=0}^d E_i = I.$$

The matrix $\frac{1}{n}J$ is a minimal idempotent (that it is an idempotent is clear, and that it is minimal follows since rk $J = 1$). We shall take $E_0 = \frac{1}{n}J$. Let P and $\frac{1}{n}Q$ be the matrices relating our two bases for $\mathscr{A}$:

$$A_j = \sum_{i=0}^{d} P_{ij} E_i, \; E_j = \frac{1}{n} \sum_{i=0}^{d} Q_{ij} A_i.$$

Then clearly

$$PQ = QP = nI.$$

It also follows that

$$A_j E_i = P_{ij} E_i,$$

which shows that the P_{ij} are the eigenvalues of A_j and that the columns of E_i are the corresponding eigenvectors. Thus $m_i = \mathrm{rk}\, E_i$ is the multiplicity of the eigenvalue P_{ij} of A_j (provided that $P_{ij} \neq P_{kj}$ for $k \neq i$). We see that $m_0 = 1$, $\sum_i m_i = n$, and $m_i = \mathrm{trace}\, E_i = n(E_i)_{jj}$ (indeed, E_i has only eigenvalues 0 and 1, so $\mathrm{rk}\, E_k$ equals the sum of the eigenvalues).

Theorem 11.2.1 *The numbers P_{ij} and Q_{ij} satisfy*

(i) $P_{i0} = Q_{i0} = 1$, $P_{0i} = n_i$, $Q_{0i} = m_i$,
(ii) $P_{ij}P_{ik} = \sum_{l=0}^{d} p_{jk}^{l} P_{il}$,
(iii) $m_i P_{ij} = n_j Q_{ji}$, $\sum_i m_i P_{ij} P_{ik} = nn_j \delta_{jk}$, $\sum_i n_i Q_{ij} Q_{ik} = nm_j \delta_{jk}$,
(iv) $|P_{ij}| \leq n_j$, $|Q_{ij}| \leq m_j$.

Proof Part (i) follows easily from $\sum_i E_i = I = A_0$, $\sum_i A_i = J = nE_0$, $A_i J = n_i J$, and $\mathrm{tr}\, E_i = m_i$. Part (ii) follows from $A_j A_k = \sum_l p_{jk}^{l} A_l$. The first equality in (iii) follows from $m_i P_{ij} = \mathrm{tr}\, A_j E_i = n_j Q_{ji}$, and the other two follow since $PQ = nI$. The first inequality of (iv) holds because the P_{ij} are eigenvalues of the n_j-regular graphs (X, R_j). The second inequality then follows from (iii). $\square$

Relations (iii) are often referred to as the *orthogonality relations*, since they state that the rows (and columns) of P (and Q) are orthogonal with respect to a suitable weight function.

An association scheme is called *primitive* if no union of the relations is a nontrivial equivalence relation. Or, equivalently, if no graph (X, R_i) with $i \neq 0$ is disconnected. For a primitive association scheme, (iv) above can be sharpened to $|P_{ij}| < n_j$ and $|Q_{ij}| < m_j$ for $j \neq 0$.

If $d = 2$, and (X, R_1) is strongly regular with parameters (v, k, λ, μ) and spectrum $k^1 \, r^f \, s^g$, the matrices P and Q are

$$P = \begin{bmatrix} 1 & k & v-k-1 \\ 1 & r & -r-1 \\ 1 & s & -s-1 \end{bmatrix}, \; Q = \begin{bmatrix} 1 & f & g \\ 1 & fr/k & gs/k \\ 1 & -f\frac{r+1}{v-k-1} & -g\frac{s+1}{v-k-1} \end{bmatrix}.$$

In general the matrices P and Q can be computed from the intersection numbers of the scheme:

Theorem 11.2.2 *For $i = 0, \ldots, d$, the intersection matrix L_j has eigenvalues P_{ij} ($0 \leq i \leq d$).*

Proof Theorem 11.2.1(ii) yields

$$\sum_{k,l} P_{il}(L_j)_{lk}(P^{-1})_{km} = P_{ij}\sum_k P_{ik}(P^{-1})_{km} = \delta_{im}P_{ij},$$

and hence $PL_jP^{-1} = \mathrm{diag}\,(P_{0j},\ldots,P_{dj})$.   □

Thanks to this theorem, it is relatively easy to compute P, Q $(=\frac{1}{n}P^{-1})$ and m_i $(=Q_{0i})$. It is also possible to express P and Q in terms of the (common) eigenvectors of the L_j. Indeed, $PL_jP^{-1} = \mathrm{diag}\,(P_{0j},\ldots,P_{dj})$ implies that the rows of P are left eigenvectors and the columns of Q are right eigenvectors. In particular, m_i can be computed from the right eigenvector u_i and the left eigenvector $v_i^\top$, normalized such that $(u_i)_0 = (v_i)_0 = 1$, by using $m_i u_i^\top v_i = n$. Clearly, each m_i must be an integer. These are the *rationality conditions* for an association scheme. As we saw in the case of a strongly regular graph, these conditions can be very powerful.

11.3 The linear programming bound

One of the main reasons association schemes have been studied is that they yield upper bounds for the size of substructures.

Let Y be a nonempty subset of X, and let its *inner distribution* be the vector a defined by $a_i = |(Y \times Y) \cap R_i|/|Y|$, the average number of elements of Y in relation R_i to a given one. Let χ be the characteristic vector of Y. Then $a_i = \frac{1}{|Y|}\chi^\top A_i\chi$.

Theorem 11.3.1 (Delsarte) $aQ \geq 0$.

Proof We have $|Y|(aQ)_j = |Y|\sum_i a_iQ_{ij} = \chi^\top \sum Q_{ij}A_i\chi = n\chi^\top E_j\chi \geq 0$ since E_j is positive semidefinite.   □

Example Consider the schemes of the triples from a 7-set, where two triples are in relation R_i when they have $3-i$ elements in common $(i=0,1,2,3)$. We find

$$P = \begin{bmatrix} 1 & 12 & 18 & 4 \\ 1 & 5 & -3 & -3 \\ 1 & 0 & -3 & 2 \\ 1 & -3 & 3 & -1 \end{bmatrix} \quad \text{and} \quad Q = \begin{bmatrix} 1 & 6 & 14 & 14 \\ 1 & 5/2 & 0 & -7/2 \\ 1 & -1 & -7/3 & 7/3 \\ 1 & -9/2 & 7 & -7/2 \end{bmatrix}.$$

How many triples can we find such that any two meet in at most one point? For the inner distribution a of such a collection Y we have $a_1 = 0$, so $a = (1,0,s,t)$, and $aQ \geq 0$ gives the three inequalities

$$6-s-\tfrac{9}{2}t \geq 0, \quad 14 - \tfrac{7}{3}s + 7t \geq 0, \quad 14 + \tfrac{7}{3}s - \tfrac{7}{2}t \geq 0.$$

The linear programming problem is to maximize $|Y| = 1+s+t$ given these inequalities, and the unique solution is $s = 6$, $t = 0$. This shows that one can have at most 7 triples that pairwise meet in at most one point in a 7-set, and if one has 7, then no two are disjoint. Of course an example is given by the Fano plane.

How many triples can we find such that any two meet in at least one point? Now $a = (1,r,s,0)$ and the optimal solution of $aQ \geq 0$ is $(1,8,6,0)$. An example of such a collection is given by the set of 15 triples containing a fixed point.

How many triples can we find such that no two meet in precisely one point? Now $a = (1,r,0,t)$ and the maximum value of $1 + r + t$ is 5. An example is given by the set of 5 triples containing two fixed points.

11.3.1 Equality

More information is available when the bound $(aQ)_j \geq 0$ is tight. Let the *outer distribution* of the set Y be the $b \times (d+1)$ matrix B, defined by

$$B_{xi} = \#\{y \in Y \mid (x,y) \in R_i\} = (A_i \chi)_x.$$

Theorem 11.3.2 *If $(aQ)_j = 0$, then $(BQ)_{xj} = 0$ for all $x \in X$. The number of nonzero $(aQ)_j$ $(0 \leq j \leq d)$ equals* $\operatorname{rk} B$.

Proof $(BQ)_{xj} = (\sum_i Q_{ij}A_i\chi)_x = n(E_j\chi)_x$ and we saw that $|Y|(aQ)_j = n\chi^\top E_j\chi = n\|E_j\chi\|^2$, so that $E_j\chi = 0$ if and only if $(aQ)_j = 0$. For the second part, note that $\operatorname{rk} B = \operatorname{rk} B^\top B$, that Q is nonsingular, and that $Q^\top B^\top BQ$ is the diagonal matrix with diagonal entries $(Q^\top B^\top BQ)_{jj} = n|Y|(aQ)_j$. □

11.3.2 The code-clique theorem

Consider a fixed association scheme with underlying set X and $d+1$ relations. For $I \subseteq \{1,\ldots,d\}$, let $\operatorname{LP}(I)$ be the linear programming upper bound for the cardinality of subsets Y of X with inner distribution a, such that $a_i = 0$ for all $i \in I$. Then $\operatorname{LP}(I)$ is (by definition) the maximum of $\sum_i a_i$ under the conditions $a_0 = 1$, $a_i = 0$ for $i \in I$, and $aQ \geq 0$. Note that $\sum_i a_i = (aQ)_0$.

Theorem 11.3.3 *Let $\{I,J\}$ be a partition of $\{1,\ldots,d\}$. Then $\operatorname{LP}(I).\operatorname{LP}(J) \leq |X|$. In particular, if Y and Z are nonempty subsets of X with inner distributions b and c, respectively, where $b_i = 0$ for $i \in I$ and $c_j = 0$ for $j \in J$, then $|Y|.|Z| \leq |X|$. Equality holds if and only if for all $i \neq 0$ we have $(bQ)_i = 0$ or $(cQ)_i = 0$.*

Proof Write $\eta = (bQ)_0$ and $\zeta = (cQ)_0$. We show that $\eta\zeta \leq n = |X|$. Define $\beta_i = \zeta^{-1}m_i^{-1}(cQ)_i$. Then $\beta_0 = 1$, $\beta_i \geq 0$ for all i, and $\zeta\sum_i \beta_i Q_{ki} = \sum_{i,j}c_j m_i^{-1}Q_{ji}Q_{ki} = \sum_{i,j}c_j n_j^{-1}P_{ij}Q_{ki} = c_k n_k^{-1}n$, so that

$$\eta = (bQ)_0 \leq \sum_i (bQ)_i \beta_i = \sum_{i,k} b_k Q_{ki}\beta_i = \frac{n}{\zeta}\sum_k n_k^{-1}b_k c_k = \frac{n}{\zeta}. \qquad \square$$

This type of result is not unexpected. For example, if Γ is any graph with transitive group, with maximal cliques and cocliques of sizes a and b, respectively, then $ab \leq n$. However, the above theorem uses not the actual sizes but the LP upper bounds for the sizes.

11.3.3 Strengthened LP bounds

One can strengthen the linear programming upper bound by adding more inequalities known to hold for a. For example, one also has $a \geq 0$.

11.4 The Krein parameters

The Bose-Mesner algebra $\mathscr{A}$ is closed not only under ordinary matrix multiplication, but also under componentwise (Hadamard, Schur) multiplication (denoted $\circ$). Clearly $\{A_0, \ldots, A_d\}$ is the basis of minimal idempotents with respect to this multiplication. Write

$$E_i \circ E_j = \frac{1}{n} \sum_{k=0}^{d} q_{ij}^k E_k.$$

The numbers q_{ij}^k thus defined are called the *Krein parameters*. (Our q_{ij}^k are those of Delsarte, but differ from SEIDEL [317]'s by a factor n.) As expected, we now have the analogue of Theorems 11.1.1 and 11.2.1.

Theorem 11.4.1 *The Krein parameters of an association scheme satisfy*

(i) $q_{0j}^k = \delta_{jk}$, $q_{ij}^0 = \delta_{ij} m_j$, $q_{ij}^k = q_{ji}^k$,

(ii) $\sum_i q_{ij}^k = m_j$, $\sum_j m_j = n$,

(iii) $q_{ij}^k m_k = q_{ik}^j m_j$,

(iv) $\sum_l q_{ij}^l q_{kl}^m = \sum_l q_{kj}^l q_{il}^m$,

(v) $Q_{ij} Q_{ik} = \sum_{l=0}^d q_{jk}^l Q_{il}$,

(vi) $n m_k q_{ij}^k = \sum_l n_l Q_{li} Q_{lj} Q_{lk}$.

Proof   Let $\sum(A)$ denote the sum of all entries of the matrix A. Then $JAJ = \sum(A)J$, $\sum(A \circ B) = \text{trace } AB^\top$ and $\sum(E_i) = 0$ if $i \neq 0$, since then $E_i J = n E_i E_0 = 0$. Now (i) follows by use of $E_i \circ E_0 = \frac{1}{n} E_i$, $q_{ij}^0 = \sum(E_i \circ E_j) = \text{trace } E_i E_j = \delta_{ij} m_j$, and $E_i \circ E_j = E_j \circ E_i$, respectively. Equation (iv) follows by evaluating $E_i \circ E_j \circ E_k$ in two ways, and (iii) follows from (iv) by taking $m = 0$. Equation (v) follows from evaluating $A_i \circ E_j \circ E_k$ in two ways, and (vi) follows from (v), using the orthogonality relation $\sum_l n_l Q_{lj} Q_{lk} = \delta_{mk} m_k n$. Finally, by use of (iii) we have

$$m_k \sum_j q_{ij}^k = \sum_j q_{ik}^j m_j = n \cdot \text{trace } (E_i \circ E_k) = n \sum_l (E_i)_{ll} (E_k)_{ll} = m_i m_k,$$

proving (ii). $\square$

The above results illustrate a dual behavior between ordinary multiplication, the numbers p_{ij}^k and the matrices A_i and P on the one hand, and Schur multiplication, the numbers q_{ij}^k and the matrices E_i and Q on the other hand. If two association schemes have the property that the intersection numbers of one are the Krein parameters of the other, then the converse is also true. Two such schemes are said to be (formally) dual to each other. One scheme may have several (formal) duals, or none at all (but when the scheme is invariant under a regular Abelian group, there is a natural way to define a dual scheme, cf. DELSARTE [139]). In fact, usually the Krein parameters are not even integers. But they cannot be negative. These important restrictions, due to SCOTT [311], are the so-called *Krein conditions*.

Theorem 11.4.2 *The Krein parameters of an association scheme satisfy $q_{ij}^k \geq 0$ for all $i, j, k \in \{0, \ldots, d\}$.*

Proof The numbers $\frac{1}{n}q_{ij}^k$ $(0 \leq k \leq d)$ are the eigenvalues of $E_i \circ E_j$ (since $(E_i \circ E_j)E_k = \frac{1}{n}q_{ij}^k E_k$). On the other hand, the Kronecker product $E_i \otimes E_j$ is positive semidefinite, since each E_i is. But $E_i \circ E_j$ is a principal submatrix of $E_i \otimes E_j$, and therefore is positive semidefinite as well, i.e., has no negative eigenvalue. $\square$

The Krein parameters can be computed by use of Theorem 11.4.1 (vi). This equation also shows that the Krein condition is equivalent to

$$\sum_l n_l Q_{li} Q_{lj} Q_{lk} \geq 0 \text{ for all } i, j, k \in \{0, \ldots, d\}.$$

In the case of a strongly regular graph we obtain

$$q_{11}^1 = \frac{f^2}{v}\left(1 + \frac{r^3}{k^2} - \frac{(r+1)^3}{(v-k-1)^2}\right) \geq 0,$$

$$q_{22}^2 = \frac{g^2}{v}\left(1 + \frac{s^3}{k^2} - \frac{(s+1)^3}{(v-k-1)^2}\right) \geq 0$$

(the other Krein conditions are trivially satisfied in this case), which is equivalent to the result mentioned in section §9.1.5.

NEUMAIER [285] generalized Seidel's absolute bound to association schemes, and obtained the following.

Theorem 11.4.3 *The multiplicities m_i $(0 \leq i \leq d)$ of an association scheme with d classes satisfy*

$$\sum_{q_{ij}^k \neq 0} m_k \leq \begin{cases} m_i m_j & \text{if } i \neq j, \\ \frac{1}{2}m_i(m_i+1) & \text{if } i = j. \end{cases}$$

Proof The left-hand side equals $\text{rk}(E_i \circ E_j)$. But $\text{rk}(E_i \circ E_j) \leq \text{rk}(E_i \otimes E_j) = \text{rk}E_i \cdot \text{rk}E_j = m_i m_j$. And if $i = j$, then $\text{rk}(E_i \circ E_i) \leq \frac{1}{2}m_i(m_i+1)$. Indeed, if the rows of E_i are linear combinations of m_i rows, then the rows of $E_i \circ E_i$ are linear combinations of the $m_i + \frac{1}{2}m_i(m_i - 1)$ rows that are the elementwise products of any two of these m_i rows. $\square$

For strongly regular graphs with $q_{11}^1 = 0$ we obtain Seidel's bound: $v \le \frac{1}{2}f(f + 3)$. But when $q_{11}^1 > 0$, Neumaier's result states that the bound can be improved to $v \le \frac{1}{2}f(f+1)$.

11.5 Automorphisms

Let π be an automorphism of an association scheme, and suppose there are N_i points x such that x and $\pi(x)$ are in relation R_i.

Theorem 11.5.1 (G. Higman) *For each j the number $\frac{1}{n}\sum_{i=0}^d N_i Q_{ij}$ is an algebraic integer.*

Proof The automorphism is represented by a permutation matrix S, where $SM = MS$ for each M in the Bose-Mesner algebra. Let $E = E_j$ be one of the idempotents. Then E has eigenvalues 0 and 1, and S has eigenvalues that are roots of unity, so ES has eigenvalues that are zero or a root of unity, and $\operatorname{tr} ES$ is an algebraic integer. But $\operatorname{tr} ES = \frac{1}{n}\sum_i N_i Q_{ij}$. ∎

If one puts $a_j = \frac{1}{n}\sum_i N_i Q_{ij}$, then $N_h = \sum_j a_j P_{jh}$ for all h.

11.5.1 The Moore graph on 3250 vertices

Let Γ be a strongly regular graph with parameters $(v,k,\lambda,\mu) = (3250,57,0,1)$ (an unknown Moore graph of diameter 2, cf. Theorem 9.1.5).

For such a graph $Q = \begin{bmatrix} 1 & 1729 & 1520 \\ 1 & \frac{637}{3} & -\frac{640}{3} \\ 1 & -\frac{13}{3} & \frac{10}{3} \end{bmatrix}$.

ASCHBACHER [12] proved that there is no such graph with a rank 3 group. G. Higman (unpublished, cf. CAMERON [82]) proved that there is no such graph with a vertex-transitive group.

Proposition 11.5.2 (G. Higman) *Γ is not vertex-transitive.*

Proof Consider any nontrivial group of automorphisms G of such a graph. The collection of points fixed by G has the properties $\lambda = 0$ and $\mu = 1$. Also, two nonadjacent fixed vertices are adjacent to the same number of fixed vertices, so the fixed subgraph is either a strongly regular Moore graph (and then has 5, 10 or 50 vertices), or all fixed vertices have distance at most 1 to some fixed vertex (so that there are at most $k+1 = 58$ of them).

Consider an involution π. If π does not interchange the endpoints of some edge, then $N_1 = 0$ and $N_0 + N_2 = 3250$. But if $\{x,y\}$ is an orbit of π, then the unique common neighbor z of x and y is fixed, and z occurs for at most 28 pairs $\{x,y\}$, so

$N_2 \le 56N_0$, so that $N_0 = 58$, $N_1 = 0$, $N_2 = 3192$ and $\frac{1}{3250}(58 \times 1729 - 3192 \times \frac{13}{3}) = \frac{133}{5}$ is not an integer, contradiction.

So, π must interchange two adjacent points x and y, and hence interchanges the remaining 56 neighbors u of x with the remaining 56 neighbors v of y. If $\{u,v\}$ is such an orbit, then the unique common neighbor of u and v is fixed, and these are all the fixed points. So $N_0 = 56$, that is, π is an odd permutation, since it is the product of 1597 transpositions. Let N be the subgroup of G consisting of the even permutations. Then N does not have any involutions, so is not transitive, and if G is transitive, N has two orbits interchanged by any element outside N. But α has fixed points and cannot interchange the two orbits of N, a contradiction, so G is not transitive.   □

11.6 *P*- and *Q*-polynomial association schemes

In many cases, the association scheme carries a distance function such that relation R_i is the relation of having distance i. Such schemes are called *metric*. They are characterized by the fact that p^i_{jk} is zero whenever one of i,j,k is larger than the sum of the other two, while p^i_{jk} is nonzero for $i = j+k$. Note that whether a scheme is metric depends on the ordering of the relations R_i. A scheme may be metric for more than one ordering. Metric association schemes are essentially the same objects as distance-regular graphs (see Chapter 12 below).

Dually, a *cometric* scheme is defined by $q^i_{jk} = 0$ for $i > j+k$ and $q^i_{jk} > 0$ for $i = j+k$.

There are several equivalent formulations of the metric (cometric) property.

An association scheme is called *P-polynomial* if there exist polynomials f_k of degree k with real coefficients, and real numbers z_i such that $P_{ik} = f_k(z_i)$. Clearly we may always take $z_i = P_{i1}$. By the orthogonality relation 11.2.1(iii) we have

$$\sum_i m_i f_j(z_i) f_k(z_i) = \sum_i m_i P_{ij} P_{ik} = nn_j \delta_{jk},$$

which shows that the f_k are orthogonal polynomials.

Dually, a scheme is called *Q-polynomial* when the same holds with Q instead of P. The following result is due to DELSARTE [139] (Theorem 5.6, p. 61).

Theorem 11.6.1 *An association scheme is metric (resp. cometric) if and only if it is P-polynomial (resp. Q-polynomial).*

Proof Let the scheme be metric. Then

$$A_1 A_i = p^{i-1}_{1i} A_{i-1} + p^i_{1i} A_i + p^{i+1}_{1i} A_{i+1}.$$

Since $p^{i+1}_{1i} \ne 0$, A_{i+1} can be expressed in terms of A_1, A_{i-1} and A_i. Hence for each j there exists a polynomial f_j of degree j such that $A_j = f_j(A_1)$, and it follows that $P_{ij} E_i = A_j E_i = f_j(A_1) E_i = f_j(P_{i1}) E_i$, and hence $P_{ij} = f_j(P_{i1})$.

Now suppose that the scheme is P-polynomial. Then the f_j are orthogonal polynomials, and therefore they satisfy a 3-term recurrence relation (see SZEGŐ [333], p. 42)

$$\alpha_{j+1} f_{j+1}(z) = (\beta_j - z) f_j(z) + \gamma_{j-1} f_{j-1}(z).$$

Hence

$$P_{i1} P_{ij} = -\alpha_{j+1} P_{i,j+1} + \beta_j P_{ij} + \gamma_{j-1} P_{i,j-1} \quad \text{for } i = 0, \ldots, d.$$

Since $P_{i1} P_{ij} = \sum_l p^l_{1j} P_{il}$ and P is nonsingular, it follows that $p^l_{1j} = 0$ for $|l - j| > 1$. Now the full metric property easily follows by induction. The proof for the cometric case is similar.   □

Given a sequence of nonzero real numbers, let its number of *sign changes* be obtained by first removing all zeros from the sequence, and then counting the number of consecutive pairs of different sign. (Thus, the number of sign changes in $1, -1, 0, 1$ is 2.)

Proposition 11.6.2 *(i) Let $(X, \mathscr{R})$ be a P-polynomial association scheme, with relations ordered according to the P-polynomial ordering and eigenspaces ordered according to descending real order on the $\theta_i := P_{i1}$. Then both row i and column i of both matrices P and Q have precisely i sign changes $(0 \le i \le d)$.*

(ii) Dually, if $(X, \mathscr{R})$ is a Q-polynomial association scheme, and the eigenspaces are ordered according to the Q-polynomial ordering and the relations are ordered according to descending real order on the $\sigma_i := Q_{i1}$, then row i and column i of the matrices P and Q have precisely i sign changes $(0 \le i \le d)$.

Proof Since $m_i P_{ij} = n_j Q_{ji}$, the statements about P and Q are equivalent. Define polynomials p_j of degree j for $0 \le j \le d + 1$ by $p_{-1}(x) = 0$, $p_0(x) = 1$, $(x - a_j) p_j(x) = b_{j-1} p_{j-1} + c_{j+1} p_{j+1}(x)$, taking $c_{d+1} = 1$. Then $A_j = p_j(A)$, and $p_{d+1}(x) = 0$ has as roots the eigenvalues of A. The numbers in row j of P are $p_i(\theta_j)$ $(0 \le i \le d)$, and by the theory of Sturm sequences the number of sign changes is the number of roots of p_{d+1} larger than θ_j, which is j. The numbers in column i of P are the values of p_i evaluated at the roots of p_{d+1}. Since p_i has degree i, and there is at least one root of p_{d+1} between any two roots of p_i there are i sign changes. The proof in the Q-polynomial case is similar.   □

Example Consider the Hamming scheme $H(4, 2)$, the association scheme on the binary vectors of length 4, where the relation is their Hamming distance. Now

$$P = Q = \begin{bmatrix} 1 & 4 & 6 & 4 & 1 \\ 1 & 2 & 0 & -2 & -1 \\ 1 & 0 & -2 & 0 & 1 \\ 1 & -2 & 0 & 2 & -1 \\ 1 & -4 & 6 & -4 & 1 \end{bmatrix}.$$

11.7 Exercises

Exercise 11.1 Show that the number of relations of valency 1 in an association scheme is 2^m for some $m \geq 0$, and $2^m | n$. (Hint: The relations of valency 1 form an elementary Abelian 2-group with operation $i \oplus j = k$ when $A_i A_j = A_k$.)

Exercise 11.2 Show that for the special case where Y is a coclique in a strongly regular graph, the linear programming bound is the Hoffman bound (Theorem 3.5.2).

Exercise 11.3 Show that if Γ is a relation of valency k in an association scheme, and θ is a negative eigenvalue of Γ, then $|S| \leq 1 - k/\theta$ for each clique S in Γ.

Exercise 11.4 Consider a primitive strongly regular graph Γ on v vertices with eigenvalues k^1, r^f, s^g ($k > r > s$) with a Hoffman coloring (that is a coloring with $1 - k/s$ colors). Consider the following relations on the vertex set of Γ:

R_0: identity,
R_1: adjacent in Γ,
R_2: nonadjacent in Γ with different colors,
R_3: nonadjacent in Γ with the same color.

Prove that these relations define an association scheme on the vertex set of Γ, and determine the matrices P and Q.

Exercise 11.5 Let (X,R) be a primitive association scheme, and let $\Gamma = (X,R_s)$ be a graph corresponding to one of the classes. Let $m > 1$ be one of the multiplicities of the scheme. Let $\eta(\ )$ denote the Haemers invariant (§3.7.2). Then $\eta(\overline{\Gamma}) \leq m+1$.

Exercise 11.6 Consider the root system of type E_6 with the five relations of having inner product 2, 1, 0, -1, -2. Show that this is an association scheme with $n = 72$ and

$$P = \begin{bmatrix} 1 & 20 & 30 & 20 & 1 \\ 1 & 10 & 0 & -10 & -1 \\ 1 & 2 & -6 & 2 & 1 \\ 1 & -2 & 0 & 2 & -1 \\ 1 & -4 & 6 & -4 & 1 \end{bmatrix} \text{ and } Q = \begin{bmatrix} 1 & 6 & 20 & 30 & 15 \\ 1 & 3 & 2 & -3 & -3 \\ 1 & 0 & -4 & 0 & 3 \\ 1 & -3 & 2 & 3 & -3 \\ 1 & -6 & 20 & -30 & 15 \end{bmatrix}.$$

Show with the notation of §11.3.2 that $LP(\{1,2\}) = 21$ and $LP(\{3,4\}) = \frac{24}{7} = n/LP(\{1,2\})$, where the latter system has unique optimal solution $a = (1,0,0,\frac{20}{7}, -\frac{3}{7})$. Adding $a \geq 0$ to the inequalities (or, in this case, just taking the integral part) improves the upper bound to 3.

Chapter 12
Distance-Regular Graphs

Consider a connected simple graph with vertex set X of diameter d. Define $R_i \subset X^2$ by $(x,y) \in R_i$ whenever x and y have graph distance i. If this defines an association scheme, then the graph (X, R_1) is called *distance-regular*. By the triangle inequality, $p_{ij}^k = 0$ if $i+j < k$ or $|i-j| > k$. Moreover, $p_{ij}^{i+j} > 0$. Conversely, if the intersection numbers of an association scheme satisfy these conditions, then (X, R_1) is easily seen to be distance-regular.

Many of the association schemes that play a role in combinatorics are metric. Families of distance-regular graphs with unbounded diameter include the Hamming graphs, the Johnson graphs, the Grassmann graphs, and graphs associated to dual polar spaces. Recently VAN DAM & KOOLEN [131] constructed a new such family, the fifteenth, and the first without transitive group.

Many constructions and results for strongly regular graphs are the $d = 2$ special case of corresponding results for distance-regular graphs.

The monograph [54] is devoted to the theory of distance-regular graphs, and gives the state of the theory in 1989.

12.1 Parameters

Conventionally, the parameters are $b_i = p_{i+1,1}^i$ and $c_i = p_{i-1,1}^i$ (and $a_i = p_{i,1}^i$). The *intersection array* of a distance-regular graph of diameter d is $\{b_0, \ldots, b_{d-1}; c_1, \ldots, c_d\}$. The valencies $p_{i,i}^0$, which were called n_i above, are usually called k_i here. We have $c_i k_i = b_{i-1} k_{i-1}$. The total number of vertices is usually called v.

It is easy to see that one has $b_0 \geq b_1 \geq \ldots \geq b_{d-1}$ and $c_1 \leq c_2 \leq \ldots \leq c_d$ and $c_j \leq b_{d-j}$ $(1 \leq j \leq d)$.

12.2 Spectrum

A distance-regular graph Γ of diameter d has $d+1$ distinct eigenvalues, and the spectrum is determined by the parameters. (Indeed, the matrices P and Q of any association scheme are determined by the parameters p^i_{jk}, and for a distance-regular graph the p^i_{jk} are determined again in terms of the b_i and c_i.)

The eigenvalues of Γ are the eigenvalues of the tridiagonal matrix $L_1 = (p^j_{1k})$ of order $d+1$ that here gets the form

$$
L_1 = \begin{pmatrix}
0 & b_0 & & & 0 \\
c_1 & a_1 & b_1 & & \\
& c_2 & a_2 & b_2 & \\
& & \cdots & \cdots & \cdots \\
0 & & & c_d & a_d
\end{pmatrix}.
$$

If $L_1 u = \theta u$ and $u_0 = 1$, then the multiplicity of θ as eigenvalue of Γ equals

$$
m(\theta) = v/(\textstyle\sum k_i u_i^2).
$$

12.3 Primitivity

A distance-regular graph Γ of diameter d is called *imprimitive* when one of the relations (X, R_i) with $i \neq 0$ is disconnected. This can happen in three cases: Γ is an n-gon (and $i|n$), or Γ is bipartite (and $i=2$), or Γ is antipodal (and $i=d$). Here Γ is called *antipodal* when having distance d is an equivalence relation on $V\Gamma$. Graphs can be both bipartite and antipodal. The $2n$-gons fall in all three cases.

12.4 Examples

12.4.1 Hamming graphs

Let Q be a set of size q. The *Hamming graph* $H(d,q)$ is the graph with vertex set Q^d, where two vertices are adjacent when they agree in $d-1$ coordinates.

This graph is distance-regular, with parameters $c_i = i$, $b_i = (q-1)(d-i)$, diameter d, and eigenvalues $(q-1)d - qi$ with multiplicity $\binom{d}{i}(q-1)^i$ $(0 \leq i \leq d)$. (Indeed, $H(d,q)$ is the Cartesian product of d copies of K_q, see §1.4.6.)

For $q = 2$, this graph is also known as the hypercube 2^d, often denoted Q_d. For $d = 2$, the graph $H(2,q)$ is also called $L_2(q)$.

Cospectral graphs

In §1.8.1 we saw that there are precisely two graphs with the spectrum of $H(4,2)$. In §9.2 we saw that there are precisely two graphs with the spectrum of $H(2,4)$. Here we give a graph cospectral with $H(3,3)$ (cf. [206]).

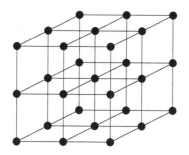

Fig. 12.1 The geometry of the Hamming graph $H(3,3)$

The graphs $H(d,q)$ have q^d vertices, and dq^{d-1} maximal cliques ("lines") of size q. Let N be the point-line incidence matrix. Then $NN^\top - dI$ is the adjacency matrix of $\Gamma = H(d,q)$, and $N^\top N - qI$ is the adjacency matrix of the graph Δ on the lines, where two lines are adjacent when they have a vertex in common. It follows that for $d = q$ the graphs Γ and Δ are cospectral. In Γ any two vertices at distance 2 have $c_2 = 2$ common neighbors. If $q \geq 3$, then two vertices at distance 2 in Δ have 1 or q common neighbors (and both occur), so that Δ is not distance-regular, and in particular not isomorphic to Γ. For $q = 3$ the geometry is displayed in Figure 12.1. See also §14.2.2.

12.4.2 Johnson graphs

Let X be a set of size n. The *Johnson graph* $J(n,m)$ is the graph with vertex set $\binom{X}{m}$, the set of all m-subsets of X, where two m-subsets are adjacent when they have $m - 1$ elements in common. For example, $J(n,0)$ has a single vertex; $J(n,1)$ is the complete graph K_n; $J(n,2)$ is the triangular graph $T(n)$.

This graph is distance-regular, with parameters $c_i = i^2$, $b_i = (m-i)(n-m-i)$, diameter $d = \min(m, n-m)$ and eigenvalues $(m-i)(n-m-i) - i$ with multiplicity $\binom{n}{i} - \binom{n}{i-1}$.

The *Kneser graph* $K(n,m)$ is the graph with vertex set $\binom{X}{m}$, where two m-subsets are adjacent when they have maximal distance in $J(n,m)$ (i.e., are disjoint when $n \geq 2m$, and have $2m - n$ elements in common otherwise). These graphs are not distance-regular in general, but the *Odd graph* O_{m+1}, which equals $K(2m+1, m)$, is.

Sending a vertex (m-set) to its complement in X is an isomorphism from $J(n,m)$ onto $J(n, n-m)$ and from $K(n,m)$ onto $K(n, n-m)$. Thus, we may always assume that $n \geq 2m$.

12.4.3 Grassmann graphs

Let V be a vector space of dimension n over the field $\mathbb{F}_q$. The *Grassmann graph* $Gr(n,m)$ is the graph with vertex set $\begin{bmatrix} V \\ m \end{bmatrix}$, the set of all m-subspaces of V, where two m-subspaces are adjacent when they intersect in an $(m-1)$-space. This graph is distance-regular, with parameters $c_i = \begin{bmatrix} i \\ 1 \end{bmatrix}^2$, $b_i = q^{2i+1} \begin{bmatrix} m-i \\ 1 \end{bmatrix} \begin{bmatrix} n-m-i \\ 1 \end{bmatrix}$, diameter $d = \min(m, n-m)$, and eigenvalues $q^{i+1} \begin{bmatrix} m-i \\ 1 \end{bmatrix} \begin{bmatrix} n-m-i \\ 1 \end{bmatrix} - \begin{bmatrix} i \\ 1 \end{bmatrix}$ with multiplicity $\begin{bmatrix} n \\ i \end{bmatrix} - \begin{bmatrix} n \\ i-1 \end{bmatrix}$. (Here $\begin{bmatrix} n \\ i \end{bmatrix} = (q^n - 1) \cdots (q^{n-i+1} - 1)/(q^i - 1) \cdots (q-1)$ is the q-binomial coefficient, the number of m-subspaces of an n-space.)

12.4.4 Van Dam-Koolen graphs

VAN DAM & KOOLEN [131] construct distance-regular graphs $vDK(m)$ with the same parameters as $Gr(2m+1, m)$. (They call them the *twisted Grassmann graphs*.) These graphs are ugly, the group of automorphisms is not transitive. The existence of such examples reinforces the idea that the parameters of distance-regular graphs of large diameter are strongly restricted, while there is some freedom for the actual structure. The construction is as follows. Let V be a vector space of dimension $2m+1$ over $\mathbb{F}_q$, and let H be a hyperplane of V. Take as vertices the $(m+1)$-subspaces of V not contained in H, and the $(m-1)$-subspaces contained in H, where two subspaces of the same dimension are adjacent when their intersection has codimension 1 in both, and two subspaces of different dimension are adjacent when one contains the other. This graph is the line graph (concurrency graph on the set of lines) of the partial linear space of which the points are the m-subspaces of V, with natural incidence, while the point graph (collinearity graph on the set of points) is $Gr(2m+1, m)$. It follows that $vDK(m)$ and $Gr(2m+1, m)$ are cospectral.

12.5 Bannai-Ito conjecture

The most famous problem about distance-regular graphs was the Bannai-Ito conjecture ([21], p. 237): show that there are only finitely many distance-regular graphs with fixed valency k larger than 2. After initial work by Bannai and Ito, the conjecture was attacked by Jack Koolen and coauthors in a long series of papers. After

25 years a complete proof was given by BANG, DUBICKAS, KOOLEN & MOULTON [19].

12.6 Connectedness

For strongly regular graphs we had Theorem 9.3.2 stating that the vertex connectivity $\kappa(\Gamma)$ equals the valency k. In [61] it was shown that the same holds for distance-regular graphs.

For strongly regular graphs we also had Proposition 9.3.1 which says that the induced subgraph on the vertices at maximal distance from a given vertex is connected. This is a very important property, but for distance-regular graphs additional hypotheses are needed. For example, there are two generalized hexagons with parameters $GH(2,2)$ (duals of each other) and in one of them the subgraphs $\Gamma_3(x)$ are disconnected.

12.7 Growth

Not surprisingly, the number of vertices of a distance-regular graph grows exponentially with the diameter. This was first proved by PYBER [296]. Currently the best bound is $d < \frac{8}{3}\log_2 v$, due to BANG et al. [18].

12.8 Degree of eigenvalues

For strongly regular graphs we saw that eigenvalues are integral, except in the "half case" where they are quadratic. Something similar happens for distance-regular graphs.

Polygons have eigenvalues of high degree: for an n-gon the degree of the i-th eigenvalue is $\phi(m)$ where $m = \gcd(i,n)$, and ϕ is the Euler totient function. But elsewhere only integral and quadratic eigenvalues seem to occur.

For the case of a P- and Q-polynomial scheme of diameter at least 34, BANNAI & ITO [21], Thm 7.11, shows that the eigenvalues are integers.

There is precisely one known distance-regular graph of valency larger than 2 with a cubic eigenvalue, namely the Biggs-Smith graph, the unique graph with intersection array $\{3,2,2,2,1,1,1; 1,1,1,1,1,1,3\}$. It has 102 vertices, and spectrum $3^1\, 2^{18}\, 0^{17}\, ((1\pm\sqrt{17})/2)^9\, \theta_j^{16}$, where the θ_j are the three roots of $\theta^3 + 3\theta^2 - 3 = 0$.

A result in this direction is

Proposition 12.8.1 *The only distance-regular graph of diameter 3 with a cubic eigenvalue is the heptagon.*

Proof Let Γ be a distance-regular graph of diameter 3 on n vertices with a cubic eigenvalue. Since algebraically conjugate eigenvalues have the same multiplicity we have three eigenvalues θ_i with multiplicity $f = (n-1)/3$. Since $\operatorname{tr} A = 0$ we find that $\theta_1 + \theta_2 + \theta_3 = -k/f$. Now k/f is rational and an algebraic integer, hence an integer, and $k \geq (n-1)/3$. The same reasoning applies to A_i for $i = 2, 3$ and hence $k_i \geq (n-1)/3$, and we must have equality. Since $k = k_2 = k_3$ we see that $b_1 = c_2 = b_2 = c_3$.

Write $\mu := c_2$. The distinct eigenvalues $k, \theta_1, \theta_2, \theta_3$ of A are the eigenvalues of the matrix L_1 (Theorem 11.2.2) and hence $k - 1 = k + \theta_1 + \theta_2 + \theta_3 = \operatorname{tr} L_1 = a_1 + a_2 + a_3 = (k - \mu - 1) + (k - 2\mu) + (k - \mu)$, so that $k = 2\mu$ and $a_2 = 0$.

Let $d(x,y) = 3$ and put $A = \Gamma(x) \cap \Gamma_2(y)$, $B = \Gamma_2(x) \cap \Gamma(y)$, so that $|A| = |B| = c_3 = \mu$. Every vertex in B is adjacent to every vertex in A, and hence two vertices in B have at least $\mu + 1$ common neighbors, so must be adjacent. Thus B is a clique, and $\mu = |B| \leq a_2 + 1$, that is, $\mu = 1$, $k = 2$. $\qquad\square$

12.9 Moore graphs and generalized polygons

Any k-regular graph of diameter d has at most

$$1 + k + k(k-1) + \ldots + k(k-1)^{d-1}$$

vertices, as is easily seen. A graph for which equality holds is called a *Moore graph*. Moore graphs are distance-regular, and those of diameter 2 were dealt with in Theorem 9.1.5. Using the rationality conditions, DAMERELL [133] and BANNAI & ITO [20] showed:

Theorem 12.9.1 *A Moore graph with diameter $d \geq 3$ is a $(2d+1)$-gon.*

A strong nonexistence result of the same nature is the theorem of FEIT & HIGMAN [154] about finite generalized polygons. We recall that a *generalized m-gon* is a point-line incidence geometry such that the incidence graph is a connected, bipartite graph of diameter m and girth $2m$. It is called *regular* of order (s,t) for certain (finite or infinite) cardinal numbers s, t if each line is incident with $s + 1$ points and each point is incident with $t + 1$ lines. From such a regular generalized m-gon of order (s,t), where s and t are finite and $m \geq 3$, we can construct a distance-regular graph with valency $s(t+1)$ and diameter $d = \lfloor \frac{m}{2} \rfloor$ by taking the collinearity graph on the points.

Theorem 12.9.2 *A finite generalized m-gon of order (s,t) with $s > 1$ and $t > 1$ satisfies $m \in \{2, 3, 4, 6, 8\}$.*

Proofs of this theorem can be found in FEIT & HIGMAN [154], BROUWER, COHEN & NEUMAIER [54] and VAN MALDEGHEM [340]; again the rationality conditions do the job. The Krein conditions yield some additional information:

Theorem 12.9.3 *A finite regular generalized m-gon with $s > 1$ and $t > 1$ satisfies $s \leq t^2$ and $t \leq s^2$ if $m = 4$ or 8; it satisfies $s \leq t^3$ and $t \leq s^3$ if $m = 6$.*

This result is due to HIGMAN [215] and HAEMERS & ROOS [204].

12.10 Euclidean representations

Let Γ be distance-regular, and let θ be a fixed eigenvalue. Let $E = E_j$ be the idempotent in the association scheme belonging to θ, so that $AE = \theta E$. Let $u_i = Q_{ij}/n$, so that $E = \sum u_i A_i$. Let $f = \mathrm{rk}E$.

The map sending vertex x of Γ to the vector $\bar{x} = Ee_x$, column x of E, provides a representation of Γ by vectors in an f-dimensional Euclidean space, namely the column span of E, where graph distances are translated into inner products: if $d(x,y) = i$ then $(\bar{x}, \bar{y}) = E_{xy} = u_i$.

If this map is not injective, and $\bar{x} = \bar{y}$ for two vertices x, y at distance $i \neq 0$, then $u_i = u_0$ and any two vertices at distance i have the same image. For $i = 1$ this happens when $\theta = k$. Otherwise, Γ is imprimitive, and either $i = 2$ and Γ is bipartite and $\theta = -k$, or $i = d$ and Γ is antipodal, or $2 < i < d$ and Γ is a polygon.

This construction allows one to translate problems about graphs into problems in Euclidean geometry. Especially when f is small, this is a very useful tool.

As an example of the use of this representation, let us prove Terwilliger's Tree Bound. Call an induced subgraph T of Γ *geodetic* when distances measured in T equal distances measured in Γ.

Proposition 12.10.1 *Let Γ be distance-regular, and let θ be an eigenvalue different from $\pm k$. Let T be a geodetic tree in Γ. Then the multiplicity f of the eigenvalue θ is at least the number of endpoints of T.*

Proof We show that the span of the vectors $\bar{x}$ for $x \in T$ has a dimension not less than the number e of endpoints of T. Induction on the size of T. If $T = \{x,y\}$ then $\bar{x} \neq \bar{y}$ since $k \neq \theta$. Assume $|T| > 2$. If $x \in T$, and S is the set of endpoints of T adjacent to x, then for $y,z \in S$ and $w \in T \setminus S$ we have $(\bar{w}, \bar{y} - \bar{z}) = 0$. Pick x such that S is nonempty, and x is an endpoint of $T' = T \setminus S$. By induction $\dim\langle \bar{w} \,|\, w \in T' \rangle \geq e - |S| + 1$. Since $\theta \neq \pm k$, we have $\dim\langle \bar{y} - \bar{z} \,|\, x, y \in S \rangle = |S| - 1$. □

Example For a distance-regular graph without triangles, $f \geq k$. Equality can hold. For example, the Higman-Sims graph is strongly regular with parameters $(v,k,\lambda,\mu) = (100, 22, 0, 6)$ and spectrum $22^1 \; 2^{77} \; (-8)^{22}$.

12.11 Extremality

This section gives a simplified account of the theory developed by Fiol and Garriga and coauthors. The gist is that among the graphs with a given spectrum with $d+1$ distinct eigenvalues the distance-regular graphs are extremal in the sense that they have a maximal number of pairs of vertices at mutual distance d.

Let Γ be a connected k-regular graph with adjacency matrix A with eigenvalues $k = \theta_1 \geq \cdots \geq \theta_n$. Suppose that A has precisely $d+1$ distinct eigenvalues (so that the diameter of Γ is at most d). Define an inner product on the $(d+1)$-dimensional vector space of real polynomials modulo the minimum polynomial of A by

$$\langle p,q \rangle = \frac{1}{n} \mathrm{tr}\, p(A)q(A) = \frac{1}{n}\sum_{i=1}^{n} p(\theta_i)q(\theta_i).$$

Note that $\langle p,p \rangle \geq 0$ for all p, and $\langle p,p \rangle = 0$ if and only if $p(A) = 0$. By applying Gram-Schmidt to the sequence of polynomials x^i $(0 \leq i \leq d)$ we find a sequence of orthogonal polynomials p_i of degree i $(0 \leq i \leq d)$ satisfying $\langle p_i, p_j \rangle = 0$ for $i \neq j$ and $\langle p_i, p_i \rangle = p_i(k)$. This latter normalization is possible since $p_i(k) \neq 0$.

(Indeed, suppose that p_i changes sign at values α_j $(0 \leq j \leq h)$ inside the interval (θ_n, k). Put $q(x) = \prod_{j=1}^{h}(x - \alpha_j)$. Then all terms in $\langle p_i, q \rangle$ have the same sign, and not all are zero, so $\langle p_i, q \rangle \neq 0$, hence $h = i$, so that all zeros of p_i are in the interval (θ_n, k), and $p_i(k) \neq 0$.)

The Hoffman polynomial (the polynomial p such that $p(A) = J$) equals $p_0 + \ldots + p_d$. Indeed, $\langle p_i, p \rangle = \frac{1}{n}\mathrm{tr}\, p_i(A)J = p_i(k) = \langle p_i, p_i \rangle$ for all i.

If Γ is distance-regular, then the p_i are the polynomials for which $A_i = p_i(A)$.

Theorem 12.11.1 ("Spectral Excess Theorem") *Let Γ be connected and regular of degree k, with $d + 1$ distinct eigenvalues. Define the polynomials p_i as above. Let $\overline{k_d} := \frac{1}{n}\sum_x k_d(x)$ be the average number of vertices at distance d from a given vertex in Γ. Then $\overline{k_d} \leq p_d(k)$, and equality holds if and only if Γ is distance-regular.*

Proof We follow FIOL, GAGO & GARRIGA [159]. Use the inner product $\langle M, N \rangle = \frac{1}{n}\mathrm{tr}\, M^{\top}N$ on the space $M_n(\mathbb{R})$ of real matrices of order n. If M, N are symmetric, then $\langle M, N \rangle = \frac{1}{n}\sum_{x,y}(M \circ N)_{xy}$. If $M = p(A)$ and $N = q(A)$ are polynomials in A, then $\langle M, N \rangle = \langle p, q \rangle$.

Since $\langle A_d, p_d(A) \rangle = \langle A_d, J \rangle = \overline{k_d}$, the orthogonal projection A'_d of A_d on the space $\langle I, A, \ldots, A^d \rangle = \langle p_0(A), \ldots, p_d(A) \rangle$ of polynomials in A equals

$$A'_d = \sum_j \frac{\langle A_d, p_j(A) \rangle}{\langle p_j, p_j \rangle} p_j(A) = \frac{\langle A_d, p_d(A) \rangle}{p_d(k)} p_d(A) = \frac{\overline{k_d}}{p_d(k)} p_d(A).$$

Now $||A'_d||^2 \leq ||A_d||^2$ gives $\overline{k_d}^2 / p_d(k) \leq \overline{k_d}$, and the inequality follows since $p_d(k) > 0$. When equality holds, $A_d = p_d(A)$.

Now it follows by downward induction on h that $A_h = p_h(A)$ $(0 \leq h \leq d)$. Indeed, from $\sum_j p_j(A) = J = \sum_j A_j$ it follows that $p_0(A) + \cdots + p_h(A) = A_0 + \cdots + A_h$. Hence $p_h(A)_{xy} = 0$ if $d(x,y) > h$, and $p_h(A)_{xy} = 1$ if $d(x,y) = h$. Since $\langle xp_{h+1}, p_j \rangle = \langle p_{h+1}, xp_j \rangle = 0$ for $j \neq h, h+1, h+2$, we have $xp_{h+1} = ap_h + bp_{h+1} + cp_{h+2}$ and hence $AA_{h+1} = ap_h(A) + bA_{h+1} + cA_{h+2}$ for certain a, b, c with $a \neq 0$. But then $p_h(A)_{xy} = 0$ if $d(x,y) < h$, so that $p_h(A) = A_h$.

Finally, the three-term recurrence for the p_h now becomes the three-term recurrence for the A_h that defines distance-regular graphs. $\qquad\square$

Noting that $p_d(k)$ depends on the spectrum only, we see that this provides a characterization of distance-regularity in terms of the spectrum and the number of pairs of vertices far apart (at mutual distance d). See [124], [158], [159], and Theorem 14.5.3 below.

12.12 Exercises

Exercise 12.1 ([244]) Show that if $c_i > c_{i-1}$, then $c_i \geq c_j + c_{i-j}$ for $1 \leq j \leq i-1$. (Hint: If $c_i > c_{i-1}$ and $d(x,y) = i$, then there is a matching between $\Gamma_{i-1}(x) \cap \Gamma(y)$ and $\Gamma(x) \cap \Gamma_{i-1}(y)$ such that corresponding vertices have distance larger than $i-2$.) Show that if $b_i > b_{i+1}$, then $b_i \geq b_{i+j} + c_j$ for $1 \leq j \leq d-i$.

Exercise 12.2 Determine the spectrum of a strongly regular graph minus a vertex. (Hint: If the strongly regular graph has characteristic polynomial $p(x) = (x-k)(x-r)^f(x-s)^g$, then the graph obtained after removing one vertex has characteristic polynomial $((x-k)(x-\lambda+\mu)+\mu)(x-r)^{f-1}(x-s)^{g-1}$.)

Determine the spectrum of a strongly regular graph minus two adjacent or non-adjacent vertices.

Show that the spectrum of a distance-regular graph minus a vertex does not depend on the vertex chosen. Give an example of two nonisomorphic cospectral graphs both obtained by removing a vertex from the same distance-regular graph.

Chapter 13
p-ranks

Designs or graphs with the same parameters can sometimes be distinguished by considering the p-rank of associated matrices. For example, there are three nonisomorphic 2-(16,6,2) designs, with point-block incidence matrices of 2-rank 6, 7 and 8, respectively.

Tight bounds on the occurrence of certain configurations are sometimes obtained by computing a rank in some suitable field, since p-ranks of integral matrices may be smaller than their ranks over $\mathbb{R}$.

Our first aim is to show that given the parameters (say, the real spectrum), only finitely many primes p are of interest.

13.1 Reduction mod p

A technical difficulty is that one would like to talk about eigenvalues that are zero or nonzero mod p for some prime p, but it is not entirely clear what that might mean when the eigenvalues are nonintegral. Necessarily some arbitrariness will be involved. For example $(5 + \sqrt{2})(5 - \sqrt{2}) \equiv 0 \bmod 23$, and one point of view is that this means that 23 is not a prime in $\mathbb{Q}(\sqrt{2})$, and one gets into algebraic number theory. But another point of view is that if one "reduces mod 23", mapping to a field of characteristic 23, then at least one factor must become 0. However, the sum of $5 + \sqrt{2}$ and $5 - \sqrt{2}$ does not become 0 upon reduction mod 23, so not both factors become 0. Since these factors are conjugate, the "reduction mod 23" cannot be defined canonically, but must involve some arbitrary choices. We follow ISAACS [232], who follows Brauer.

Let R be the ring of algebraic integers in $\mathbb{C}$, and let p be a prime. Let M be a maximal ideal in R containing the ideal pR. Put $F = R/M$. Then F is a field of characteristic p. Let $r \mapsto \bar{r}$ be the quotient map $R \to R/M = F$. This will be our "reduction mod p". (It is not canonical because M is not determined uniquely.)

Lemma 13.1.1 (ISAACS [232], (15.1)) *Let $U = \{z \in \mathbb{C} \mid z^m = 1 \text{ for some integer } m$ not divisible by $p\}$. Then the quotient map $R \to R/M = F$ induces an isomorphism* *of groups* $U \to F^*$ *from* U *onto the multiplicative group* F^* *of* F. *Moreover,* F *is algebraically closed, and is algebraic over its prime field.*

One consequence is that on integers "reduction mod p" has the usual meaning: if m is an integer not divisible by p, then some power is 1 (mod p) and it follows that $\bar{m} \neq 0$. More generally, if $\bar{\theta} = 0$, then $p | N(\theta)$, where $N(\theta)$ is the *norm* of θ, the product of its conjugates, up to sign the constant term of its minimal polynomial.

13.2 The minimal polynomial

Let M be a matrix of order n over a field F. For each eigenvalue θ of M in F, let $m(\theta)$ be the geometric multiplicity of θ, so that $\mathrm{rk}(M - \theta I) = n - m(\theta)$.

Let $e(\theta)$ be the algebraic multiplicity of the eigenvalue θ, so that the characteristic polynomial of M factors as $c(x) := \det(xI - M) = \prod(x - \theta)^{e(\theta)} c_0(x)$, where $c_0(x)$ has no roots in F. Then $m(\theta) \leq e(\theta)$.

The *minimal polynomial* $p(x)$ of M is the unique monic polynomial over F of minimal degree such that $p(M) = 0$. The numbers $\theta \in F$ for which $p(\theta) = 0$ are precisely the eigenvalues of M (in F). By the Cayley-Hamilton theorem, $p(x)$ divides $c(x)$. It follows that if $p(x) = \prod(x - \theta)^{h(\theta)} p_0(x)$, where $p_0(x)$ has no roots in F, then $1 \leq h(\theta) \leq e(\theta)$.

In terms of the Jordan decomposition of M, $m(\theta)$ is the number of Jordan blocks for θ, $h(\theta)$ is the size of the largest block, and $e(\theta)$ is the sum of the sizes of all Jordan blocks for θ.

We see that $n - e(\theta) + h(\theta) - 1 \leq \mathrm{rk}(M - \theta I) \leq n - e(\theta)/h(\theta)$, and also that $1 \leq \mathrm{rk}((M - \theta I)^i) - \mathrm{rk}((M - \theta I)^{i+1}) \leq m(\theta)$ for $1 \leq i \leq h - 1$.

13.3 Bounds for the *p*-rank

Let M be a square matrix of order n, and let $\mathrm{rk}_p(M)$ be its p-rank. Let R and F be as above in §13.1. Use a suffix F or p to denote rank or multiplicity over the field F or $\mathbb{F}_p$ (instead of $\mathbb{C}$).

Proposition 13.3.1
Let M be an integral square matrix. Then $\mathrm{rk}_p(M) \leq \mathrm{rk}(M)$.
Let M be a square matrix with entries in R. Then $\mathrm{rk}_F(\overline{M}) \leq \mathrm{rk}(M)$.

Proof The rank of a matrix is the size of the largest submatrix with nonzero determinant.   □

Proposition 13.3.2 *Let M be an integral square matrix. Then*

$$\mathrm{rk}_p(M) \geq \sum \{m(\theta) \mid \bar{\theta} \neq 0\}.$$

Proof   Let M have order n. Then $\mathrm{rk}_p(M) = n - m_p(0) \geq n - e_p(0) = n - e_F(0) = \Sigma\{e_F(t) \mid t \neq 0\} = \Sigma\{e(\theta) \mid \bar{\theta} \neq 0\} \geq \Sigma\{m(\theta) \mid \bar{\theta} \neq 0\}.$ $\square$

Proposition 13.3.3 *Let the integral square matrix M be diagonalizable. Then we have $\mathrm{rk}_F(\overline{M} - \bar{\theta}I) \leq n - e(\theta)$ for each eigenvalue θ of M.*

Proof $\mathrm{rk}_F(\overline{M} - \bar{\theta}I) \leq \mathrm{rk}(M - \theta I) = n - m(\theta) = n - e(\theta).$ $\square$

It follows that if $\bar{\theta} = 0$ for a unique θ, then $\mathrm{rk}_p(M) = n - e(\theta)$. We can still say something when $\bar{\theta} = 0$ for two eigenvalues θ, when one has multiplicity 1:

Proposition 13.3.4 *Let the integral square matrix M be diagonalizable, and suppose that $\bar{\theta} = 0$ for only two eigenvalues θ, say θ_0 and θ_1, where $e(\theta_0) = 1$. Let M have minimal polynomial $p(x) = (x - \theta_0)f(x)$. Then $\mathrm{rk}_F(\overline{M}) = n - e(\theta_1) - \varepsilon$, where $\varepsilon = 1$ if $\overline{f(M)} = 0$ and $\varepsilon = 0$ otherwise.*

Proof   By the above $n - e(\theta_1) - 1 \leq \mathrm{rk}_F(\overline{M}) \leq n - e(\theta_1)$. By the previous section $n - e_F(0) + h_F(0) - 1 \leq \mathrm{rk}_F(\overline{M}) \leq n - e_F(0)/h_F(0)$. Since $e_F(0) = e(\theta_1) + 1$ we find $\mathrm{rk}_F(\overline{M}) = n - e(\theta_1) - \varepsilon$, where $\varepsilon = 1$ if $h_F(0) = 1$ and $\varepsilon = 0$ otherwise. But $h_F(0) = 1$ iff $\overline{f(M)} = 0$. $\square$

If M is a matrix with integral entries, then the minimal polynomial $p(x)$ and its factor $f(x)$ have integral coefficients. In particular, if M is an integral symmetric matrix with constant row sums k, and the eigenvalue k of M has multiplicity 1, then $f(M) = (f(k)/n)J$ and the condition $\overline{f(M)} = 0$ becomes $\bar{c} = 0$, where $c = \frac{1}{n}\prod_{\theta \neq k}(k - \theta)$ is an integer.

13.4 Interesting primes p

Let A be an integral matrix of order n, and let $M = A - aI$ for some integer a. If θ is an eigenvalue of A, then $\theta - a$ is an eigenvalue of M.

If $\bar{\theta} = \bar{a}$ for no θ, then $\mathrm{rk}_p(M) = n$.

If $\bar{\theta} = \bar{a}$ for a unique θ, then $\mathrm{rk}_p(M) = \mathrm{rk}_F(\overline{M}) = \mathrm{rk}_F(M - (\bar{\theta} - a)I) \leq \mathrm{rk}(A - \theta I) = n - m(\theta)$ by Proposition 13.3.1, but also $\mathrm{rk}_p(M) \geq n - m(\theta)$ by Proposition 13.3.2, so that $\mathrm{rk}_p(M) = n - m(\theta)$.

So, if the p-rank of M is interesting, if it gives information not derivable from the spectrum of A and the value a, then at least two eigenvalues of M become zero upon reduction mod p. But if $\overline{\theta - a} = \overline{\eta - a} = 0$, then $\overline{\theta - \eta} = 0$, and in particular $p \mid N(\theta - \eta)$, which happens for finitely many p only.

Example The unique distance-regular graph with intersection array $\{4, 3, 2; 1, 2, 4\}$ has 14 vertices and spectrum $4, \sqrt{2}^6, (-\sqrt{2})^6, -4$ (with multiplicities written as exponents).

Let A be the adjacency matrix of this graph, and consider the p-rank of $M = A - aI$ for integers a. The norms of $\theta - a$ are $4 - a$, $a^2 - 2$, $-4 - a$, and if these are all nonzero mod p, then the p-rank of M is 14. If p is not 2 or 7, then at most one of these norms can be 0 mod p, and for $a \equiv 4 \pmod{p}$ or $a \equiv -4 \pmod{p}$ the p-rank of M is 13. If $a^2 \equiv 2 \pmod{p}$, then precisely one of the eigenvalues $\sqrt{2} - a$ and $-\sqrt{2} - a$ reduces to 0, and the p-rank of M is 8. Finally, for $p = 2$ and $p = 7$ we need to look at the matrix M itself, and find $\mathrm{rk}_2(A) = 6$ and $\mathrm{rk}_7(A \pm 3I) = 8$.

13.5 Adding a multiple of J

Let A be an integral matrix of order n with row and column sums k, and consider the rank and p-rank of $M = M_b = A + bJ$. Since J has rank 1, all these matrices differ in rank by at most 1, so either all have the same rank r, or two ranks $r, r + 1$ occur, and in the latter case rank $r + 1$ occurs whenever the row space of M contains the vector $\mathbf{1}$.

The matrix M has row sums $k + bn$.

If $p \nmid n$, then the row space of M over $\mathbb{F}_p$ contains $\mathbf{1}$ when $k + bn \not\equiv 0 \pmod{p}$. On the other hand, if $k + bn \equiv 0 \pmod{p}$, then all rows have zero row sum (mod p) while $\mathbf{1}$ does not, so $\mathbf{1}$ is not in the row space over $\mathbb{F}_p$. Thus, we are in the second case, where the smaller p-rank occurs for $b = -k/n$ only.

If $p \mid n$ and $p \nmid k$, then all row sums are nonzero (mod p) for all b, and we are in the former case: the rank is independent of b, and the row space over $\mathbb{F}_p$ always contains $\mathbf{1}$.

Finally, if $p \mid n$ and also $p \mid k$, then further inspection is required.

Example (PEETERS [293]). According to [206], there are precisely ten graphs with the spectrum $7^1 \sqrt{7}^8 (-1)^7 (-\sqrt{7})^8$, one of which is the Klein graph, the unique distance-regular graph with intersection array $\{7, 4, 1; 1, 2, 7\}$. It turns out that the p-ranks of $A - aI + bJ$ for these graphs depend on the graph only for $p = 2$ ([293]). Here $n = 24$ and $k = 7 - a + 24b$.

graph	$\mathrm{rk}_2(A+I)$	$\mathrm{rk}_2(A+I+J)$
#1,2	14	14
#3,8,9	15	14
#4,7	13	12
#5	12	12
#6	11	10
#10	9	8

$\mathrm{rk}_3(A - aI + bJ)$			
$a\backslash b$	0	1	2
0	24	24	24
1	16	15	16
2	16	16	16

$\mathrm{rk}_7(A - aI)$	
$a\backslash b$	0
0	15
1–5	24
6	17

Interesting primes (dividing the norm of the difference of two eigenvalues) are 2, 3, and 7. All p-ranks follow from the parameters except possibly $\mathrm{rk}_2(A + I + bJ)$, $\mathrm{rk}_3(A - I + bJ)$, $\mathrm{rk}_3(A + I)$, and $\mathrm{rk}_7(A)$.

The interesting 2-rank is $\mathrm{rk}_2(A + I)$, and inspection of the graphs involved shows that this takes the values 9, 11, 12, 13, 14, 15 where 9 occurs only for the Klein graph. The value of $\mathrm{rk}_2(A + I + J)$ follows, since a symmetric matrix with zero diagonal has even 2-rank, and the diagonal of a symmetric matrix lies in the $\mathbb{F}_2$-

space of its rows. Hence if $\mathrm{rk}_2(A+I)$ is even, then $\mathrm{rk}_2(A+I+J) = \mathrm{rk}_2(A+I)$, and if $\mathrm{rk}_2(A+I)$ is odd, then $\mathrm{rk}_2(A+I+J) = \mathrm{rk}_2(A+I) - 1$.

The 3-rank of $A - I + bJ$ is given by Proposition 13.3.4. Here $f(x) = (x+2)((x+1)^2 - 7)$ and $k = 6 + 24b$, so that $f(k)/n \equiv 0 \pmod 3$ is equivalent to $b \equiv 1 \pmod 3$. One has $\mathrm{rk}_3(A+I) = 16$ in all ten cases.

The value of $\mathrm{rk}_7(A)$ can be predicted: We have $\det(A+J) = -7^8.31$, so the Smith normal form (§13.8) of $A+J$ has at most 8 entries divisible by 7 and $\mathrm{rk}_7(A+J) \geq 16$. By Proposition 13.3.3, $\mathrm{rk}_7(A+J) = 16$. Since $7 \nmid n$ and $\mathbf{1}$ is in the row space of $A+J$ but not in that of A, $\mathrm{rk}_7(A) = 15$.

13.6 Paley graphs

Let q be a prime power, $q \equiv 1 \pmod 4$, and let Γ be the graph with vertex set $\mathbb{F}_q$ where two vertices are adjacent whenever their difference is a nonzero square. (Then Γ is called the *Paley graph* of order q.) In order to compute the p-rank of the Paley graphs, we first need a lemma.

Lemma 13.6.1 *Let $p(x,y) = \sum_{i=0}^{d-1} \sum_{j=0}^{e-1} c_{ij} x^i y^j$ be a polynomial with coefficients in a field F. Let $A, B \subseteq F$, with $m := |A| \geq d$ and $n := |B| \geq e$. Consider the $m \times n$ matrix $P = (p(a,b))_{a \in A, b \in B}$ and the $d \times e$ matrix $C = (c_{ij})$. Then $\mathrm{rk}_F(P) = \mathrm{rk}_F(C)$.*

Proof For any integer s and subset X of F, let $Z(s,X)$ be the $|X| \times s$ matrix $(x^i)_{x \in X, 0 \leq i \leq s-1}$. Note that if $|X| = s$ then this is a Vandermonde matrix and hence invertible. We have $P = Z(d,A) C Z(e,B)^\top$, so $\mathrm{rk}_F(P) \leq \mathrm{rk}_F(C)$, but P contains a submatrix $Z(d,A') C Z(e,B')$ with $A' \subseteq A$, $B' \subseteq B$, $|A'| = d$, $|B'| = e$, and this submatrix has the same rank as C. □

For odd prime powers $q = p^e$, p prime, let Q be the $\{0,\pm1\}$-matrix of order q with entries $Q_{xy} = \chi(y - x)$ $(x,y \in \mathbb{F}_q$, χ the quadratic residue character, $\chi(0) = 0)$.

Proposition 13.6.2 ([55]) $\mathrm{rk}_p Q = ((p+1)/2)^e$.

Proof Applying the above lemma with $p(x,y) = \chi(y - x) = (y - x)^{(q-1)/2} = \sum_i (-1)^i \binom{(q-1)/2}{i} x^i y^{(q-1)/2-i}$, we see that $\mathrm{rk}_p Q$ equals the number of binomial coefficients $\binom{(q-1)/2}{i}$ with $0 \leq i \leq (q-1)/2$ not divisible by p. Now Lucas' Theorem says that if $l = \sum_i l_i p^i$ and $k = \sum_i k_i p^i$ are the p-ary expansions of l and k, then $\binom{l}{k} \equiv \prod_i \binom{l_i}{k_i} \pmod p$. Since $\frac{1}{2}(q-1) = \sum_i \frac{1}{2}(p-1)p^i$, this means that for each p-ary digit of i there are $(p+1)/2$ possibilities and the result follows. □

For Lucas' Theorem, cf. MACWILLIAMS & SLOANE [267], §13.5, p. 404 (and references given there). Note that this proof shows that each submatrix of Q of order at least $(q+1)/2$ has the same rank as Q.

The relation between Q here and the adjacency matrix A of the Paley graph is $Q = 2A + I - J$. From $Q^2 = qI - J \equiv -J \pmod p$ and $(2A+I)^2 = qI + (q-1)J \equiv -J \pmod p$ it follows that both $\langle Q \rangle$ and $\langle 2A + I \rangle$ contain $\mathbf{1}$, so $\mathrm{rk}_p(A + \frac{1}{2}I) = \mathrm{rk}_p(2A + I) = \mathrm{rk}_p(Q) = ((p+1)/2)^e$.

13.7 Strongly regular graphs

Let Γ be a strongly regular graph with adjacency matrix A, and assume that A has integral eigenvalues k, r, s with multiplicities $1, f, g$, respectively. We investigate the p-rank of a linear combination of A, I and J.

The following proposition shows that only the case $p|(r-s)$ is interesting.

Proposition 13.7.1 *Let $M = A + bJ + cI$. Then M has eigenvalues $\theta_0 = k + bv + c$, $\theta_1 = r + c$, $\theta_2 = s + c$, with multiplicities $m_0 = 1$, $m_1 = f$, $m_2 = g$, respectively.*

(i) If none of the θ_i vanishes (mod p), then $\mathrm{rk}_p M = v$.

(ii) If precisely one θ_i vanishes (mod p), then M has p-rank $v - m_i$.

Put $e := \mu + b^2 v + 2bk + b(\mu - \lambda)$.

(iii) If $\theta_0 \equiv \theta_1 \equiv 0$ (mod p), $\theta_2 \not\equiv 0$ (mod p), then $\mathrm{rk}_p M = g$ if and only if $p|e$, and $\mathrm{rk}_p M = g + 1$ otherwise.

(iii)' If $\theta_0 \equiv \theta_2 \equiv 0$ (mod p), $\theta_1 \not\equiv 0$ (mod p), then $\mathrm{rk}_p M = f$ if and only if $p|e$, and $\mathrm{rk}_p M = f + 1$ otherwise.

(iv) In particular, if $k \equiv r \equiv 0$ (mod p) and $s \not\equiv 0$ (mod p), then $\mathrm{rk}_p A = g$. And if $k \equiv s \equiv 0$ (mod p) and $r \not\equiv 0$ (mod p), then $\mathrm{rk}_p A = f$.

(v) If $\theta_1 \equiv \theta_2 \equiv 0$ (mod p), then $\mathrm{rk}_p M \leq \min(f+1, g+1)$.

Proof Parts (i), (ii), and (v) are immediate from Propositions 13.3.1 and 13.3.2. Suppose $\theta_0 \equiv \theta_1 \equiv 0$ (mod p), $\theta_2 \not\equiv 0$ (mod p). Then we can apply Proposition 13.3.4 with the two eigenvalues 0 and θ_2. Since $\mathrm{rk}_p(M - \theta_2 I) = v - g$, and $g \leq \mathrm{rk}_p M \leq g + 1$, it follows that $\mathrm{rk}_p M = g$ if and only if $M(M - \theta_2 I) \equiv 0$ (mod p). But using $(A - rI)(A - sI) = \mu J$ and $r + s = \lambda - \mu$, we find $M(M - \theta_2 I) \equiv (A + bJ - rI)(A + bJ - sI) = eJ$. Part (iii)' is similar. $\qquad\square$

Thus, the only interesting case (where the structure of Γ plays a role) is that where p divides both θ_1 and θ_2, so that $p \mid (r - s)$. In particular, only finitely many primes are of interest. In this case we only have the upper bound (v).

Looking at the idempotents sometimes improves this bound by 1: We have $E_1 = (r-s)^{-1}(A - sI - (k-s)v^{-1}J)$ and $E_2 = (s-r)^{-1}(A - rI - (k-r)v^{-1}J)$. Thus, if $k - s$ and v are divisible by the same power of p (so that $(k-s)/v$ can be interpreted in $\mathbb{F}_p$), then $\mathrm{rk}_p(A - sI - (k-s)v^{-1}J) \leq \mathrm{rk} E_1 = f$, and, similarly, if $k - r$ and v are divisible by the same power of p then $\mathrm{rk}_p(A - rI - (k-r)v^{-1}J) \leq \mathrm{rk} E_2 = g$.

For $M = A + bJ + cI$ and $p|(r+c)$, $p|(s+c)$ we have $ME_1 = JE_1 = 0$ (over $\mathbb{F}_p$) so that $\mathrm{rk}_p \langle M, \mathbf{1} \rangle \leq g + 1$, and hence $\mathrm{rk}_p M \leq g$ (and similarly $\mathrm{rk}_p M \leq f$) when $\mathbf{1} \notin \langle M \rangle$.

Much more detail is given in [55] and [292].

In the table below we give for a few strongly regular graphs for each prime p dividing $r - s$ the p-rank of $A - sI$ and the unique b_0 such that $\mathrm{rk}_p(A - sI - b_0 J) = \mathrm{rk}_p(A - sI - bJ) - 1$ for all $b \neq b_0$, or "-" in case $\mathrm{rk}_p(A - sI - bJ)$ is independent of b. (When $p \nmid v$ we are in the former case, and b_0 follows from the parameters. When $p|v$ and $p \nmid \mu$, we are in the latter case.)

For a description of most of these graphs, see [62].

Name	v	k	λ	μ	r^f	s^g	p	$\mathrm{rk}_p(A-sI)$	b_0
Folded 5-cube	16	5	0	2	1^{10}	$(-3)^5$	2	6	-
Schläfli	27	16	10	8	4^6	$(-2)^{20}$	2	6	0
							3	7	-
$T(8)$	28	12	6	4	4^7	$(-2)^{20}$	2	6	0
							3	8	2
3 Chang graphs	28	12	6	4	4^7	$(-2)^{20}$	2	8	-
							3	8	2
$G_2(2)$	36	14	4	6	2^{21}	$(-4)^{14}$	2	8	-
							3	14	-
$Sp_4(3)$	40	12	2	4	2^{24}	$(-4)^{15}$	2	16	-
							3	11	1
$O_5(3)$	40	12	2	4	2^{24}	$(-4)^{15}$	2	10	-
							3	15	1
Hoffman-Singleton	50	7	0	1	2^{28}	$(-3)^{21}$	5	21	-
Gewirtz	56	10	0	2	2^{35}	$(-4)^{20}$	2	20	-
							3	20	1
M_{22}	77	16	0	4	2^{55}	$(-6)^{21}$	2	20	0
Brouwer-Haemers	81	20	1	6	2^{60}	$(-7)^{20}$	3	19	-
Higman-Sims	100	22	0	6	2^{77}	$(-8)^{22}$	2	22	-
							5	23	-
Hall-Janko	100	36	14	12	6^{36}	$(-4)^{63}$	2	36	0
							5	23	-
$GQ(3,9)$	112	30	2	10	2^{90}	$(-10)^{21}$	2	22	-
							3	20	1
001... in $S(5,8,24)$	120	42	8	18	2^{99}	$(-12)^{20}$	2	20	-
							7	20	5
$Sp_4(5)$	156	30	4	6	4^{90}	$(-6)^{65}$	2	66	-
							5	36	1
Sub McL	162	56	10	24	2^{140}	$(-16)^{21}$	2	20	0
							3	21	-
Edges of Ho-Si	175	72	20	36	2^{153}	$(-18)^{21}$	2	20	0
							5	21	-
01... in $S(5,8,24)$	176	70	18	34	2^{154}	$(-18)^{21}$	2	22	-
							5	22	3
a switched version	176	90	38	54	2^{153}	$(-18)^{22}$	2	22	-
of the previous graph							5	22	3
Cameron	231	30	9	3	9^{55}	$(-3)^{175}$	2	55	1
							3	56	1
Berlekamp-van Lint-Seidel	243	22	1	2	4^{132}	$(-5)^{110}$	3	67	-
Delsarte	243	110	37	60	2^{220}	$(-25)^{22}$	3	22	-
$S(4,7,23)$	253	112	36	60	2^{230}	$(-26)^{22}$	2	22	0
							7	23	5
McLaughlin	275	112	30	56	2^{252}	$(-28)^{22}$	2	22	0

continued...

Name	v	k	λ	μ	r^f	s^g	p	$\mathrm{rk}_p(A-sI)$	b_0
							3	22	1
							5	23	-
a switched version	276	140	58	84	2^{252}	$(-28)^{23}$	2	24	-
of previous plus							3	23	2
isolated point							5	24	3
$G_2(4)$	416	100	36	20	20^{65}	$(-4)^{350}$	2	38	-
							3	65	1
Dodecads mod **1**	1288	792	476	504	8^{1035}	$(-36)^{252}$	2	22	0
							11	230	3

Table 13.1 *p*-ranks of some strongly regular graphs ([55])

13.8 Smith normal form

The *Smith normal form* $S(M)$ of an integral matrix M is a diagonal matrix $S(M) = PMQ = \mathrm{diag}(s_1,\ldots,s_n)$, where P and Q are integral with determinant ± 1 and $s_1|s_2|\cdots|s_n$. It exists and is uniquely determined up to the signs of the s_i. The s_i are called the *elementary divisors* or *invariant factors*. For example, if $M = \begin{bmatrix} 1111 \\ 3111 \end{bmatrix}$, then $S(M) = \begin{bmatrix} 1000 \\ 0200 \end{bmatrix}$.

Let $\langle M \rangle$ denote the row space of M over $\mathbb{Z}$. By the fundamental theorem for finitely generated Abelian groups, the group $\mathbb{Z}^n/\langle M \rangle$ is isomorphic to a direct sum $\mathbb{Z}_{s_1} \oplus \cdots \oplus \mathbb{Z}_{s_m} \oplus \mathbb{Z}^s$ for certain $s_1,\ldots,s_m,s$, where $s_1|\cdots|s_m$. Since $\mathbb{Z}^n/\langle M \rangle \cong \mathbb{Z}^n/\langle S(M) \rangle$, we see that $\mathrm{diag}(s_1,\ldots,s_m,0^t)$ is the Smith normal form of M, when M has r rows and $n = m+s$ columns, and $t = \min(r,n) - m$.

If M is square, then $\prod s_i = \det S(M) = \pm \det M$. More generally, $\prod_{i=1}^{t} s_i$ is the g.c.d. of all minors of M of order t.

The Smith normal form is a finer invariant than the p-rank: the p-rank is just the number of s_i not divisible by p. (It follows that if M is square and $p^e \| \det M$, then $\mathrm{rk}_p M \geq n - e$.)

We give some examples of graphs distinguished by Smith normal form or p-rank.

Example Let A and B be the adjacency matrices of the lattice graph $K_4 \square K_4$ and the Shrikhande graph. Then $S(A) = S(B) = \mathrm{diag}(1^6, 2^4, 4^5, 12)$, but $S(A + 2I) = \mathrm{diag}(1^6, 8^1, 0^9)$ and $S(B + 2I) = \mathrm{diag}(1^6, 2^1, 0^9)$. All have 2-rank equal to 6.

Example An example where the p-rank suffices to distinguish, is given by the Chang graphs, strongly regular graphs with the same parameters as the triangular graph $T(8)$, with $(v,k,\lambda,\mu) = (28,12,6,4)$ and spectrum $12^1\, 4^7\, (-2)^{20}$. If A is the adjacency matrix of the triangular graph and B that of one of the Chang graphs, then $S(A) = \mathrm{diag}(1^6, 2^{15}, 8^6, 24^1)$ and $S(B) = \mathrm{diag}(1^8, 2^{12}, 8^7, 24^1)$, so that A and B have different 2-rank.

Example Another example is given by the point graph and the line graph of the $GQ(3,3)$ constructed in §9.6.2. The 2-ranks of the adjacency matrices are 10 and 16, respectively.

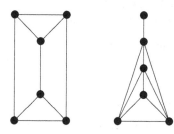

Fig. 13.1 Graphs with the same Laplacian Smith normal form $(1^3, 5, 15, 0)$

GRONE, MERRIS & WATKINS [187] gave the pair of graphs in Figure 13.1 that have $S(L) = \mathrm{diag}(1^3, 5, 15, 0)$, where L is the Laplacian. The Laplace spectrum of the left one (which is $K_2 \,\Box\, K_3$) is 0, 2, 3^2, 5^2. That of the right one is 0, 0.914, 3.572, 5^2, 5.514, where the three nonintegers are roots of $\lambda^3 - 10\lambda^2 + 28\lambda - 18 = 0$.

13.8.1 Smith normal form and spectrum

There is no very direct connection between Smith normal form and spectrum. For example, the matrix $\begin{bmatrix} 3 & 1 \\ 1 & 3 \end{bmatrix}$ has eigenvalues 2 and 4, and invariant factors 1 and 8.

Proposition 13.8.1 *Let M be an integral matrix of order n, with invariant factors $s_1, \ldots, s_n$.*

(i) If a is an integral eigenvalue of M, then $a|s_n$.

(ii) If a is an integral eigenvalue of M with geometric multiplicity m, then $a|s_{n-m+1}$.

(iii) If M is diagonalizable with distinct eigenvalues $a_1, \ldots, a_m$, all integral, then we have $s_n|a_1 a_2 \cdots a_m$.

Proof Part (i) is a special case of (ii). Part (ii) is Proposition 13.8.4 below. For (iii) we may assume that all a_i are nonzero. It suffices to show that every element in $\mathbb{Z}^n / \langle M \rangle$ has an order dividing $a_1 a_2 \cdots a_m$. We show by induction on k that if $u = \sum u_i$ is integral and is sum of k left eigenvectors u_i of M, with $u_i M = a_i u_i$, then $a_1 \cdots a_k u \in \langle M \rangle$. Indeed, since $uM = \sum a_i u_i \in \langle M \rangle$ and $a_k u - uM = \sum (a_k - a_i) u_i$ is integral and sum of at most $k - 1$ eigenvectors, we find by induction that $a_1 \cdots a_{k-1}(a_k u - uM) \in \langle M \rangle$, and hence $a_1 \cdots a_k u \in \langle M \rangle$. $\square$

The invariant factors are determined when we know for each prime p and each $i \geq 0$ how many invariant factors are divisible by p^i, and the following proposition tells us.

Proposition 13.8.2 *Let A be an integral matrix of order n, p a prime number and i a nonnegative integer. Put $M_i := M_i(A) := \{x \in \mathbb{Z}^n \mid p^{-i}Ax \in \mathbb{Z}^n\}$. Let $\overline{M}_i \subseteq \mathbb{F}_p^n$ be the mod p reduction of M_i. Then $\overline{M}_i$ is an $\mathbb{F}_p$-vectorspace, and the number of invariant factors of A divisible by p^i equals $\dim_p \overline{M}_i$.*

Proof $\dim_p \overline{M}_i$ does not change when A is replaced by PAQ, where P and Q are integral matrices of determinant 1. So we may assume that A is already in Smith normal form. Now the statement is obvious. $\square$

There is a dual statement:

Proposition 13.8.3 *Let A be an integral matrix of order n, p a prime number, and i a nonnegative integer. Put $N_i := N_i(A) := \{p^{-i}Ax \mid x \in M_i\}$. Then the number of invariant factors of A not divisible by p^{i+1} equals $\dim_p \overline{N}_i$.*

Proof $\dim_p \overline{N}_i$ does not change when A is replaced by PAQ, where P and Q are integral matrices of determinant 1. So we may assume that A is already in Smith normal form. Now the statement is obvious. $\square$

Proposition 13.8.4 *Let A be a square integral matrix with integral eigenvalue a of (geometric) multiplicity m. Then the number of invariant factors of A divisible by a is at least m.*

Proof Let $W = \{x \in \mathbb{Q}^n \mid Ax = ax\}$ be the a-eigenspace of A over $\mathbb{Q}$, so that $\dim_\mathbb{Q}(W) = m$. By Proposition 13.8.2, it suffices to show that $\dim_p \overline{W} = m$ for all primes p, where $\overline{W}$ is the mod p reduction of $W \cap \mathbb{Z}^n$. Pick a basis $x_1, \ldots, x_m$ of W consisting of m integral vectors, chosen in such a way that the $n \times m$ matrix X that has columns x_j has a (nonzero) minor of order m with the minimum possible number of factors p. If upon reduction mod p these vectors become dependent, that is, if $\sum c_j \overline{x}_j = 0$ where not all c_j vanish, then $\sum c_j x_j$ has coefficients divisible by p, so that $y := \frac{1}{p} \sum c_j x_j \in W \cap \mathbb{Z}^n$, and we can replace some x_j (with nonzero c_j) by y and get a matrix X' where the minors have fewer factors p, contrary to assumption. So, the x_i remain independent upon reduction mod p, and $\dim_p \overline{W} = m$. $\square$

Example Let $q = p^t$ for some prime p. Consider the adjacency matrix A of the graph Γ of which the vertices are the lines of $PG(3,q)$, where two lines are adjacent when they are disjoint. This graph is strongly regular, with eigenvalues $k = q^4$, $r = q$, $s = -q^2$ and multiplicities 1, $f = q^4 + q^2$, $g = q^3 + q^2 + q$, respectively. Since $\det A$ is a power of p, all invariant factors are powers of p. Let p^i occur as invariant factor with multiplicity e_i.

Claim. *We have $e_0 + e_1 + \cdots + e_t = f$ and $e_{2t} + \cdots + e_{3t} = g$ and $e_{4t} = 1$ and $e_i = 0$ for $t < i < 2t$ and $3t < i < 4t$ and $i > 4t$. Moreover, $e_{3t-i} = e_i$ for $0 \le i < t$.*

Proof The total number of invariant factors is the size of the matrix, so $\sum_i e_i = f + g + 1$. The number of factors p in $\det A$ is $\sum_i i e_i = t(f + 2g + 4)$. Hence $\sum_i (i-t)e_i = t(g+3)$.

Let $m_i := \sum_{j \ge i} e_j$. By (the proof of) Proposition 13.8.4 we have $m_{4t} \ge 1$ and $m_{2t} \ge g + 1$. (The $+1$ follows because $\mathbf{1}$ is orthogonal to eigenvectors with eigenvalue other than k, but has a nonzero (mod p) inner product with itself, so that $\mathbf{1} \notin \overline{W}$ for an eigenspace W with $\mathbf{1} \notin W$.)

The matrix A satisfies the equation $(A - rI)(A - sI) = \mu J$, that is, $A(A + q(q - 1)I) = q^3 I + q^3(q - 1)J$, and the right-hand side is divisible by p^{3t}. If $x \in \mathbb{Z}^n$ and $p^{-i}(A + q(q - 1)I)x \in \mathbb{Z}^n$, then $p^{-i}(A + q(q - 1)I)x \in M_{3t-i}(A)$ for $0 \le i \le 3t$. If $0 \le i < t$, then $p^{-i}q(q - 1)x = 0 \pmod{p}$, so that $\overline{N}_i \subseteq \overline{M}_{3t-i}$. Also $\mathbf{1} \in \overline{M}_{3t-i}$, while $\mathbf{1} \notin \overline{N}_i$ because $\mathbf{1}^\top p^{-i}Ax = p^{4t-i}\mathbf{1}^\top x$ reduces to 0 (mod p) for integral x, unlike $\mathbf{1}^\top \mathbf{1}$. By Proposition 13.8.3 we find $m_{3t-i} \ge e_0 + \cdots + e_i + 1$ $(0 \le i < t)$.

Adding the inequalities $-\sum_{0 \le i \le h} e_i + \sum_{i \ge 3t-h} e_i \ge 1$ $(0 \le h < t)$, and $t \sum_{i \ge 2t} e_i \ge t(g + 1)$ and $t \sum_{i \ge 4t} e_i \ge t$ yields

$$\sum_{0 \le i < t} (i - t)e_i + \sum_{2t \le i \le 3t} (i - t)e_i + 2t \sum_{3t+1 \le i < 4t} e_i + 3t \sum_{i \ge 4t} e_i \ge t(g + 3)$$

and equality must hold everywhere since $\sum_i (i - t)e_i = t(g + 3)$.   □

Note that our conclusion also holds for any strongly regular graph with the same parameters as this graph on the lines of $PG(3, q)$.

In the particular case $q = p$, the invariant factors are $1, p, p^2, p^3, p^4$ with multiplicities $e, f - e, g - e, e, 1$, respectively, where $e = \frac{1}{3}p(2p^2 + 1)$ in the case of the lines of $PG(3, p)$ (cf. [148]). Indeed, the number e of invariant factors not divisible by p is the p-rank of A, determined in SIN [326].

For $p = 2$, there are 3854 strongly regular graphs with parameters (35,16,6,8) ([277]), and the 2-ranks occurring are 6, 8, 10, 12, 14 (with frequencies 1, 3, 44, 574, 3232, respectively)—they must be even because A is alternating (mod 2).

The invariant factors of the disjointness graph of the lines of $PG(3, 4)$ are $1^{36} 2^{16}$ 4^{220} 16^{32} 32^{16} 64^{36} 256^1, with multiplicities written as exponents.

One can generalize the above observations, and show for example that if p is a prime, and A is the adjacency matrix of a strongly regular graph, and $p^a||k$, $p^b||r$, $p^c||s$, where $a \ge b + c$ and $p \nmid v$, and A has e_i invariant factors s_j with $p^i||s_j$, then $e_i = 0$ for $\min(b, c) < i < \max(b, c)$ and $b + c < i < a$ and $i > a$. Moreover, $e_{b+c-i} = e_i$ for $0 \le i < \min(b, c)$.

13.9 Exercises

Exercise 13.1 ([55]) Let A be the adjacency matrix of the $n \times n$ grid graph ($n \ge 2$). Then A has Smith normal form $S(A) = \mathrm{diag}(1^{2n-2}, 2^{(n-2)^2}, (2n - 4)^{2n-3}, 2(n - 1)(n - 2))$ and 2-rank $2n - 2$.

Exercise 13.2 ([55]) Let A be the adjacency matrix of the triangular graph $T(n)$ ($n \ge 4$). Then A has Smith normal form

$$S(A) = \begin{cases} \mathrm{diag}(1^{n-2}, 2^{(n-2)(n-3)/2}, (2n - 8)^{n-2}, (n - 2)(n - 4)) & \text{if } n \text{ is even,} \\ \mathrm{diag}(1^{n-1}, 2^{(n-1)(n-4)/2}, (2n - 8)^{n-2}, 2(n - 2)(n - 4)) & \text{if } n \text{ is odd.} \end{cases}$$

The 2-rank of A is $n - 2$ if n is even, and $n - 1$ if n is odd.

Chapter 14
Spectral Characterizations

In this chapter, we consider the question to what extent graphs are determined by their spectrum. First we give several constructions of families of cospectral graphs and then give cases in which it has been shown that the graph is determined by its spectrum.

Let us abbreviate "determined by the spectrum" to DS.[1] Here, of course, "spectrum" (and DS) depends on the type of adjacency matrix. If the matrix is not specified, we mean the ordinary adjacency matrix.

Large parts of this chapter were taken from VAN DAM & HAEMERS [126, 127, 128].

14.1 Generalized adjacency matrices

Let $A = A_\Gamma$ be the adjacency matrix of a graph Γ. The choice of 0, 1, 0 in A to represent equality, adjacency, and nonadjacency was rather arbitrary, and one can more generally consider a matrix $xI + yA + z(J - I - A)$ that uses x, y, z instead. Any such matrix with $y \neq z$ is called a *generalized adjacency matrix* of Γ. The spectrum of any such matrix is obtained by scaling and shifting from that of a matrix of the form $A + yJ$, so for matters of cospectrality we can restrict ourselves to this case.

Call two graphs Γ and Δ *y-cospectral* (for some real y) when $A_\Gamma - yJ$ and $A_\Delta - yJ$ have the same spectrum. Then 0-cospectral is what we called cospectral, $\frac{1}{2}$-cospectral is Seidel-cospectral, and 1-cospectrality is cospectrality for the complementary graphs. Call two graphs *just y-cospectral* when they are y-cospectral but not z-cospectral for any $z \neq y$.

The graphs $K_{1,4}$ and $K_1 + C_4$ are just 0-cospectral. The graphs $2K_3$ and $2K_1 + K_4$ are just $\frac{1}{3}$-cospectral. The graphs $K_1 + C_6$ and $\hat{E}_6$ (cf. §1.3.7) are y-cospectral for all y.

Proposition 14.1.1 *(i)* (JOHNSON & NEWMAN [235])

---

[1] We shall use the somewhat ugly "(non-)DS graph" for "graph (not) determined by the spectrum".

If two graphs are y-cospectral for two distinct values of y, then they are for all y.

(ii) (VAN DAM, HAEMERS & KOOLEN [129]) If two graphs are y-cospectral for an irrational value of y, then they are for all y.

Proof Define $p(x,y) = \det(A_\Gamma - xI - yJ)$. Thus, for fixed y, $p(x,y)$ is the characteristic polynomial of $A_\Gamma - yJ$. Since J has rank 1, the degree in y of $p(x,y)$ is 1 (this follows from Gaussian elimination in $A_\Gamma - xI - yJ$), so there exist integers $a_0,\ldots,a_n$ and $b_0,\ldots,b_n$ such that

$$p(x,y) = \sum_{i=0}^{n}(a_i + b_i y)x^i .$$

Suppose Γ and Γ' are y-cospectral for some $y = y_0$ but not for all y. Then the corresponding polynomials $p(x,y)$ and $p'(x,y)$ are not identical, while $p(x,y_0) = p'(x,y_0)$. This implies that $a_i + b_i y_0 = a_i' + b_i' y_0$ with $b_i \neq b_i'$ for some i. So $y_0 = (a_i' - a_i)/(b_i - b_i')$ is unique and rational.   □

VAN DAM, HAEMERS & KOOLEN [129] show that there is a pair of nonisomorphic just y-cospectral graphs if and only if y is rational. Values of y other than $0, \frac{1}{2}, 1$ occur naturally when studying subgraphs of strongly regular graphs.

Proposition 14.1.2 *Let Γ be strongly regular with vertex set X of size n, and let θ be an eigenvalue other than the valency k. Let $y = (k - \theta)/n$. Then for each subset S of X, the spectrum of Γ and the y-spectrum of the graph induced on S determines the y-spectrum of the graph induced on $X \setminus S$.*

Proof Since $A - yJ$ has only two eigenvalues, this follows immediately from Lemma 2.11.1.   □

This can be used to produce cospectral pairs. For example, let Γ be the Petersen graph, and let S induce a 3-coclique. Then the y-spectrum of the graph induced on $X \setminus S$ is determined by that on S, and does not depend on the coclique chosen. Since θ can take two values, the graphs induced on the complement of a 3-coclique ($\hat{E}_6$ and $K_1 + C_6$) are y-cospectral for all y.

14.2 Constructing cospectral graphs

Many constructions of cospectral graphs are known. Most constructions from before 1988 can be found in [115], §6.1, and [114], §1.3; see also [172], §4.6. More recent constructions of cospectral graphs are presented by SERESS [321], who gives an infinite family of cospectral 8-regular graphs. Graphs cospectral to distance-regular graphs can be found in [54], [126], [206], and in §14.2.2. Notice that the graphs mentioned are regular, so they are cospectral with respect to any generalized adjacency matrix, which in this case includes the Laplace matrix.

There exist many more papers on cospectral graphs. On regular as well as non-regular graphs, and with respect to the Laplace matrix as well as the adjacency matrix. We mention [47], [164], [210], [264], [278] and [304], but don't claim to be complete.

Here we discuss four construction methods for cospectral graphs. One used by Schwenk to construct cospectral trees, one from incidence geometry to construct graphs cospectral with distance-regular graphs, one presented by Godsil and McKay, which seems to be the most productive one, and finally one due to Sunada.

14.2.1 Trees

Let Γ and Δ be two graphs, with vertices x and y, respectively. SCHWENK [308] examined the spectrum of what he called the *coalescence* of these graphs at x and y, namely, the result $\Gamma +_{x,y} \Delta$ of identifying x and y in the disjoint union $\Gamma + \Delta$. He proved the following (see also [115], p. 159, and [172], p. 65).

Lemma 14.2.1 *Let Γ and Γ' be cospectral graphs and let x and x' be vertices of Γ and Γ' respectively. Suppose that $\Gamma - x$ (that is the subgraph of Γ obtained by deleting x) and $\Gamma' - x'$ are cospectral too. Let Δ be an arbitrary graph with a fixed vertex y. Then $\Gamma +_{x,y} \Delta$ is cospectral with $\Gamma' +_{x',y} \Delta$.*

Proof Let z be the vertex of $Z := \Gamma +_{x,y} \Delta$ that is the result of identifying x and y. A directed cycle in Z cannot meet both $\Gamma - x$ and $\Delta - y$. By §1.2.1, the characteristic polynomial $p(t)$ of Z can be expressed in the numbers of unions of directed cycles with given number of vertices and of components. We find $p(t) = p_{\Gamma-x}(t)p_{\Delta}(t) + p_{\Gamma}(t)p_{\Delta-y}(t) - tp_{\Gamma-x}(t)p_{\Delta-y}(t)$. □

For example, let $\Gamma = \Gamma'$ be as given below. Then $\Gamma - x$ and $\Gamma - x'$ are cospectral, because they are isomorphic.

Suppose $\Delta = P_3$ and let y be the vertex of degree 2. Then Lemma 14.2.1 shows that the graphs in Figure 14.1 are cospectral.

Fig. 14.1 Cospectral trees

It is clear that Schwenk's method is very suitable for constructing cospectral trees. In fact, the lemma above enabled him to prove his famous theorem:

Theorem 14.2.2 *With respect to the adjacency matrix, almost all trees are non-DS.*

After Schwenk's result, trees were proved to be almost always non-DS with respect to all kinds of matrices. GODSIL & MCKAY [175] proved that almost all trees are non-DS with respect to the adjacency matrix of the complement $\overline{A}$, while MCKAY [276] proved it for the Laplace matrix L and for the distance matrix D.

14.2.2 Partial linear spaces

A *partial linear space* consists of a (finite) set of points $\mathscr{P}$, and a collection $\mathscr{L}$ of subsets of $\mathscr{P}$ called lines, such that two lines intersect in at most one point (and consequently, two points are on at most one line). Let $(\mathscr{P}, \mathscr{L})$ be such a partial linear space and assume that each line has exactly q points, and each point is on q lines. Then clearly $|\mathscr{P}| = |\mathscr{L}|$. Let N be the point-line incidence matrix of $(\mathscr{P}, \mathscr{L})$. Then $NN^\top - qI$ and $N^\top N - qI$ both are the adjacency matrix of a graph, called the *point graph* (also known as *collinearity graph*) and *line graph* of $(\mathscr{P}, \mathscr{L})$, respectively. These graphs are cospectral, since $NN^\top$ and $N^\top N$ are. But in many examples they are nonisomorphic. An example was given in §12.4.1.

14.2.3 GM switching

Seidel switching was discussed above in §1.8.2. No graph with more than one vertex is DS for the Seidel adjacency matrix. In some cases Seidel switching also leads to cospectral graphs for the adjacency spectrum, for example when graph and switched graph are regular of the same degree.

GODSIL & MCKAY [176] consider a different kind of switching and give conditions under which the adjacency spectrum is unchanged by this operation. We will refer to their method as GM switching. (See also §1.8.3.) Though GM switching was invented to make cospectral graphs with respect to the adjacency matrix, the idea also works for the Laplace matrix and the signless Laplace matrix, as will be clear from the following formulation.

Theorem 14.2.3 *Let N be a $(0,1)$-matrix of size $b \times c$ (say) whose column sums are 0, b or $b/2$. Define $\widetilde{N}$ to be the matrix obtained from N by replacing each column v with $b/2$ ones by its complement $\mathbf{1} - v$. Let B be a symmetric $b \times b$ matrix with constant row (and column) sums, and let C be a symmetric $c \times c$ matrix. Put*

$$M = \begin{bmatrix} B & N \\ N^\top & C \end{bmatrix} \text{ and } \widetilde{M} = \begin{bmatrix} B & \widetilde{N} \\ \widetilde{N}^\top & C \end{bmatrix}.$$

Then M and $\widetilde{M}$ are cospectral.

Proof Define $Q = \begin{bmatrix} \frac{2}{b}J - I_b & 0 \\ 0 & I_c \end{bmatrix}$. Then $Q^{-1} = Q$ and $QMQ^{-1} = \widetilde{M}$. $\square$

The matrix partition used in [176] (and in §1.8.3) is more general than the one presented here. But this simplified version suffices to show that GM switching produces many cospectral graphs.

If M and $\widetilde{M}$ are adjacency matrices of graphs, then GM switching gives cospectral graphs with cospectral complements and hence, by the result of Johnson and Newman quoted in §14.1, it produces cospectral graphs with respect to any generalized adjacency matrix.

If one wants to apply GM switching to the Laplace matrix L of a graph Γ, take $M = -L$ and let B and C (also) denote the sets of vertices indexing the rows and columns of the matrices B and C, respectively. The requirement that the matrix B has constant row sums means that N has constant row sums, that is, the vertices of B all have the same number of neighbors in C.

For the signless Laplace matrix, take $M = Q$. Now all vertices in B must have the same number of neighbors in C, and, in addition, the subgraph of Γ induced by B must be regular.

When Seidel switching preserves the valency of a graph, it is a special case of GM switching, where all columns of N have $b/2$ ones. So the above theorem also gives sufficient conditions for Seidel switching to produce cospectral graphs with respect to the adjacency matrix A and the Laplace matrix L.

If $b = 2$, GM switching just interchanges the two vertices of B, and we call it trivial. But if $b \geq 4$, GM switching almost always produces nonisomorphic graphs.

Fig. 14.2 Two graphs cospectral w.r.t. any generalized adjacency matrix

Fig. 14.3 Two graphs cospectral w.r.t. the Laplace matrix

In Figures 14.2 and 14.3 we have two examples of pairs of cospectral graphs produced by GM switching. In both cases $b = c = 4$ and the upper vertices correspond to B and the lower vertices to C. In the example of Figure 14.2, B induces a regular subgraph and so the graphs are cospectral with respect to every generalized adjacency matrix. In the example of Figure 14.3 all vertices of B have the same number of neighbors in C, so the graphs are cospectral with respect to the Laplace matrix L.

14.2.4 Sunada's method

As a corollary of the discussion in §6.4 we have:

Proposition 14.2.4 *Let Γ be a finite graph, and G a group of automorphisms. If H_1 and H_2 are subgroups of G such that Γ is a cover of Γ/H_i $(i = 1,2)$ and such that each conjugacy class of G meets H_1 and H_2 in the same number of elements, then the quotients Γ/H_i $(i = 1,2)$ have the same spectrum and the same Laplace spectrum.*

Proof The condition given just means that the induced characters $1_{H_i}^G$ $(i = 1,2)$ are the same. Now apply Lemma 6.4.1 with $M = A$ and $M = L$. □

SUNADA [331] did this for manifolds, and the special case of graphs was discussed in [210]. BROOKS [47] shows a converse: any pair of regular connected cospectral graphs arises from this construction.

14.3 Enumeration

14.3.1 Lower bounds

GM switching gives lower bounds for the number of pairs of cospectral graphs with respect to several types of matrices.

Let Γ be a graph on $n - 1$ vertices and fix a set X of three vertices. There is a unique way to extend Γ by one vertex x to a graph Γ' such that $X \cup \{x\}$ induces a regular graph in Γ' and that every other vertex in Γ' has an even number of neighbors in $X \cup \{x\}$. Thus the adjacency matrix of Γ' admits the structure of Theorem 14.2.3, where B corresponds to $X \cup \{x\}$. This implies that from a graph Γ on $n - 1$ vertices one can make $\binom{n-1}{3}$ graphs with a cospectral mate on n vertices (with respect to any generalized adjacency matrix) and every such n-vertex graph can be obtained in four ways from a graph on $n - 1$ vertices. Of course some of these graphs may be isomorphic, but the probability of such a coincidence tends to zero as $n \to \infty$ (see [208] for details). So, if g_n denotes the number of nonisomorphic graphs on n vertices, then:

Theorem 14.3.1 *The number of graphs on n vertices that are non-DS with respect to any generalized adjacency matrix is at least*

$$(\tfrac{1}{24} - o(1))\, n^3 g_{n-1}.$$

The fraction of graphs with the required condition with $b = 4$ for the Laplace matrix is roughly $2^{-n} n\sqrt{n}$. This leads to the following lower bound (again see [208] for details):

Theorem 14.3.2 *The number of non-DS graphs on n vertices with respect to the Laplace matrix is at least*

$$rn\sqrt{n}g_{n-1},$$

for some constant r > 0.

In fact, a lower bound like the one in Theorem 14.3.2 can be obtained for any matrix of the form $A + \alpha D$, including the signless Laplace matrix Q.

14.3.2 Computer results

GODSIL & MCKAY [175, 176] give interesting computer results for small cospectral graphs. In [176] all graphs up to 9 vertices are generated and checked on cospectrality. This enumeration has been extended to 11 vertices by HAEMERS & SPENCE [208], and cospectrality was tested with respect to the adjacency matrix A, the set of generalized adjacency matrices ($A \& \overline{A}$), the Laplace matrix L, and the signless Laplace matrix Q. The results are given in Table 14.1, where we list the fractions of non-DS graphs for each of the four cases. The last three columns give the fractions of graphs for which GM switching gives cospectral nonisomorphic graphs with respect to A, L and Q, respectively. Since GM switching gives cospectral graphs with cospectral complements, column GM-A gives a lower bound for column $A \& \overline{A}$.

n	# graphs	A	$A \& \overline{A}$	L	Q	GM-A	GM-L	GM-Q
2	2	0	0	0	0	0	0	0
3	4	0	0	0	0	0	0	0
4	11	0	0	0	0.182	0	0	0
5	34	0.059	0	0	0.118	0	0	0
6	156	0.064	0	0.026	0.103	0	0	0
7	1044	0.105	0.038	0.125	0.098	0.038	0.069	0
8	12346	0.139	0.094	0.143	0.097	0.085	0.088	0
9	274668	0.186	0.160	0.155	0.069	0.139	0.110	0
10	12005168	0.213	0.201	0.118	0.053	0.171	0.080	0.001
11	1018997864	0.211	0.208	0.090	0.038	0.174	0.060	0.001
12	165091172592	0.188		0.060	0.027			

Table 14.1 Fractions of non-DS graphs

Notice that for $n \leq 4$ there are no cospectral graphs with respect to A or to L, but there is one such pair with respect to Q, namely $K_{1,3}$ and $K_1 + K_3$. For $n = 5$ there is just one pair with respect to A. This is of course the Saltire pair ($K_{1,4}$ and $K_1 + C_4$).

An interesting result from the table is that the fraction of non-DS graphs is non-decreasing for small n, but starts to decrease at $n = 10$ for A, at $n = 9$ for L, and at $n = 6$ for Q. Especially for the Laplace matrix and the signless Laplace matrix, these data suggest that the fraction of DS graphs might tend to 1 as $n \to \infty$. In addition, the table shows that the majority of non-DS graphs with respect to $A\&\overline{A}$ and L comes from GM switching (at least for $n \geq 7$). If this tendency continues, the lower bounds given in Theorems 14.3.1 and 14.3.2 will be asymptotically tight (with maybe another constant) and almost all graphs will be DS for all three cases. Indeed, the fraction of graphs that admit a nontrivial GM switching tends to zero as n tends to infinity, and the partitions with $b = 4$ account for most of these switchings (see also [176]). Results for the normalized Laplacian, and for trees, are given in [351]. For the data for $n = 12$, see [66] and [329].

14.4 DS graphs

In §14.2 we saw that many constructions for non-DS graphs are known, and in the previous section we remarked that it seems more likely that almost all graphs are DS, than that almost all graphs are non-DS. Yet much less is known about DS graphs than about non-DS graphs. For example, we do not know of a satisfying counterpart to the lower bounds for non-DS graphs given in §14.3.1. The reason is that it is not easy to prove that a given graph is DS. Below we discuss the graphs known to be DS. The approach is via structural properties of a graph that follow from the spectrum. So let us start with a short survey of such properties.

14.4.1 Spectrum and structure

Let us first investigate for which matrices one can see from the spectrum whether the graph is regular.

Proposition 14.4.1 *Let D denote the diagonal matrix of degrees. If a regular graph is cospectral with a nonregular one with respect to the matrix $R = A + \beta J + \gamma D + \delta I$, then $\gamma = 0$ and $-1 < \beta < 0$.*

Proof Without loss of generality, $\delta = 0$. Let Γ be a graph with the given spectrum, and suppose that Γ has n vertices and vertex degrees d_i $(1 \leq i \leq n)$.

First suppose that $\gamma \neq 0$. Then $\sum_i d_i$ is determined by $\text{tr}(R)$ and hence by the spectrum of R. Since $\text{tr}(R^2) = \beta^2 n^2 + (1 + 2\beta + 2\beta\gamma)\sum_i d_i + \gamma^2 \sum_i d_i^2$, it follows that also $\sum_i d_i^2$ is determined by the spectrum. Now Cauchy's inequality states that $(\sum_i d_i)^2 \leq n \sum_i d_i^2$ with equality if and only if $d_1 = \ldots = d_n$. This shows that regularity of the graph can be seen from the spectrum of R.

Now suppose $\gamma = 0$ and $\beta \neq -1/2$. By considering $\text{tr}(R^2)$ we see that $\sum_i d_i$ is determined by the spectrum of R. The matrix $R = A + \beta J$ has average row sum $r = \beta n + \sum_i d_i/n$ determined by its spectrum. Let R have eigenvalues $\theta_1 \geq \ldots \geq \theta_n$. By interlacing, $\theta_1 \geq r \geq \theta_n$, and equality on either side implies that R has constant row sums, and Γ is regular. On the other hand, if $\beta \geq 0$ (resp. $\beta \leq -1$), then R

(resp. $-R$) is a nonnegative matrix, hence if Γ is regular, then **1** is an eigenvector for eigenvalue $r = \theta_1$ (resp. $r = -\theta_n$). Thus also here regularity of the graph can be seen from the spectrum. □

It remains to see whether one can see from the spectrum of $A - yJ$ (with $0 < y < 1$) whether the graph is regular. For $y = \frac{1}{2}$ the answer is clearly no: The Seidel adjacency matrix is $S = J - I - 2A$, and for S a regular graph can be cospectral with a nonregular one (e.g. K_3 and $K_1 + K_2$), or with another regular one with different valency (e.g. $4K_1$ and C_4). CHESNOKOV & HAEMERS [89] constructed pairs of y-cospectral graphs where one is regular and the other not for all rational y, $0 < y < 1$. Finally, if y is irrational, then one can deduce regularity from the spectrum of $A - yJ$ by Proposition 14.1.1(ii).

Corollary 14.4.2 *For regular graphs, being DS (or not DS) is equivalent for the adjacency matrix, the adjacency matrix of the complement, the Laplace matrix, and the signless Laplace matrix.*

Proof For each of these matrices the above proposition says that regularity can be recognized. It remains to find the valency k. For $A, \overline{A}, Q$, the largest eigenvalue is k, $n - 1 - k$, $2k$, respectively. For L, the trace is nk. □

Lemma 14.4.3 *For the adjacency matrix, the Laplace matrix and the signless Laplace matrix of a graph Γ, the following can be deduced from the spectrum:*

   (i) *the number of vertices,*
   (ii) *the number of edges,*
   (iii) *whether Γ is regular,*
   (iv) *whether Γ is regular with any fixed girth.*

For the adjacency matrix the following follows from the spectrum:

   (v) *the number of closed walks of any fixed length,* ·
   (vi) *whether Γ is bipartite.*

For the Laplace matrix the following follows from the spectrum:

   (vii) *the number of components,*
   (viii) *the number of spanning trees.*

Proof Part (i) is clear. For L and Q the number of edges is twice the trace of the matrix, while parts (ii) and (v) for A were shown in Proposition 1.3.1. Part (vi) follows from (v), since Γ is bipartite if and only if Γ has no closed walks of odd length. Part (iii) follows from Proposition 14.4.1, and (iv) follows from (iii) and the fact that in a regular graph the number of closed walks of length less than the girth depends on the degree only. Parts (vii) and (viii) follow from Propositions 1.3.7 and 1.3.4. □

The Saltire pair shows that (vii) and (viii) do not hold for the adjacency matrix. The two graphs of Figure 14.4 have cospectral Laplace matrices. They illustrate that (v) and (vi) do not follow from the Laplace spectrum. The graphs $K_1 + K_3$ and $K_{1,3}$ show that (v)–(viii) are false for the signless Laplace matrix.

Fig. 14.4 Two graphs cospectral w.r.t. the Laplace matrix
(Laplace spectrum: $0, 3 - \sqrt{5}, 2, 3, 3, 3 + \sqrt{5}$)

14.4.2 Some DS graphs

Lemma 14.4.3 immediately leads to some DS graphs.

Proposition 14.4.4 *The graphs K_n and $K_{m,m}$ and C_n and their complements are DS for any matrix $R = A + \beta J + \gamma D + \delta I$ for which regularity follows from the spectrum of R. In particular this holds for the matrices A, $\bar{A}$, L and R.*

Proof Since these graphs are regular, we only need to show that they are DS with respect to the adjacency matrix. A graph cospectral with K_n has n vertices and $n(n-1)/2$ edges and therefore equals K_n. A graph cospectral with $K_{m,m}$ is regular and bipartite with $2m$ vertices and m^2 edges, so it is isomorphic to $K_{m,m}$. A graph cospectral with C_n is 2-regular with girth n, so it equals C_n. $\qquad\square$

Proposition 14.4.5 *The disjoint union of k complete graphs, $K_{m_1} + \ldots + K_{m_k}$, is DS with respect to the adjacency matrix.*

Proof The spectrum of the adjacency matrix A of any graph cospectral with $K_{m_1} + \ldots + K_{m_k}$ equals $\{[m_1 - 1]^1, \ldots, [m_k - 1]^1, [-1]^{n-k}\}$, where $n = m_1 + \ldots + m_k$. This implies that $A + I$ is positive semidefinite of rank k, and hence $A + I$ is the matrix of inner products of n vectors in $\mathbb{R}^k$. All these vectors are unit vectors, and the inner products are 1 or 0. So two such vectors coincide or are orthogonal. This clearly implies that the vertices can be ordered in such a way that $A + I$ is a block diagonal matrix with all-1 diagonal blocks. The sizes of these blocks are nonzero eigenvalues of $A + I$. $\qquad\square$

This proposition shows that a complete multipartite graph is DS with respect to $\bar{A}$. In general, the disjoint union of complete graphs is not DS with respect to $\bar{A}$ and L. The Saltire pair shows that $K_1 + K_4$ is not DS for $\bar{A}$, and $K_5 + 5K_2$ is not DS for L, because it is cospectral with the Petersen graph extended by five isolated vertices (both graphs have Laplace spectrum $[0]^6 [2]^5 [5]^4$). See also BOULET [40].

Proposition 14.4.6 *The path with n vertices is determined by the spectrum of its adjacency matrix. More generally, each connected graph with largest eigenvalue less than 2 is determined by its spectrum.*

Proof Let Γ be connected with n vertices and have largest eigenvalue less than 2, and let the graph Δ be cospectral. Then Δ does not contain a cycle, and has $n-1$ edges, so is a tree. By Theorem 3.1.3 (and following remarks) we find that Δ is one of $A_n = P_n$, D_n, E_6, E_7, E_8, and has largest eigenvalue $2\cos\frac{\pi}{h}$, where h is the Coxeter number. Now Δ is determined by n and h, that is, by its number of vertices and its largest eigenvalue.   □

In fact, P_n is also DS with respect to $\overline{A}$, L, and Q. The result for $\overline{A}$, however, is nontrivial and the subject of [147]. The hypothesis "connected" here is needed, but we can describe precisely which pairs of graphs with largest eigenvalue less than 2 are cospectral.

Proposition 14.4.7

> (i) $D_{n+2} + P_n$ is cospectral with $P_{2n+1} + P_1$ for $n \geq 2$.
> (ii) $D_7 + P_2$ is cospectral with $E_6 + P_3$.
> (iii) $D_{10} + P_2$ is cospectral with $E_7 + P_5$.
> (iv) $D_{16} + P_4 + P_2$ is cospectral with $E_8 + P_9 + P_5$.
> (v) If two graphs Γ and Δ with largest eigenvalue less than 2 are cospectral, then there exist integers a, b, c such that $\Delta + aP_4 + bP_2 + cP_1$ arises from $\Gamma + aP_4 + bP_2 + cP_1$ by (possibly repeatedly) replacing some connected components by some others cospectral with the replaced ones according to (i)–(iv).

For example, $P_{11} + P_2 + P_1$ is cospectral with $E_6 + P_5 + P_3$, and $P_{17} + P_2 + P_1$ is cospectral with $E_7 + P_8 + P_5$, and $P_{29} + P_4 + P_2 + P_1$ is cospectral with $E_8 + P_{14} + P_9 + P_5$, and $E_6 + D_{10} + P_7$ is cospectral with $E_7 + D_5 + P_{11}$, and $E_7 + D_4$ is cospectral with $D_{10} + P_1$, and $E_8 + D_6 + D_4$ is cospectral with $D_{16} + 2P_1$.

It follows that $P_{n_1} + \ldots + P_{n_k}$ (with $n_i > 1$ for all i) and $D_{n_1} + \ldots + D_{n_k}$ (with $n_i > 3$ for all i) are DS.

We do not know whether $P_{n_1} + \ldots + P_{n_k}$ is DS with respect to $\overline{A}$. But it is easy to show that this graph is DS for L and for Q.

Proposition 14.4.8 *The union of k disjoint paths, $P_{n_1} + \ldots + P_{n_k}$ each having at least one edge, is DS with respect to the Laplace matrix L and the signless Laplace matrix Q.*

Proof The Laplace eigenvalues of P_n are $2 + 2\cos\frac{\pi i}{n}$, $i = 1, \ldots, n$ (see §1.4.4). Since P_n is bipartite, the signless Laplace eigenvalues are the same (see Proposition 1.3.10).

Suppose Γ is a graph cospectral with $P_{n_1} + \ldots + P_{n_k}$ with respect to L. Then all eigenvalues of L are less than 4. Lemma 14.4.3 implies that Γ has k components and $n_1 + \ldots + n_k - k$ edges, so Γ is a forest. Let L' be the Laplace matrix of $K_{1,3}$. The spectrum of L' equals $[0]^1 [1]^2 [4]^1$. If degree 3 (or more) occurs in Γ, then $L' + D$ is a principal submatrix of L for some diagonal matrix D with nonnegative entries. But then $L' + D$ has largest eigenvalue at least 4, a contradiction. So the degrees in Γ are at most two and hence Γ is the disjoint union of paths. The length m (say) of the longest path follows from the largest eigenvalue. Then the other lengths follow recursively by deleting P_m from the graph and the eigenvalues of P_m from the spectrum.

For a graph Γ' cospectral with $P_{n_1} + \ldots + P_{n_k}$ with respect to Q, the first step is to see that Γ' is a forest. But a circuit in Γ' gives a submatrix L' in Q with all row sums at least 4. So L' has an eigenvalue at least 4, a contradiction (by Corollary 2.5.2), and it follows that Γ' is a forest and hence bipartite. Since for bipartite graphs L and Q have the same spectrum, Γ' is also cospectral with $P_{n_1} + \ldots + P_{n_k}$ with respect to L, and we are done.   □

The above two propositions show that for A, $\overline{A}$, L, and Q the number of DS graphs on n vertices is bounded below by the number of partitions of n, which is asymptotically equal to $2^{\alpha\sqrt{n}}$ for some constant α. This is clearly a very poor lower bound, but we know of no better one.

We have seen that the disjoint union of some DS graphs is not necessarily DS. One might wonder whether the disjoint union of regular DS graphs with the same degree is always DS. The disjoint union of cycles is DS, as can be shown by an argument similar to that in the proof of Proposition 14.4.8. Also the disjoint union of some copies of a strongly regular DS graph is DS. In general we expect a negative answer, however.

14.4.3 Line graphs

The smallest adjacency eigenvalue of a line graph is at least -2 (see §1.4.5). Other graphs with least adjacency eigenvalue -2 are the cocktail party graphs ($\overline{mK_2}$, the complement of m disjoint edges) and the so-called generalized line graphs, which are common generalizations of line graphs and cocktail party graphs (see [114], Ch. 1). We will not need the definition of a generalized line graph, but only use the fact that if a generalized line graph is regular, it is a line graph or a cocktail party graph. Graphs with least eigenvalue -2 have been characterised by CAMERON, GOETHALS, SEIDEL & SHULT [84] (cf. §8.4). They prove that such a graph is a generalized line graph or is in a finite list of exceptions that comes from root systems. Graphs in this list are called *exceptional graphs*. A consequence of the above characterization is the following result of CVETKOVIĆ & DOOB [113], Thm. 5.1; see also [114], Thm. 1.8.

Theorem 14.4.9 *Suppose a regular graph Δ has the adjacency spectrum of the line graph $L(\Gamma)$ of a connected graph Γ. Suppose Γ is not one of the fifteen regular 3-connected graphs on 8 vertices, or $K_{3,6}$, or the semiregular bipartite graph with 9 vertices and 12 edges. Then Δ is the line graph $L(\Gamma')$ of a graph Γ'.*

It does not follow that the line graph of a connected regular DS graph, which is not one of the exceptions mentioned, is DS itself. The reason is that it can happen that two noncospectral graphs Γ and Γ' have cospectral line graphs. For example, both $L(K_6)$ and $K_{6,10}$ have a line graph with spectrum $14^1\ 8^5\ 4^9\ -2^{45}$, and both L(Petersen) and the incidence graph of the 2-$(6,3,2)$ design have a line graph with spectrum $6^1\ 4^5\ 1^4\ 0^5\ -2^{15}$. The following lemma gives necessary conditions for this phenomenon (cf. [76], Thm. 1.7).

Lemma 14.4.10 *Let Γ be a k-regular connected graph on n vertices and let Γ' be a connected graph such that $L(\Gamma)$ is cospectral with $L(\Gamma')$. Then either Γ' is* *cospectral with Γ, or Γ' is a semiregular bipartite graph with $n+1$ vertices and $nk/2$ edges, where $(n,k) = (b^2-1, ab)$ for integers a and b with $a \leq \frac{1}{2}b$.*

Proof Suppose that Γ has m edges. Then $L(\Gamma)$ has m vertices.

If N is the point-edge incidence matrix of Γ, then $NN^\top$ is the signless Laplace matrix of Γ, and $NN^\top - kI$ is the adjacency matrix of Γ, and $N^\top N - 2I$ is the adjacency matrix of $L(\Gamma)$. Since Γ is connected, the matrix N has eigenvalue 0 with multiplicity 1 if Γ is bipartite, and does not have eigenvalue 0 otherwise. Consequently, $L(\Gamma)$ has eigenvalue -2 with multiplicity $m-n+1$ if Γ is bipartite, and with multiplicity $m-n$ otherwise. If $\eta \neq 0$, then the multiplicity of $\eta - 2$ as eigenvalue of $L(\Gamma)$ equals the multiplicity of $\eta - k$ as eigenvalue of Γ.

We see that for a regular connected graph Γ, the spectrum of $L(\Gamma)$ determines that of Γ (since $L(G)$ is regular of valency $2k-2$ and n is determined by $m = \frac{1}{2}nk$).

Since $L(\Gamma')$ is cospectral with $L(\Gamma)$, also Γ' has m edges. $L(\Gamma')$ is regular and hence Γ' is regular or semiregular bipartite. Suppose that Γ' is not cospectral with Γ. Then Γ' is semiregular bipartite with parameters (n_1, n_2, k_1, k_2) (say), and

$$m = \tfrac{1}{2}nk = n_1 k_1 = n_2 k_2.$$

Since the signless Laplace matrices Q and Q' of Γ and Γ' have the same nonzero eigenvalues, their largest eigenvalues are equal:

$$2k = k_1 + k_2.$$

If $n = n_1 + n_2$, then $k_1 = k_2$, a contradiction. So

$$n = n_1 + n_2 - 1.$$

Write $k_1 = k - a$ and $k_2 = k + a$. Then $nk = n_1 k_1 + n_2 k_2$ yields

$$k = (n_1 - n_2)a.$$

Now $n_1 k_1 = n_2 k_2$ gives

$$(n_1 - n_2)^2 = n_1 + n_2.$$

Put $b = n_1 - n_2$. Then $(n,k) = (b^2 - 1, ab)$. Since $2ab = k_1 + k_2 \leq n_2 + n_1 = b^2$, it follows that $a \leq \frac{1}{2}b$. $\square$

Now the following can be concluded from Theorem 14.4.9 and Lemma 14.4.10.

Theorem 14.4.11 *Suppose Γ is a connected regular DS graph, which is not a 3-connected graph with 8 vertices or a regular graph with $b^2 - 1$ vertices and degree ab for some integers a and b, with $a \leq \frac{1}{2}b$. Then also the line graph $L(\Gamma)$ of Γ is DS.*

BUSSEMAKER, CVETKOVIĆ & SEIDEL [76] determined all connected regular exceptional graphs (see also [119]). There are exactly 187 such graphs, of which 32 are DS. This leads to the following characterization.

Theorem 14.4.12 *Suppose Γ is a connected regular DS graph with all its adjacency eigenvalues at least -2. Then one of the following occurs.*

(i) Γ is the line graph of a connected regular DS graph.
(ii) Γ is the line graph of a connected semiregular bipartite graph, which is DS with respect to the signless Laplace matrix.
(iii) Γ is a cocktail party graph.
(iv) Γ is one of the 32 connected regular exceptional DS graphs.

Proof Suppose Γ is not an exceptional graph or a cocktail party graph. Then Γ is the line graph of a connected graph Δ, say. WHITNEY [348] has proved that Δ is uniquely determined from Γ, unless $\Gamma = K_3$. If this is the case, then $\Gamma = L(K_3) = L(K_{1,3})$, so (i) holds. Suppose Δ' is cospectral with Δ with respect to the signless Laplace matrix Q. Then Γ and $L(\Delta')$ are cospectral with respect to the adjacency matrix, so $\Gamma = L(\Delta')$ (since Γ is DS). Hence $\Delta = \Delta'$. Because Γ is regular, Δ must be regular, or semiregular bipartite. If Δ is regular, DS with respect to Q is the same as DS. $\square$

All four cases from Theorem 14.4.12 do occur. For (i) and (iv) this is obvious, and (iii) occurs because the cocktail party graphs $\overline{mK_2}$ are DS (since they are regular and $\overline{A}$-cospectral by Proposition 14.4.5). Examples for Case (ii) are the complete graphs $K_n = L(K_{1,n})$ with $n \neq 3$. Thus, the fact that K_n is DS implies that $K_{1,n}$ is DS with respect to Q if $n \neq 3$.

14.5 Distance-regular graphs

All regular DS graphs constructed so far have the property that either the adjacency matrix A or the adjacency matrix $\overline{A}$ of the complement has smallest eigenvalue at least -2. In this section we present other examples.

Recall that a distance-regular graph with diameter d has $d+1$ distinct eigenvalues and that its (adjacency) spectrum can be obtained from the intersection array. Conversely, the spectrum of a distance-regular graph determines the intersection array (see, e.g., [126]). However, in general the spectrum of a graph doesn't tell you whether it is distance-regular or not.

For $d \geq 3$ we have constructed graphs cospectral with, but nonisomorphic to $H(d,d)$ in §14.2.2. Many more examples are given in [206] and [130].

In the theory of distance-regular graphs an important question is: "Which graphs are determined by their intersection array?" For many distance-regular graphs this is known to be the case. Here we investigate in the cases where the graph is known to be determined by its intersection array, whether it is in fact already determined by its spectrum.

14.5.1 Strongly regular DS graphs

The spectrum of a graph Γ determines whether Γ is strongly regular. Indeed, by Proposition 3.3.1 we can see whether Γ is regular. And a regular graph with spectrum $\theta_1 \geq \ldots \geq \theta_n$ is strongly regular if and only if $|\{\theta_i \mid 2 \leq i \leq n\}| = 2$.

(That is, a regular graph is strongly regular if and only if either it is connected, and then has precisely three distinct eigenvalues: its valency and two others, or it is the disjoint union aK_ℓ ($a \geq 2$, $\ell \geq 2$) of a complete graphs of size ℓ.)

Indeed, if Γ has valency k and all eigenvalues θ_i with $i > 1$ are in $\{r,s\}$, then $(A - rI)(A - sI) = cJ$ so that A^2 is a linear combination of A, I and J, and Γ is strongly regular.

By Propositions 14.4.4 and 14.4.5 and Theorem 14.4.11, we find the following infinite families of strongly regular DS graphs.

Proposition 14.5.1 *If $n \neq 8$ and $m \neq 4$, the graphs aK_ℓ, $L(K_n)$, $L(K_{m,m})$, and their complements are strongly regular DS graphs.*

Note that $L(K_n)$ is the triangular graph $T(n)$, and $L(K_{m,m})$ is the lattice graph $L_2(n)$. For $n = 8$ and $m = 4$ cospectral graphs exist. There is exactly one graph cospectral with $L(K_{4,4})$, the Shrikhande graph ([325]), and there are three graphs cospectral with $L(K_8)$, the so-called Chang graphs ([87]). See also §9.2.

Besides the graphs of Proposition 14.5.1, only a few strongly regular DS graphs are known; these are surveyed in Table 14.2. (Here a *local* graph of a graph Γ is the subgraph induced by the neighbors of a vertex of Γ.)

v	Spectrum			Name	Reference
5	2	$[(-1 \pm \sqrt{5})/2]^2$		pentagon	
13	6	$[(-1 \pm \sqrt{13})/2]^6$		Paley	[316]
17	8	$[(-1 \pm \sqrt{17})/2]^8$		Paley	[316]
16	5	1^{10}	$(-3)^5$	folded 5-cube	[315]
27	10	1^{20}	$(-5)^6$	GQ(2,4)	[315]
50	7	2^{28}	$(-3)^{21}$	Hoffman-Singleton	[206]
56	10	2^{35}	$(-4)^{20}$	Gewirtz	[170], [58]
77	16	2^{55}	$(-6)^{21}$	M_{22}	[48]
81	20	2^{60}	$(-7)^{20}$	Brouwer-Haemers	[57], §9.7
100	22	2^{77}	$(-8)^{22}$	Higman-Sims	[170]
105	32	2^{84}	$(-10)^{20}$	flags of PG(2,4)	[137]
112	30	2^{90}	$(-10)^{21}$	GQ(3,9)	[83]
120	42	2^{99}	$(-12)^{20}$	001... in $S(5,8,24)$	[137]
126	50	2^{105}	$(-13)^{20}$	Goethals	[105]
162	56	2^{140}	$(-16)^{21}$	local McLaughlin	[83]
176	70	2^{154}	$(-18)^{21}$	01... in $S(5,8,24)$	[137]
275	112	2^{252}	$(-28)^{22}$	McLaughlin	[180]

Table 14.2 The known sporadic strongly regular DS graphs (up to complements)

Being DS seems to be a very strong property for strongly regular graphs. Most strongly regular graphs have (many) cospectral mates. For example, there are exactly 32548 nonisomorphic strongly regular graphs with spectrum $15, 3^{15}, (-3)^{20}$ ([277]). Other examples can be found in the survey [51]. FON-DER-FLAASS [161] showed that the number of nonisomorphic cospectral strongly regular graphs on at most n vertices grows exponentially in n. This implies that almost all strongly regular graphs are non-DS. One might be tempted to conjecture that there are only finitely many strongly regular DS graphs besides the ones from Proposition 14.5.1.

14.5.2 Distance-regularity from the spectrum

If $d \geq 3$, only in some special cases does it follow from the spectrum of a graph that it is distance-regular. The following result surveys the cases known to us.

Theorem 14.5.2 *If Γ is a distance-regular graph with diameter d and girth g satisfying one of the following properties, then every graph cospectral with Γ is also distance-regular, with the same parameters as Γ.*

(i) $g \geq 2d - 1$.
(ii) $g \geq 2d - 2$ and Γ is bipartite.
(iii) $g \geq 2d - 2$ and $c_{d-1}c_d < -(c_{d-1}+1)(\theta_1 + \ldots + \theta_d)$.
(iv) Γ is a generalized Odd graph, that is, $a_1 = \ldots = a_{d-1} = 0, a_d \neq 0$.
(v) $c_1 = \ldots = c_{d-1} = 1$.
(vi) Γ is the dodecahedron, or the icosahedron.
(vii) Γ is the coset graph of the extended ternary Golay code.
(viii) Γ is the Ivanov-Ivanov-Faradjev graph.

For parts (i), (iv) and (vi), see [58] (and also [200]), [228], and [206], respectively. Parts (ii), (iii), (v), (vii) are proved in [126] (in fact, (ii) is a special case of (iii)) and (viii) is proved in [130]. Notice that the polygons C_n and the strongly regular graphs are special cases of (i), while bipartite distance-regular graphs with $d = 3$ (these are the incidence graphs of symmetric block designs, see also [115], Thm. 6.9) are a special case of (ii).

An important result on spectral characterizations of distance-regular graphs is the following theorem of FIOL & GARRIGA [158], a direct consequence of Theorem 12.11.1.

Theorem 14.5.3 *Let Γ be a distance-regular graph with diameter d and $k_d = |\Gamma_d(u)|$ vertices at distance d from any given vertex u. If Γ' is cospectral with Γ and $|\Gamma'_d(x)| = k_d$ for every vertex x of Γ', then Γ' is distance-regular.*

Let us illustrate the use of this theorem by proving Case (i) of Theorem 14.5.2. Since the girth and the degree follow from the spectrum, any graph Γ' cospectral with Γ also has girth g and degree k_1. Fix a vertex x in Γ'. Clearly, $c_{x,y} = 1$ for every vertex y at distance at most $(g-1)/2$ from x, and $a_{x,y} = 0$ (where $a_{x,y}$ is the number of neighbors of y at distance $d(x,y)$ from x) if the distance between x and y is at most $(g-2)/2$. This implies that the number k_i' of vertices at distance i from x equals $k_1(k_1-1)^{i-1}$ for $i = 1,\ldots,d-1$. Hence $k_i' = k_i$ for these i. But then also $k_d' = k_d$ and Γ' is distance-regular by Theorem 14.5.3.

14.5.3 Distance-regular DS graphs

BROUWER, COHEN & NEUMAIER [54] gives many distance-regular graphs determined by their intersection array. We only need to check which ones satisfy one of the properties of Theorem 14.5.2. First we give the known infinite families:

Proposition 14.5.4 *The following distance-regular graphs are DS:*
(i) the polygons C_n,
(ii) the complete bipartite graphs minus a perfect matching,
(iii) the Odd graphs O_{d+1},
(iv) the folded $(2d+1)$-cubes.

Proof Part (i) follows from Theorem 14.5.2 (i) (and Proposition 14.4.4). Part (ii) follows from Theorem 14.5.2 (ii). The graphs of parts (iii) and (iv) are generalized Odd graphs, so the result follows from Theorem 14.5.2 (iv). □

Next, there are the infinite families where the spectrum determines the combinatorial or geometric structure, where the graphs are DS if and only if the corresponding structure is determined by its parameters.

Proposition 14.5.5 *A graph cospectral with the incidence graph of a symmetric block design with parameters* 2-(v,k,λ) *is itself the incidence graph of a symmetric block design with these same parameters.*

The designs known to be uniquely determined by their parameters are the six projective planes $PG(2,q)$ for $q = 2,3,4,5,7,8$, and the biplane 2-$(11,5,2)$, and their complementary designs with parameters 2-$(v, v-k, v-2k+\lambda)$.

The remaining known distance-regular DS graphs are presented in Tables 14.3, 14.4, and 14.5. For all but one graph, the fact that they are unique (that is, determined by their parameters) can be found in [54]. Uniqueness of the Perkel graph has been proved only recently [103]. The last columns in the tables refer to the relevant theorems by which distance-regularity follows from the spectrum. In these tables we denote by $IG(v,k,\lambda)$ the point-block incidence graph of a 2-(v,k,λ) design, and by GH, GO, and GD the point graph of a generalized hexagon, generalized octagon, and generalized dodecagon, respectively.

Recall that the point graph of a $GH(1,q)$ ($GO(1,q)$, $GD(1,q)$) is the point-line incidence graph of a projective plane (generalized quadrangle, generalized hexagon) of order q. Recall that the point graph of a $GH(q,1)$ is the line graph of the dual $GH(1,q)$, that is, the line graph of the point-line incidence graph (also known as the *flag graph*) of a projective plane of order q.

Finally, $\mathcal{G}_{23}$, $\mathcal{G}_{21}$, and $\mathcal{G}_{12}$ denote the binary Golay code, the doubly truncated binary Golay code, and the extended ternary Golay code, and HoSi is the Hoffman-Singleton graph.

n		Spectrum			g	Name	Theorem
12	5	$\sqrt{5}^3$	$(-1)^5$	$(-\sqrt{5})^3$	3	icosahedron	14.5.2vi
14	3	$\sqrt{2}^6$	$(-\sqrt{2})^6$	$(-3)^1$	6	Heawood; $GH(1,2)$	14.5.2i
14	4	$\sqrt{2}^6$	$(-\sqrt{2})^6$	$(-4)^1$	4	$IG(7,4,2)$	14.5.2ii
15	4	2^5	$(-1)^4$	$(-2)^5$	3	L(Petersen)	14.4.11
21	4	$(1+\sqrt{2})^6$	$(1-\sqrt{2})^6$	$(-2)^8$	3	$GH(2,1)$	14.5.2v
22	5	$\sqrt{3}^{10}$	$(-\sqrt{3})^{10}$	$(-5)^1$	4	$IG(11,5,2)$	14.5.2ii
22	6	$\sqrt{3}^{10}$	$(-\sqrt{3})^{10}$	$(-6)^1$	4	$IG(11,6,3)$	14.5.2ii
26	4	$\sqrt{3}^{12}$	$(-\sqrt{3})^{12}$	$(-4)^1$	6	$GH(1,3)$	14.5.2i
26	9	$\sqrt{3}^{12}$	$(-\sqrt{3})^{12}$	$(-9)^1$	4	$IG(13,9,6)$	14.5.2ii
36	5	2^{16}	$(-1)^{10}$	$(-3)^9$	5	Sylvester	14.5.2i
42	6	2^{21}	$(-1)^6$	$(-3)^{14}$	5	antipodal 6-cover of K_7	14.5.2i
42	5	2^{20}	$(-2)^{20}$	$(-5)^1$	6	$GH(1,4)$	14.5.2i
42	16	2^{20}	$(-2)^{20}$	$(-16)^1$	4	$IG(21,16,12)$	14.5.2ii
52	6	$(2+\sqrt{3})^{12}$	$(2-\sqrt{3})^{12}$	$(-2)^{27}$	3	$GH(3,1)$	14.5.2v
57	6	$(\frac{3+\sqrt{5}}{2})^{18}$	$(\frac{3-\sqrt{5}}{2})^{18}$	$(-3)^{20}$	5	Perkel	14.5.2i
62	6	$\sqrt{5}^{30}$	$(-\sqrt{5})^{30}$	$(-6)^1$	6	$GH(1,5)$	14.5.2i
62	25	$\sqrt{5}^{30}$	$(-\sqrt{5})^{30}$	$(-25)^1$	4	$IG(31,25,20)$	14.5.2ii
63	8	$\sqrt{8}^{27}$	$(-1)^8$	$(-\sqrt{8})^{27}$	4	antipodal 7-cover of K_9	14.5.2v
105	8	5^{20}	1^{20}	$(-2)^{64}$	3	$GH(4,1)$	14.5.2v
114	8	$\sqrt{7}^{56}$	$(-\sqrt{7})^{56}$	$(-8)^1$	6	$GH(1,7)$	14.5.2i
114	49	$\sqrt{7}^{56}$	$(-\sqrt{7})^{56}$	$(-49)^1$	4	$IG(57,49,42)$	14.5.2ii
146	9	$\sqrt{8}^{72}$	$(-\sqrt{8})^{72}$	$(-9)^1$	6	$GH(1,8)$	14.5.2i
146	64	$\sqrt{8}^{72}$	$(-\sqrt{8})^{72}$	$(-64)^1$	4	$IG(73,64,56)$	14.5.2ii
175	21	7^{28}	2^{21}	$(-2)^{125}$	3	L(HoSi)	14.4.11
186	10	$(4+\sqrt{5})^{30}$	$(4-\sqrt{5})^{30}$	$(-2)^{125}$	3	$GH(5,1)$	14.5.2v
456	14	$(6+\sqrt{7})^{56}$	$(6-\sqrt{7})^{56}$	$(-2)^{343}$	3	$GH(7,1)$	14.5.2v
506	15	4^{230}	$(-3)^{253}$	$(-8)^{22}$	5	M_{23} graph	14.5.2i
512	21	5^{210}	$(-3)^{280}$	$(-11)^{21}$	4	Coset graph of $\mathcal{G}_{21}$	14.5.2iii
657	16	$(7+\sqrt{8})^{72}$	$(7-\sqrt{8})^{72}$	$(-2)^{512}$	3	$GH(8,1)$	14.5.2v
729	24	6^{264}	$(-3)^{440}$	$(-12)^{24}$	3	Coset graph of $\mathcal{G}_{12}$	14.5.2vii
819	18	5^{324}	$(-3)^{468}$	$(-9)^{26}$	3	$GH(2,8)$	14.5.2v
2048	23	7^{506}	$(-1)^{1288}$	$(-9)^{253}$	4	Coset graph of $\mathcal{G}_{23}$	14.5.2iii,iv
2457	24	11^{324}	3^{468}	$(-3)^{1664}$	3	$GH(8,2)$	14.5.2v

Table 14.3 Sporadic distance-regular DS graphs with diameter 3

By Biaff(q), we denote the point-line incidence graph of an affine plane of order q minus a parallel class of lines (sometimes called a *biaffine plane*). Any graph cospectral with a graph Biaff(q) is also such a graph. For prime powers $q < 9$, there is a unique affine plane of order q. (Biaff(2) is the 8-gon.)

n	Nonnegative spectrum	d	g	Name	Theorem
18	$3^1 \ \sqrt{3}^6 \ 0^4$	4	6	Pappus; Biaff(3)	14.5.2ii
30	$3^1 \ 2^9 \ 0^{10}$	4	8	Tutte's 8-cage; $GO(1,2)$	14.5.2i
32	$4^1 \ 2^{12} \ 0^6$	4	6	Biaff(4)	14.5.2ii
50	$5^1 \ \sqrt{5}^{20} \ 0^8$	4	6	Biaff(5)	14.5.2ii
80	$4^1 \ \sqrt{6}^{24} \ 0^{30}$	4	8	$GO(1,3)$	14.5.2i
98	$7^1 \ \sqrt{7}^{42} \ 0^{12}$	4	6	Biaff(7)	14.5.2ii
126	$3^1 \ \sqrt{6}^{21} \ \sqrt{2}^{27} \ 0^{28}$	6	12	$GD(1,2)$	14.5.2i
128	$8^1 \ \sqrt{8}^{56} \ 0^{14}$	4	6	Biaff(8)	14.5.2ii
170	$5^1 \ \sqrt{8}^{50} \ 0^{68}$	4	8	$GO(1,4)$	14.5.2i

Table 14.4 Sporadic bipartite distance-regular DS graphs with $d \geq 4$

n	Spectrum	d	g	Name	Theorem
20	$3^1 \ \sqrt{5}^3 \ 1^5 \ 0^4 \ (-2)^4 \ (-\sqrt{5})^3$	5	5	dodecahedron	14.5.2vi
28	$3^1 \ 2^8 \ (-1+\sqrt{2})^6 \ (-1)^7 \ (-1-\sqrt{2})^6$	4	7	Coxeter	14.5.2i
45	$4^1 \ 3^9 \ 1^{10} \ (-1)^9 \ (-2)^{16}$	4	3	$GO(2,1)$	14.5.2v
102	$3^1 \ (\frac{1+\sqrt{17}}{2})^9 \ 2^{18} \ \theta_1^{16} \ 0^{17}$ $\theta_2^{16} \ (\frac{1-\sqrt{17}}{2})^9 \ \theta_3^{16}$ $(\theta_1, \theta_2, \theta_3 \text{ roots of } \theta^3 + 3\theta^2 - 3 = 0)$	7	9	Biggs-Smith	14.5.2v
160	$6^1 \ (2+\sqrt{6})^{24} \ 2^{30} \ (2-\sqrt{6})^{24} \ (-2)^{81}$	4	3	$GO(3,1)$	14.5.2v
189	$4^1 \ (1+\sqrt{6})^{21} \ (1+\sqrt{2})^{27} \ 1^{28}$ $(1-\sqrt{2})^{27} \ (1-\sqrt{6})^{21} \ (-2)^{64}$	6	3	$GD(2,1)$	14.5.2v
330	$7^1 \ 4^{55} \ 1^{154} \ (-3)^{99} \ (-4)^{21}$	4	5	M_{22} graph	14.5.2v
425	$8^1 \ (3+\sqrt{8})^{50} \ 3^{68} \ (3-\sqrt{8})^{50} \ (-2)^{256}$	4	3	$GO(4,1)$	14.5.2v
990	$7^1 \ 5^{42} \ 4^{55} \ (\frac{-1+\sqrt{33}}{2})^{154} \ 1^{154} \ 0^{198}$ $(-3)^{99} \ (\frac{-1-\sqrt{33}}{2})^{154} \ (-4)^{21}$	8	5	Ivanov-Ivanov-Faradjev	14.5.2$viii$

Table 14.5 Sporadic nonbipartite distance-regular DS graphs with $d \geq 4$

We finally remark that also the complements of distance-regular DS graphs are DS (but not distance-regular, unless $d = 2$).

14.6 The method of Wang and Xu

WANG & XU [344] invented a method to show that relatively many graphs are determined by their spectrum and the spectrum of their complement. We give a brief sketch.

Let Γ be a graph on n vertices with adjacency matrix A. The *walk matrix* W of Γ is the square matrix of order n with i-th column $A^{i-1}\mathbf{1}$ ($1 \leq i \leq n$). It is nonsingular if and only if A does not have an eigenvector orthogonal to $\mathbf{1}$. (Indeed, let $u^\top A = \theta u^\top$. Then $u^\top W = (1, \theta, \ldots, \theta^{n-1}) u^\top \mathbf{1}$. If $u^\top \mathbf{1} = 0$, then this shows that the rows of W are dependent. If for no eigenvector $u^\top$ we have $u^\top \mathbf{1} = 0$, then all eigenvalues have multiplicity 1, and by Vandermonde W is nonsingular.)

Let $p(t) = \sum c_i t^i = \det(tI - A)$ be the characteristic polynomial of A. Let the *companion matrix* $C = (c_{ij})$ be given by $c_{in} = -c_i$ and $c_{ij} = \delta_{i,j+1}$ for $1 \le j \le n-1$. Then $AW = WC$. (Indeed, this follows from $p(A) = 0$.)

Assume that Γ and Γ' are cospectral with cospectral complements. Call their walk matrices W and W'. Then $W^\top W = W'^\top W'$. (Indeed, $(W^\top W)_{i,j} = \mathbf{1}^\top A^{i+j-2} \mathbf{1}$, and we saw in the proof of Proposition 14.1.1 that if Γ and Γ', with adjacency matrices A and A', are y-cospectral for two distinct y, then $\mathbf{1}^\top A^m \mathbf{1} = \mathbf{1}^\top A'^m \mathbf{1}$ for all m.)

Suppose that W is nonsingular. Then W' is nonsingular, and $Q = W'W^{-1}$ is the unique orthogonal matrix such that $A' = QAQ^\top$ and $Q\mathbf{1} = \mathbf{1}$. (Indeed, since $W^\top W = W'^\top W'$ also W' is nonsingular, and $Q\mathbf{1} = \mathbf{1}$ since $QW = W'$, and $QQ^\top = W'(W^\top W)^{-1} W'^\top = I$. Since Γ and Γ' are cospectral, their companion matrices are equal and $QAQ^\top = QWCW^{-1}Q^\top = W'CW'^{-1} = A'$. If Q is arbitrary with $QQ^\top = I$, $Q\mathbf{1} = \mathbf{1}$ (hence also $Q^\top \mathbf{1} = \mathbf{1}$) and $QAQ^\top = A'$, then $QA^m \mathbf{1} = QA^m Q^\top \mathbf{1} = A'^m \mathbf{1}$ for all m, and $QW = W'$.)

Forget about Γ' and study rational matrices Q with $QQ^\top = I$, $Q\mathbf{1} = \mathbf{1}$ and $QAQ^\top$ a $(0,1)$-matrix with zero diagonal. Let the *level* of Q be the smallest integer ℓ such that ℓQ is integral. The matrices Q of level 1 are permutation matrices leading to isomorphic graphs. So the graph Γ (without eigenvector orthogonal to $\mathbf{1}$) is determined by its spectrum and the spectrum of its complement when all such matrices Q have level 1.

If Q has level ℓ, then clearly $\ell \mid \det W$. A tighter restriction on ℓ is found by looking at the Smith normal form S of W. Let $S = UWV$ with unimodular integral U and V, where $S = \mathrm{diag}(s_1, \ldots, s_n)$ with $s_1 | s_2 | \ldots | s_n$. Then $W^{-1} = VS^{-1}U$ so that $s_n W^{-1}$ is integral, and $\ell | s_n$.

Let p be prime, $p | \ell$. There is an integral row vector z, $z \not\equiv 0 \pmod{p}$ such that $zW \equiv 0 \pmod{p}$ and $zz^\top \equiv 0 \pmod{p}$. (Indeed, let z be a row of ℓQ, nonzero mod p. Now $QW = W'$ is integral and hence $zW \equiv 0 \pmod{p}$. And $QQ^\top = I$, so $zz^\top = \ell^2 \equiv 0 \pmod{p}$.)

This observation can be used to rule out odd prime divisors of ℓ in some cases. Suppose that all numbers s_i are powers of 2, except possibly the last one s_n. Let p be an odd prime divisor of s_n, and suppose that $uu^\top \not\equiv 0 \pmod{p}$, where u is the last row of U. Then $p \nmid \ell$. (Indeed, $zW \equiv 0 \pmod{p}$ and $W = U^{-1}SV^{-1}$ with unimodular V implies $zU^{-1}S \equiv 0 \pmod{p}$. Assume $p | \ell$, so that $p | s_n$. Let $y = zU^{-1}$. Then all coordinates of y except for the last one are $0 \pmod{p}$. And $z = yU$ is a nonzero constant times $u \pmod{p}$. This contradicts $uu^\top \not\equiv 0 \pmod{p}$.)

It remains to worry about $p = 2$. Assume that $s_n \equiv 2 \pmod 4$, so that (with all of the above assumptions) $\ell = 2$. For z we now have $z \not\equiv 0 \pmod 2$, $zW \equiv 0 \pmod 2$, $zz^\top = 4$, $z\mathbf{1} = 2$, so that z has precisely four nonzero entries, three 1 and one -1.

We proved the following:

Theorem 14.6.1 *Let Γ be a graph on n vertices without eigenvector orthogonal to $\mathbf{1}$, and let $S = \mathrm{diag}(s_1, \ldots, s_n) = UWV$ be the Smith normal form of its walk matrix W, where U and V are unimodular. Let u be the last row of U. If $s_n \equiv 2 \pmod 4$,* and $\gcd(uu^\top, s_n/2) = 1$, and $zW \neq 0$ (mod 2) *for every* $(0,1)$*-vector z with weight* *4, then* Γ *is determined by its spectrum and the spectrum of its complement.* □

Wang and Xu generate a number of random graphs where this method applies.

Let us abbreviate the condition "determined by its spectrum and the spectrum of its complement" by DGS (determined by the generalized spectrum). WANG & XU [345] used their approach to find conditions for which a DGS graph remains DGS if an isolated vertex is added.

Theorem 14.6.2 *Let* Γ *be a graph without eigenvector orthogonal to* **1**. *If we have* $\gcd(\det A, \det W) = 1$, *then the graph obtained from* Γ *by adding an isolated vertex is DGS if and only if* Γ *is.*

There is experimental evidence that in most cases where a cospectral mate exists, the level ℓ is 2.

14.7 Exercises

Exercise 14.1 Show for the adjacency matrix A
   (i) that there is no pair of cospectral graphs on fewer than 5 vertices,
   (ii) that the Saltire pair is the only cospectral pair on 5 vertices,
   (iii) that there are precisely 5 cospectral pairs on 6 vertices.

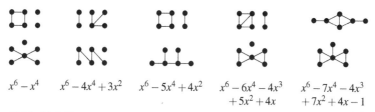

$x^6 - x^4$	$x^6 - 4x^4 + 3x^2$	$x^6 - 5x^4 + 4x^2$	$x^6 - 6x^4 - 4x^3$ $+ 5x^2 + 4x$	$x^6 - 7x^4 - 4x^3$ $+ 7x^2 + 4x - 1$

Table 14.6 The cospectral graphs on 6 vertices (with characteristic polynomial)

Exercise 14.2 Let Γ have spectrum 4^1, $(-2)^{10}$, $(-1 \pm \sqrt{3})^{10}$, $((3 \pm \sqrt{5})/2)^{12}$. Show that Γ has no m-cycles, for $m = 3, 4, 6, 7$. Show that every 2-path is contained in a unique pentagon. In fact, there is a unique such graph (Blokhuis and Brouwer).

Exercise 14.3 ([203]) Let Γ be the Kneser graph $K(m,k)$ with vertex set $V = \binom{X}{k}$, where $|X| = m = 3k - 1$ ($k \geq 2$). Fix $Y \subset X$ with $|Y| = k - 1$ and consider the subset W of the vertices of Γ consisting of the k-subsets of X containing Y. Prove that W satisfies the conditions for GM switching, and that the switching produces a graph nonisomorphic to Γ, provided $k \geq 3$.

Chapter 15
Graphs with Few Eigenvalues

Graphs with few distinct eigenvalues tend to have some kind of regularity. A graph with only one eigenvalue (for A, L, or Q) is edgeless, and a connected graph with two distinct adjacency eigenvalues (for A, L, or Q) is complete. A connected regular graph Γ has three eigenvalues if and only if Γ is connected and strongly regular. Two obvious next cases are connected regular graphs with four eigenvalues and general graphs with three eigenvalues. In the latter case, the graphs need not be regular, so it matters which type of matrix we consider. For the Laplace matrix, there is an elegant characterization in terms of the structure, which gives a natural generalization of the spectral characterization of strongly regular graphs.

15.1 Regular graphs with four eigenvalues

Suppose Γ is regular with r distinct (adjacency) eigenvalues $k = \lambda_1 > \ldots > \lambda_r$. Then the Laplace matrix has eigenvalues $0 = k - \lambda_1 < \ldots < k - \lambda_r$, and the signless Laplacian has eigenvalues $k + \lambda_1 > \ldots > k + \lambda_r$. So for regular graphs these three matrices have the same number of distinct eigenvalues. If, in addition, both Γ and its complement $\overline{\Gamma}$ are connected, then $\overline{\Gamma}$ also has r distinct eigenvalues, being $n - k - 1 > -\lambda_r - 1 > \ldots > -\lambda_2 - 1$. However, for the Seidel matrix the eigenvalues become $-2\lambda_r - 1 > \ldots > -2\lambda_2 - 1$ and $n - 2k - 1$. But $n - 2k - 1$ may be equal to one of the other eigenvalues, in which case S has $r - 1$ distinct eigenvalues. For example, the Petersen graph has three distinct adjacency eigenvalues but only two distinct Seidel eigenvalues, being ± 3.

Connected regular graphs with four distinct (adjacency) eigenvalues have been studied by DOOB [145, 146], VAN DAM [121], and VAN DAM & SPENCE [132]. Many such graphs are known, for example the line graphs of primitive strongly regular graphs and distance-regular graphs of diameter 3. More generally, most graphs defined by a relation of a three-class association scheme have four eigenvalues. There is no nice characterization as for regular graphs with three eigenvalues, but they do possess an interesting regularity property. A graph is *walk-regular* whenever for every $\ell \geq 2$ the number of closed walks of length ℓ at a vertex v is independent of the choice of v. Note that walk-regularity implies regularity (take $\ell = 2$). Examples of walk-regular graphs are distance-regular graphs and vertex-transitive graphs, but there is more.

Proposition 15.1.1 *Let Γ be a connected graph whose adjacency matrix A has $r \geq 4$ distinct eigenvalues. Then Γ is walk-regular if and only if A^ℓ has constant diagonal for $2 \leq \ell \leq r - 2$.*

Proof. We know that the number of closed walks of length ℓ at vertex v equals $(A^\ell)_{v,v}$. Therefore, Γ is walk-regular if and only if A^ℓ has constant diagonal for all $\ell \geq 2$. Suppose A^ℓ has constant diagonal for $2 \leq \ell \leq r - 2$. Then A^2 has constant diagonal, so Γ is regular. The Hoffman polynomial of Γ has degree $r - 1$ and hence $A^{r-1} \in \langle A^{r-2}, \ldots, A^2, A, I, J \rangle$. This implies $A^\ell \in \langle A^{r-2}, \ldots, A^2, A, I, J \rangle$ for all $\ell \geq 0$. Therefore A^ℓ has constant diagonal for all $\ell \geq 0$.   □

Corollary 15.1.2 *If Γ is connected and regular with four distinct eigenvalues, then Γ is walk-regular.*   □

For a graph Γ with adjacency matrix A, the average number of triangles through a vertex equals $\frac{1}{2n} \operatorname{tr} A^3$. Suppose Γ is walk-regular. Then this number must be an integer. Similarly, $\frac{1}{2n} \operatorname{tr} A^\ell$ is an integer if ℓ is odd, and $\frac{1}{n} \operatorname{tr} A^\ell$ is an integer if n is even. VAN DAM & SPENCE [132] have used these (and other) conditions in their computer generation of feasible spectra for connected regular graphs with four eigenvalues. For constructions, characterizations, and other results on regular graphs with four eigenvalues we refer to VAN DAM [121, 122]. Here we finish with the bipartite case, which can be characterized in terms of block designs (see §4.9).

Proposition 15.1.3 *A connected bipartite regular graph Γ with four eigenvalues is the incidence graph of a symmetric 2-design (and therefore distance-regular).*

Proof. Since Γ is connected, bipartite, and regular, the spectrum is

$$\{k, \; \lambda_2^{v-1}, \; (-\lambda_2)^{v-1}, \; -k\},$$

where $2v$ is the number of vertices. For the adjacency matrix A of Γ, we have

$$A = \begin{bmatrix} O & N \\ N^\top & O \end{bmatrix} \quad \text{and} \quad A^2 = \begin{bmatrix} NN^\top & O \\ O & N^\top N \end{bmatrix}$$

for some square $(0,1)$-matrix N satisfying $N\mathbf{1} = N^\top\mathbf{1} = k\mathbf{1}$. It follows that $NN^\top$ has spectrum $\{(k^2)^1, (\lambda_2^2)^{v-1}\}$, where k^2 corresponds to the row and column sum of $NN^\top$. This implies that $NN^\top \in \langle J, I \rangle$, and hence N is the incidence matrix of a symmetric design.   □

15.2 Three Laplace eigenvalues

If a connected graph Γ has three distinct Laplace eigenvalues $0 < v < v'$ (say), the complement $\overline{\Gamma}$ has eigenvalues $0 \leq n - v' < n - v$, so if $\overline{\Gamma}$ is connected, it also has three distinct eigenvalues. To avoid the disconnected exceptions, it is convenient to use the notion of restricted eigenvalues (recall that an eigenvalue is *restricted* if it has an eigenvector orthogonal to the all-1 vector $\mathbf{1}$) and consider graphs with two distinct restricted Laplace eigenvalues.

We say that a graph Γ has constant $\mu(\Gamma)$ if Γ is not complete and any two distinct nonadjacent vertices of Γ have the same number of common neighbors (equal to $\mu(\Gamma)$).

Theorem 15.2.1 *A graph Γ has two distinct restricted Laplace eigenvalues v and v' if and only if Γ has constant $\mu(\Gamma)$ and its complement $\overline{\Gamma}$ has constant $\mu(\overline{\Gamma})$. If Γ is such a graph, only two vertex degrees, d and d', occur, and*

$$v + v' = d + d' + 1 = \mu(\Gamma) + n - \mu(\overline{\Gamma}), \ vv' = dd' + \mu(\Gamma) = \mu(\Gamma)n.$$

Proof. Suppose Γ has just two restricted Laplace eigenvalues v and v'. Then $(L - vI)(L - v'I)$ has rank 1 and row sum vv', so

$$(L - vI)(L - v'I) = \frac{vv'}{n}J.$$

If u and v are nonadjacent vertices, then $(L)_{uv} = 0$, so $(L^2)_{uv} = vv'/n$, and $\mu(\Gamma) = vv'/n$ is constant. Similarly, $\overline{\Gamma}$ has constant $\mu(\overline{\Gamma}) = (n - v)(n - v')/n$.

Next, suppose $\mu = \mu(\Gamma)$ and $\overline{\mu} = \mu(\overline{\Gamma})$ are constant. If u and v are adjacent vertices, then $((nI - J - L)^2)_{uv} = \overline{\mu}$, so $\overline{\mu} = (L^2)_{uv} + n$, and if u and v are nonadjacent, then $(L^2)_{uv} = \mu$. Furthermore, $(L^2)_{uu} = d_u^2 + d_u$, where d_u is the degree of u. Writing $D = \mathrm{diag}(d_1, \ldots, d_n)$, we obtain

$$L^2 = (\overline{\mu} - n)(D - L) + \mu(J - I - D + L) + D^2 + D =$$
$$(\mu + n - \overline{\mu})L + D^2 - (\mu + n - \overline{\mu} - 1)D - \mu I + \mu J.$$

Since L and L^2 have zero row sums, it follows that $d_u^2 - d_u(\mu + n - \overline{\mu} - 1) - \mu + \mu n = 0$ for every vertex u, so $L^2 - (\mu + n - \overline{\mu})L + \mu nI = \mu J$. Now let v and v' be such that $v + v' = \mu + n - \overline{\mu}$ and $vv' = \mu n$. Then $(L - vI)(L - v'I) = \frac{vv'}{n}J$, so L has distinct restricted eigenvalues v and v'. As a side result, we obtained that all vertex degrees d_u satisfy the same quadratic equation, so d_u can only take two values d and d', and the formulas readily follow. $\square$

Regular graphs with constant $\mu(\Gamma)$ and $\mu(\overline{\Gamma})$ are strongly regular, so Theorem 15.2.1 generalizes the spectral characterization of strongly regular graphs. Several nonregular graphs with two restricted Laplace eigenvalues are known. A geodetic graph of diameter 3 with connected complement provides an example with $\mu(\Gamma) = 1$ (see [54], Theorem 1.17.1). Here we give two other constructions. Both constructions use symmetric block designs (see §4.9). Correctness easily follows by using Theorem 15.2.1.

Proposition 15.2.2 *Let N be the incidence matrix of a symmetric 2-(n,k,λ) design. Suppose that N is symmetric (which means that the design has a polarity). Then $L = kI - N$ is the Laplace matrix of a graph with two restricted eigenvalues, being $k \pm \sqrt{k - \lambda}$. The possible degrees are k and $k - 1$.*   □

If all diagonal elements of N are 0, then the graph Γ is an (n,k,λ)-graph (a strongly regular graph with $\lambda = \mu$), and if all diagonal elements of N are 1, then $\overline{\Gamma}$ is such a graph. Otherwise both degrees k and $k-1$ occur. For example, the Fano plane admits a symmetric matrix with three ones on the diagonal. The corresponding graph has restricted Laplace eigenvalues $3 \pm \sqrt{2}$ and vertex degrees 2 and 3. See also §4.10.

Proposition 15.2.3 *Let N be the incidence matrix of a symmetric block design. Write*

$$N = \begin{bmatrix} 1 & N_1 \\ 0 & N_2 \end{bmatrix}, \text{ and define } L = \begin{bmatrix} vI - J & O & N_1 - J \\ O & vI - J & -N_2 \\ N_1^\top - J & -N_2^\top & 2(k-\lambda)I \end{bmatrix}.$$

Then L has two restricted eigenvalues.   □

Other examples, characterizations, and a table of feasible spectra can be found in [125] and [122] (see also Exercise 15.1). See [346] for some more recent results on graphs with three Laplace eigenvalues.

15.3 Other matrices with at most three eigenvalues

No characterization of nonregular graphs with three M-eigenvalues is known, for a matrix M other than the Laplacian. However, several examples and properties are known. Some of these will be discussed below.

15.3.1 Few Seidel eigenvalues

Seidel switching (see §1.8.2) doesn't change the Seidel spectrum, so having few Seidel eigenvalues is actually a property of the switching class of a graph. For example the switching class of $\overline{K_n}$, the edgeless graph on n vertices, consists of the complete bipartite graphs $K_{m,n-m}$, and all of them have Seidel spectrum $\{(-1)^{n-1}, n-1\}$. Only the one-vertex graph K_1 has one Seidel eigenvalue. Graphs with two Seidel eigenvalues are strong (see §10.1). To be precise, they are the graphs whose associated two-graph is regular (Theorem 10.3.1). The Seidel matrix is a special case of a generalized adjacency matrix. These are matrices of the form $M(x,y,z) = xI + yA + z(J - I - A)$ with $y \neq z$, where A is the adjacency matrix; see also Chapter 14. If A is the adjacency matrix of a strongly regular graph with eigenvalues $k \geq r > s$, then both $nA - (k - r)J$ and $nA - (k - s)J$ (these are basically the nontrivial idempotents of the association scheme) are generalized adjacency matrices with two eigenvalues. We recall that a strong graph either has two Seidel eigenvalues or is strongly regular. Thus, for every strong graph there exist numbers x, y, and z such that $M(x, y, z)$ has two eigenvalues.

Proposition 15.3.1 *A graph is strong if and only if at least one generalized adjacency matrix has two eigenvalues.*

Proof Correctness of the "only if" part of the statement has been established already. Without loss of generality we assume that the eigenvalues of $M = M(x, y, z)$ are 0 and 1. Then $M^2 = M$. Let d_i be the degree of vertex i. Then $x = M_{ii} = (M^2)_{ii} = x^2 + d_i y^2 + (n - 1 - d_i)z^2$, which gives $d_i(y^2 - z^2) = x - x^2 - (n - 1)z^2$. So $y = -z$ or Γ is regular. In the first case, $S = \frac{1}{z}(M - xI)$ is the Seidel matrix of Γ with two eigenvalues, so Γ is strong. If Γ is regular, the adjacency matrix $A = \frac{1}{y-z}(M + (z - x)I - zJ)$ has three eigenvalues, so Γ is strongly regular and therefore strong. $\square$

So, if a generalized adjacency matrix $M(x, y, z)$ of a nonregular graph has two eigenvalues, then $y = -z$ (and we basically deal with the Seidel matrix).

A strongly regular graph Γ on n vertices with adjacency eigenvalues k, r, s ($k \geq r > s$) has Seidel eigenvalues $\rho_0 = n - 1 - 2k$, $\rho_1 = -2s - 1$, and $\rho_2 = -2r - 1$. If $\rho_0 = \rho_1$ or $\rho_0 = \rho_2$, then Γ has two eigenvalues; otherwise Γ, and all graphs switching equivalent to Γ, have three eigenvalues. For example, the (switching class of the) Petersen graph has two Seidel eigenvalues, 3 and -3, while the pentagon C_5 has three Seidel eigenvalues 0 and $\pm\sqrt{5}$. However, not every graph with three Seidel eigenvalues is switching equivalent to a strongly regular graph. Not even if the graph is regular. Indeed, consider a graph Γ whose Seidel matrix S has two eigenvalues, ρ_1 and ρ_2. Then $(S + I) \otimes (S + I) - I$ represents a graph Γ^2 with eigenvalues $(\rho_1 + 1)^2 - 1$, $(\rho_1 + 1)(\rho_2 + 1) - 1$, and $(\rho_2 + 1)^2 - 1$. Moreover, Γ^2 is regular if Γ is.

15.3.2 Three adjacency eigenvalues

Connected regular graphs with three adjacency eigenvalues are strongly regular. The complete bipartite graphs $K_{\ell,m}$ have spectrum $\{-\sqrt{\ell m}, 0^{n-2}, \sqrt{\ell m}\}$. If $\ell \neq m$ they are nonregular with three adjacency eigenvalues. Other nonregular graphs with three adjacency eigenvalues have been constructed by BRIDGES & MENA [45], KLIN & MUZYCHUK [243], and VAN DAM [122, 123]. CHUANG & OMIDI [91] characterized all such graphs with largest eigenvalue at most 8. Many nonregular graphs with three eigenvalues can be made from a strongly regular graph by introducing one new vertex adjacent to all other vertices. Such a graph is called a *cone* over a strongly regular graph.

Proposition 15.3.2 *Let Γ be a strongly regular graph on n vertices with eigenvalues $k > r > s$. Then the cone $\hat{\Gamma}$ over Γ has three eigenvalues if and only if $n = s(s - k)$.*

Proof. If $\hat{A}$ is the adjacency matrix of $\hat{\Gamma}$, then $\hat{A}$ admits an equitable partition with quotient matrix

$$\begin{bmatrix} 0 & n \\ 1 & k \end{bmatrix}$$

with eigenvalues $(k \pm \sqrt{k^2+4n})/2$, which are also eigenvalues of $\hat{A}$. The other eigenvalues of $\hat{A}$ have eigenvectors orthogonal to the characteristic vectors of the partition, so they remain eigenvalues if the all-1 blocks of the equitable partition are replaced by all-zero blocks. Therefore they are precisely the restricted eigenvalues r and s of Γ. So the eigenvalues of $\hat{A}$ are $(k \pm \sqrt{k^2+4n})/2$, r, and s. Two of these values coincide if and only if $s = (k - \sqrt{k^2+4n})/2$. □

There exist infinitely many strongly regular graphs for which $n = s(s-k)$, the smallest of which is the Petersen graph. The cone over the Petersen graph has eigenvalues 5, 1, and -2. If a cone over a strongly regular graph has three eigenvalues, then these eigenvalues are integers (see Exercise 15.3). The complete bipartite graphs provide many examples with nonintegral eigenvalues. In fact:

Proposition 15.3.3 *If Γ is a connected graph with three distinct adjacency eigenvalues, of which the largest is not an integer, then Γ is a complete bipartite graph.*

Proof. Assume Γ has $n \geq 4$ vertices. Since the largest eigenvalue ρ is nonintegral with multiplicity 1, one of the other two eigenvalues $\overline{\rho}$ (say) also has this property, and the third eigenvalue has multiplicity $n - 2 \geq 2$, so it cannot be irrational. Thus the spectrum of Γ is

$$\{\rho = \tfrac{1}{2}(a+\sqrt{b}), \ \overline{\rho} = \tfrac{1}{2}(a-\sqrt{b}), \ c^{n-2}\}$$

for integers a, b, and c. Now $\mathrm{tr}\,A = 0$ gives $c = -a/(n-2)$. By the Perron-Frobenius theorem, $\rho \geq |\overline{\rho}|$, and therefore $a \geq 0$ and $c \leq 0$. If $c = 0$, the eigenvalues of Γ are $\pm\sqrt{b}/2$ and 0, and Γ is bipartite of diameter at most 2, and hence Γ is complete bipartite. If $c \leq -2$, then $\mathrm{tr}\,A^2 \geq 4(n-2)^2$, so Γ has at least $2(n-2)^2$ edges, which is ridiculous. If $c = -1$, then $\rho = \tfrac{1}{2}(n-2+\sqrt{b}) \leq n-1$, and hence $\sqrt{b} \leq n$ and $\overline{\rho} > -1$. This implies that $A + I$ is positive semidefinite (of rank 2), so $A + I$ is the Gram matrix of a set of unit vectors (in $\mathbb{R}^2$) with angles 0 and $\pi/2$. This implies that being adjacent is an equivalence relation, so $\Gamma = K_n$, a contradiction. □

The conference graphs are examples of regular graphs where only the largest eigenvalue is an integer. VAN DAM & SPENCE [79] found a number of nonregular graphs on 43 vertices with eigenvalues 21, $-\tfrac{1}{2} \pm \tfrac{1}{2}\sqrt{41}$. It turns out that all these graphs have three distinct vertex degrees: 19, 26, and 35 (which was impossible in the case of the Laplace spectrum).

15.3.3 Three signless Laplace eigenvalues

Recently, AYOOBI, OMIDI & TAYFEH-REZAIE [13] started to investigate nonregular graphs whose signless Laplace matrix Q has three distinct eigenvalues. They found three infinite families.

(i) The complete K_n with one edge deleted has Q-spectrum

$$\{\frac{1}{2}(3n - 6 + \sqrt{n^2 + 4n - 12}),\ (n - 2)^{n-2},\ \frac{1}{2}(3n - 6 - \sqrt{n^2 + 4n - 12})\}.$$

(ii) The star $K_{1,n-1}$ has Q-spectrum $0^1, 1^{n-2}, n^1$.

(iii) The complement of $K_{m,m} + mK_1$ has Q-spectrum

$$(5m - 2)^1,\ (3m - 2)^m,\ (2m - 2)^{2m-2}.$$

In addition, there are some sporadic examples (see also Exercise 15.4). As in Proposition 15.3.3, the case in which the spectral radius is nonintegral can be characterized.

Proposition 15.3.4 ([13]) *Let Γ be a connected graph on at least four vertices, of which the signless Laplace matrix has three distinct eigenvalues. Then the largest of these eigenvalues is nonintegral if and only if Γ is the complete graph minus one edge.*

It is not known if there exist other nonregular examples with a nonintegral eigenvalue. We expect that the list above is far from complete.

15.4 Exercises

Exercise 15.1 Prove that a graph with two restricted Laplace eigenvalues whose degrees d and d' differ by 1 comes from a symmetric design with a polarity as described in Proposition 15.2.2.

Exercise 15.2 Let Γ be a strongly regular graph with a coclique C whose size meets Hoffman's bound (3.5.2). Prove that the subgraph of Γ induced by the vertices outside C is regular with at most four distinct eigenvalues. Can it have fewer than four eigenvalues?

Exercise 15.3 Suppose $\hat{\Gamma}$ is a cone over a strongly regular graph. Show that if $\hat{\Gamma}$ has three distinct eigenvalues, then all three are integral.

Exercise 15.4 Show that the cone over the Petersen graph has three signless Laplace eigenvalues. Find a necessary and sufficient condition on the parameters (n, k, λ, μ) of a strongly regular graph Γ under which the cone over Γ has three signless Laplace eigenvalues.

References

Page numbers at the end of each citation indicate the page(s) on which the work is cited in the text.

1. O. Ahmadi, N. Alon, I. F. Blake & I. E. Shparlinski, *Graphs with integral spectrum*, Lin. Alg. Appl. **430** (2009) 547–552. (p. 50)
2. AIM minimum rank—special graphs work group (18 authors), *Zero forcing sets and the minimum rank of graphs*, Lin. Alg. Appl. **428** (2008) 1628–1648. (p. 65)
3. M. Ajtai, J. Komlós & E. Szemerédi, *Sorting in c log n parallel steps*, Combinatorica **3** (1983) 1–19. (p. 70)
4. N. Alon, *Eigenvalues and expanders*, Combinatorica **6** (1986) 83–96. (p. 67)
5. N. Alon, *The Shannon capacity of a union*, Combinatorica **18** (1998) 301–310. (p. 43)
6. N. Alon, *Large sets in finite fields are sumsets*, J. Number Theory **126** (2007) 110–118. (p. 97)
7. N. Alon & F. R. K. Chung, *Explicit constructions of linear sized tolerant networks*, Discr. Math. **2** (1988) 15–19. (p. 69)
8. N. Alon & V. D. Milman, λ_1, *isoperimetric inequalities for graphs and superconcentrators*, J. Combin. Theory (B) **38** (1985) 73–88. (p. 74)
9. N. Alon & P. Seymour, *A counterexample to the rank-coloring conjecture*, J. Graph Theory **13** (1989) 523–525. (p. 42)
10. W. N. Anderson & T. D. Morley, *Eigenvalues of the Laplacian of a graph*, Lin. Multilin. Alg. **18** (1985) 141–145. (p. 51)
11. D. M. Appleby, *Symmetric informationally complete-positive operator valued measures and the extended Clifford group*, J. Math. Phys. **46** 052107 (2005). (p. 162)
12. M. Aschbacher, *The non-existence of rank three permutation groups of degree 3250 and subdegree 57*, J. Alg. **19** (1971) 538–540. (p. 172)
13. F. Ayoobi, G. R. Omidi & B. Tayfeh-Rezaie, *A note on graphs whose signless Laplacian has three distinct eigenvalues*, Lin. Multilin. Alg. **59** (2011) 701–706. (p. 227)
14. L. Babai, D. Yu. Grigoryev & D. M. Mount, *Isomorphism of graphs with bounded eigenvalue multiplicity*, pp. 310–324 in: Proc. 14th ACM Symp. on Theory of Computing, 1982. (pp. 62, 63)
15. Bhaskar Bagchi, *On strongly regular graphs with $\mu \leq 2$*, Discr. Math. **306** (2006) 1502–1504. (p. 121)
16. Hua Bai, *The Grone-Merris conjecture*, Trans. Amer. Math. Soc. **363** (2011) 4463–4474. (p. 53)
17. G. A. Baker, *Drum shapes and isospectral graphs*, J. Math. Phys. **7** (1966) 2238–2242. (p. 19)

18. S. Bang, A. Hiraki & J. H. Koolen, *Improving diameter bounds for distance-regular graphs*, Europ. J. Combin. **27** (2006) 79–89. (p. 181)

19. S. Bang, A. Dubickas, J. H. Koolen & V. Moulton, *There are only finitely many distance-regular graphs of fixed valency greater than two*, arXiv.org:0909.5253, 29 Sep 2009. (p. 181)

20. E. Bannai & T. Ito, *On finite Moore graphs*, J. Fac. Sci. Univ. Tokyo, Sect IA **20** (1973) 191–208. (p. 182)

21. E. Bannai & T. Ito, *Algebraic Combinatorics I: Association Schemes*, Benjamin/Cummings, London, 1984. (pp. vi, 180, 181)

22. D. Bauer, H. J. Broersma, J. van den Heuvel & H. J. Veldman, *Long cycles in graphs with prescribed toughness and minimum degree*, Discr. Math. **141** (1995) 1–10. (p. 71)

23. D. Bauer, H. J. Broersma & H. J. Veldman, *Not every 2-tough graph is Hamiltonian*, Discr. Appl. Math. **99** (2000) 317–321. (p. 71)

24. V. Belevitch, *Theory of 2n-terminal networks with application to conference telephony*, Electron. Comm. **27** (1950) 231–244. (p. 158)

25. E. R. Berlekamp, J. H. van Lint & J. J. Seidel, *A strongly regular graph derived from the perfect ternary Golay code*, pp. 25–30 in: A survey of combinatorial theory, Symp. Colorado State Univ. 1971 (J. N. Srivastava et al., eds.), North Holland, Amsterdam, 1973. (p. 148)

26. A. Berman & R. J. Plemmons, *Nonnegative Matrices in the Mathematical Sciences*, Acad. Press, New York, 1979. (p. 24)

27. Jochem Berndsen & Mayank, *The Brouwer conjecture for split graphs*, preprint, 2010. (p. 53)

28. Monica Bianchini, Marco Gori & Franco Scarselli, *Inside PageRank*, ACM Trans. Internet Techno. **5** (2005) 92–128. (p. 60)

29. A. Bigalke & H. A. Jung, *Über Hamiltonsche Kreise und unabhängige Ecken in Graphen*, Monatsh. Math. **88** (1979) 195–210. (p. 127)

30. N. L. Biggs, *Algebraic Graph Theory*, Cambridge Tracts in Mathematics 67, Cambridge Univ. Press, Cambridge, 1974. (Second ed., 1993.) (p. vi)

31. A. Blokhuis, *Few-distance sets*, Ph.D. thesis, Eindhoven, 1983. (p. 164)

32. A. Blokhuis & A. E. Brouwer, *Uniqueness of a Zara graph on 126 points and non-existence of a completely regular two-graph on 288 points*, pp. 6–19 in: Papers Dedicated to J. J. Seidel (P. J. de Doelder et al., eds.), Eindhoven Univ. Techn. Report 84-WSK-03, Aug 1984. (p. 157)

33. A. Blokhuis, A. E. Brouwer & W. H. Haemers, *The graph with spectrum* $14^1 \ 2^{40} \ (-4)^{10} \ (-6)^9$, Designs, Codes and Cryptography, to appear. Preprint 2011. (p. 136)

34. J. A. Bondy, *A graph reconstruction manual*, pp. 221–252 in: Surveys in Combinatorics, London Math. Soc. Lecture Note Ser. 166, Cambridge Univ. Press, Cambridge, 1991. (p. 16)

35. J. Bosák, A. Rosa & Š. Znám, *On decompositions of complete graphs into factors with given diameters*, pp. 37–56 in: Theory of Graphs (Proc. Colloq. Tihany, 1966), Academic Press, New York, 1968. (p. 12)

36. R. C. Bose, *Strongly regular graphs, partial geometries and partially balanced designs*, Pacific J. Math **13** (1963) 389–419. (p. 124)

37. R. C. Bose & T. A. Dowling, *A generalization of Moore graphs of diameter two*, J. Combin. Theory **11** (1971) 213–226. (p. 149)

38. R. C. Bose & D. M. Mesner, *On linear associative algebras corresponding to association schemes of partially balanced designs*, Ann. Math. Statist. **30** (1959) 21–38. (p. 166)

39. R. C. Bose & T. Shimamoto, *Classification and analysis of partially balanced incomplete block designs with two associate classes*, J. Amer. Statist. Assoc. **47** (1952) 151–184. (p. 165)

40. R. Boulet, *Disjoint unions of complete graphs characterized by their Laplacian spectrum*, Electron. J. Lin. Alg. **18** (2009) 773–783. (p. 208)

41. N. Bourbaki, *Groupes et algèbres de Lie*, Chap. 4, 5 et 6, Hermann, Paris, 1968. (p. 35)

42. V. Brankov, P. Hansen & D. Stevanović, *Automated conjectures on upper bounds for the largest Laplacian eigenvalue of graphs*, Lin. Alg. Appl. **414** (2006) 407–424. (p. 51)

43. A. Brauer & I. C. Gentry, *On the characteristic roots of tournament matrices*, Bull. Amer. Math. Soc. **74** (1968) 1133–1135. (p. 18)

44. W. G. Bridges & R. A. Mena, *Rational circulants with rational spectra and cyclic strongly regular graphs*, Ars. Combin. **8** (1979) 143–161. (p. 119)

45. W. G. Bridges & R. A. Mena, *Multiplicative cones—a family of three eigenvalue graphs*, Aequat. Math. **22** (1981) 208–214. (p. 225)
46. Sergey Brin & Lawrence Page, *The anatomy of a large-scale hypertextual web search engine*, Proceedings of the Seventh International World Wide Web Conference, Computer Networks and ISDN Systems **30** (1998) 107–117. (p. 59)
47. R. Brooks, *Non-Sunada graphs*, Ann. Inst. Fourier (Grenoble) **49** (1999) 707–725. (pp. 201, 204)
48. A. E. Brouwer, *The uniqueness of the strongly regular graph on 77 points*, J. Graph Theory **7** (1983) 455–461. (pp. 145, 213)
49. A. E. Brouwer, *A non-degenerate generalized quadrangle with lines of size four is finite*, pp. 47–49 in: Advances in Finite Geometries and Designs, Proc. 3rd Isle of Thorns Conf., Chelwood Gate, UK, 1990. (p. 130)
50. A. E. Brouwer, *Toughness and spectrum of a graph*, Lin. Alg. Appl. **226–228** (1995) 267–271. (p. 72)
51. A. E. Brouwer, *Strongly regular graphs*, pp. 667–685 in: The CRC handbook of combinatorial designs (C. J. Colbourn & J. H. Dinitz, eds.), CRC Press, Boca Raton, 1996. (p. 214)
52. A. E. Brouwer, *Small integral trees*, Electron. J. Combin. **15** (2008) N1. (pp. 89, 90)
53. A. E. Brouwer & A. M. Cohen, *Some remarks on Tits geometries*, with an appendix by J. Tits, Indag. Math. **45** (1983) 393–402. (p. 129)
54. A. E. Brouwer, A. M. Cohen & A. Neumaier, *Distance-Regular Graphs*, Springer-Verlag, Berlin, 1989. (pp. v, vi, 35, 119, 120, 132, 140, 177, 182, 200, 215, 223)
55. A. E. Brouwer & C. A. van Eijl, *On the p-rank of the adjacency matrices of strongly regular graphs*, J. Alg. Combin. **1** (1992) 329–346. (pp. 191, 192, 194, 197)
56. A. E. Brouwer & M. van Eupen, *The correspondence between projective codes and 2-weight codes*, Designs, Codes and Cryptography **11** (1997) 261–266. (p. 140)
57. A. E. Brouwer & W. H. Haemers, *Structure and uniqueness of the (81,20,1,6) strongly regular graph*, Discr. Math. **106/107** (1992) 77–82. (pp. 132, 145, 213)
58. A. E. Brouwer & W. H. Haemers, *The Gewirtz graph: An exercise in the theory of graph spectra*, Europ. J. Combin. **14** (1993) 397–407. (pp. 144, 213, 214)
59. A. E. Brouwer & W. H. Haemers, *Eigenvalues and perfect matchings*, Lin. Alg. Appl. **395** (2005) 155–162. (pp. 75, 76)
60. A. E. Brouwer & W. H. Haemers, *A lower bound for the Laplacian eigenvalues of a graph—proof of a conjecture by Guo*, Lin. Alg. Appl. **429** (2008) 2131–2135. (p. 52)
61. A. E. Brouwer & J. H. Koolen, *The vertex connectivity of a distance-regular graph*, Europ. J. Combin. **30** (2009) 668–673. (p. 181)
62. A. E. Brouwer & J. H. van Lint, *Strongly regular graphs and partial geometries*, pp. 85–122 in: Enumeration and Design, Proc. Silver Jubilee Conf. on Combinatorics, Waterloo, 1982 (D. M. Jackson & S. A. Vanstone, eds.), Academic Press, Toronto, 1984. (pp. 121, 127, 147, 161, 192)
63. A. E. Brouwer & D. M. Mesner, *The connectivity of strongly regular graphs*, Europ. J. Combin. **6** (1985) 215–216. (p. 126)
64. A. E. Brouwer & A. Neumaier, *A remark on partial linear spaces of girth 5 with an application to strongly regular graphs*, Combinatorica **8** (1988) 57–61. (p. 121)
65. A. E. Brouwer & A. Neumaier, *The graphs with spectral radius between 2 and $\sqrt{2+\sqrt{5}}$*, Lin. Alg. Appl. **114/115** (1989) 273–276. (p. 35)
66. A. E. Brouwer & E. Spence, *Cospectral graphs on 12 vertices*, Electron. J. Combin. **16** (2009) N20. (p. 206)
67. A. E. Brouwer & H. A. Wilbrink, *Block designs*, pp. 349–382 in: Handbook of Incidence Geometry (F. Buekenhout, ed.), Elsevier, Amsterdam, 1995. (p. 79)
68. A. E. Brouwer, R. M. Wilson & Qing Xiang, *Cyclotomy and Strongly Regular Graphs*, J. Alg. Combin. **10** (1999) 25–28. (p. 143)
69. R. A. Brualdi & A. J. Hoffman, *On the spectral radius of (0,1) matrices*, Lin. Alg. Appl. **65** (1985) 133–146. (p. 36)
70. N. G. de Bruijn, *A combinatorial problem*, Nederl. Akad. Wetensch. **49** (1946) 758–764 = Indag. Math. **8** (1946) 461–467. (p. 18)

71. N. G. de Bruijn, *Acknowledgement of priority to C. Flye sainte-Marie on the counting of circular arrangements of 2^n zeros and ones that show each n-letter word exactly once*, Report 75-WSK-06, Techn. Hogeschool Eindhoven, 1975. (p. 18)

72. Kurt Bryan & Tanya Leise, *The $25,000,000,000 eigenvector—the linear algebra behind Google*, SIAM Review **48** (2006) 569–581. (p. 60)

73. F. Buekenhout & H. Van Maldeghem, *A characterization of some rank 2 incidence geometries by their automorphism group*, Mitt. Math. Sem. Giessen **218** (1994), i+70 pp. (p. 121)

74. M. Burrow, *Representation Theory of Finite Groups*, Academic Press, New York, 1965. (p. 166)

75. F. C. Bussemaker & D. M. Cvetković, *There are exactly 13 connected, cubic, integral graphs*, Univ. Beograd Publ. Elektrotehn. Fak. Ser. Mat. Fiz. **544–576** (1976) 43–48. (p. 46)

76. F. C. Bussemaker, D. M. Cvetković & J. J. Seidel, *Graphs related to exceptional root systems*, T.H.-Report 76-WSK-05, Eindhoven University of Technology, 1976. (pp. 210, 211)

77. F. C. Bussemaker, W. H. Haemers, R. Mathon & H. A. Wilbrink, *A (49,16,3,6) strongly regular graph does not exist*, Europ. J. Combin. **10** (1989) 413–418. (p. 144)

78. F. C. Bussemaker, R. A. Mathon & J. J. Seidel, *Tables of two-graphs*, pp. 70–112 in: Combinatorics and Graph Theory, Proc. Calcutta 1980 (S. B. Rao, ed.), Lecture Notes in Math. 885, Springer-Verlag, Berlin, 1981. (p. 153)

79. D. de Caen, E. R. van Dam & E. Spence, *A nonregular analogue of conference graphs*, J. Combin. Theory (A) **88** (1999) 194–204. (p. 226)

80. A. R. Calderbank & W. M. Kantor, *The geometry of two-weight codes*, Bull. London Math. Soc. **18** (1986) 97–122. (p. 136)

81. P. J. Cameron, *Orbits of permutation groups on unordered sets. II*, J. London Math. Soc., II. Ser. **23** (1981) 249–264. (p. 130)

82. P. J. Cameron, *Automorphism groups of graphs*, pp. 89–127 in: Selected Topics in Graph Theory 2 (L. W. Beineke & R. J. Wilson, eds.), Acad. Press, London, 1983. (p. 172)

83. P. J. Cameron, J.-M. Goethals & J. J. Seidel, *Strongly regular graphs having strongly regular subconstituents*, J. Alg. **55** (1978) 257–280. (pp. 133, 213)

84. P. J. Cameron, J.-M. Goethals, J. J. Seidel & E. E. Shult, *Line graphs, root systems, and elliptic geometry*, J. Alg. **43** (1976) 305–327. (pp. 105, 210)

85. A. Cayley, *A theorem on trees*, Quart. J. Pure Appl. Math. **23** (1889) 376–378. (p. 6)

86. L. C. Chang, *The uniqueness and nonuniqueness of the triangular association scheme*, Sci. Record **3** (1959) 604–613. (p. 125)

87. L. C. Chang, *Association schemes of partially balanced block designs with parameters $v = 28$, $n_1 = 12$, $n_2 = 15$ and $p_{11}^2 = 4$*, Sci. Record **4** (1960) 12–18. (pp. 125, 143, 213)

88. Gregory Cherlin, *Locally finite generalized quadrangles with at most five points per line*, Discr. Math. **291** (2005) 73–79. (p. 130)

89. A. A. Chesnokov & W. H. Haemers, *Regularity and the generalized adjacency spectra of graphs*, Lin. Alg. Appl. **416** (2006) 1033–1037. (p. 207)

90. B. Cheyne, V. Gupta & C. Wheeler, *Hamilton cycles in addition graphs*, Rose-Hulman Undergraduate Math. J. **1** (2003) (electronic). (p. 97)

91. H. Chuang & G. R. Omidi, *Graphs with three distinct eigenvalues and largest eigenvalue less than 8*, Lin. Alg. Appl. **430** (2009) 2053–2062. (p. 225)

92. F. R. K. Chung, *Diameters and eigenvalues*, J. Amer. Math. Soc. **2** (1989) 187–196. (pp. 72, 97)

93. F. R. K. Chung, *Spectral Graph Theory*, CBMS Lecture Notes 92, Amer. Math. Soc., Providence, 1997. (p. vi)

94. V. Chvátal, *Tough graphs and hamiltonian circuits*, Discr. Math. **5** (1973) 215–228. (p. 71)

95. V. Chvátal & P. Erdős, *A note on Hamiltonian circuits*, Discr. Math. **2** (1972) 111–113. (p. 127)

96. S. M. Cioabă, *Perfect matchings, eigenvalues and expansion*, C. R. Math. Acad. Sci. Soc. R. Can. **27** (4) (2005) 101–104. (p. 76)

97. S. M. Cioabă, D. A. Gregory & W. H. Haemers, *Matchings in regular graphs from eigenvalues*, J. Combin. Theory (B) **99** (2009) 287–297. (p. 77)

98. Sebastian M. Cioabă, Edwin R. van Dam, Jack H. Koolen & Jae-Ho Lee, *Asymptotic results on the spectral radius and the diameter of graphs*, Lin. Alg. Appl. **432** (2010) 722–737. (p. 35)

99. W. H. Clatworthy, *Partially balanced incomplete block designs with two associate classes and two treatments per block*, J. Res. Nat. Bur. Standards **54** (1955) 177–190. (p. 143)

100. A. Clebsch, *Ueber die Flächen vierter Ordnung, welche eine Doppelcurve zweiten Grades besitzen*, J. Math. **69** (1868) 142–184. (p. 143)

101. Y. Colin de Verdière, *Sur un nouvel invariant des graphs et un critère de planarité*, J. Combin. Theory (B) **50** (1990) 11–21. (pp. 102, 103)

102. L. Collatz & U. Sinogowitz, *Spektren endlicher Grafen*, Abh. Math. Sem. Univ. Hamburg **21** (1957) 63–77. (p. 18)

103. K. Coolsaet & J. Degraer, *A computer assisted proof of the uniqueness of the Perkel graph*, Designs, Codes and Cryptography **34** (2005) 155–171. (p. 215)

104. K. Coolsaet, J. Degraer & E. Spence, *The strongly Regular (45,12,3,3) graphs*, Electron. J. Combin. **13** (2006) R32. (p. 144)

105. K. Coolsaet & J. Degraer, *Using algebraic properties of minimal idempotents for exhaustive computer generation of association schemes*, Electron. J. Combin. **15** (2008) R30. (p. 213)

106. C. A. Coulson, *On the calculation of the energy in unsaturated hydrocarbon molecules*, Proc. Cambridge Philos. Soc. **36** (1940) 201–203. (p. 91)

107. R. Courant & D. Hilbert, *Methoden der Mathematischen Physik*, Springer-Verlag, Berlin, 1924. (p. 27)

108. H. S. M. Coxeter, *Self-dual configurations and regular graphs*, Bull. Amer. Math. Soc. **56** (1950) 413–455. (p. 143)

109. R. Craigen & H. Kharaghani, *Hadamard matrices and Hadamard designs*, Chapter V.1., pp. 273–280 in: Handbook of Combinatorial Designs, 2nd ed. (C. J. Colbourn & J. H. Dinitz, eds.), Chapman & Hall / CRC Press, Boca Raton, 2007. (p. 160)

110. P. Csikvári, *On a conjecture of V. Nikiforov*, Discr. Math. **309** (2009) 4522–4526. (p. 36)

111. P. Csikvári, *Integral trees of arbitrarily large diameters*, J. Alg. Combin. **32** (2010) 371–377. (p. 90)

112. D. M. Cvetković, *Graphs and their spectra*, Univ. Beograd Publ. Elektrotehn. Fak. Ser. Mat. Fiz. **354–356** (1971) 1–50. (p. 39)

113. D. Cvetković & M. Doob, *Root systems, forbidden subgraphs and spectral characterizations of line graphs*, pp. 69–99 in: Graph Theory, Proc. 4th Yugoslav Sem. Novi Sad 1983 (D. Cvetković et al., eds.), Univ. Novi Sad, 1984. (p. 210)

114. D. M. Cvetković, M. Doob, I. Gutman & A. Torgašev, *Recent Results in the Theory of Graph Spectra*, Annals of Discr. Math. 36, North Holland, Amsterdam, 1988. (pp. 200, 210)

115. D. M. Cvetković, M. Doob & H. Sachs, *Spectra of Graphs: Theory and Applications*, V.E.B. Deutscher Verlag der Wissenschaften, Berlin, 1979. (Also: Academic Press, New York, 1980. Third ed., Johann Abrosius Barth Verlag, Heidelberg-Leipzig, 1995. Third rev. enl. ed., Wiley, New York, 1998.) (pp. vi, 13, 200, 201, 214)

116. D. Cvetković, M. Lepović, P. Rowlinson & S. K. Simić, *The maximal exceptional graphs*, J. Combin. Theory (B) **86** (2002) 347–363. (p. 64)

117. D. Cvetković, P. Rowlinson & S. Simić, *Eigenspaces of Graphs*, Cambridge Univ. Press, Cambridge, 1997. (p. 64)

118. D. Cvetković, P. Rowlinson & S. K. Simić, *Graphs with least eigenvalue −2: The star complement technique*, J. Alg. Combin. **14** (2001) 5–16. (p. 64)

119. D. Cvetković, P. Rowlinson & S. Simić, *Spectral Generalizations of Line Graphs; on Graphs with Least Eigenvalue −2*, London Math. Soc. Lecture Note Ser. 314, Cambridge Univ. Press, Cambridge, 2004. (pp. 64, 211)

120. D. Cvetković, P. Rowlinson & S. Simić, *An Introduction to the Theory of Graph Spectra*, Cambridge Univ. Press, Cambridge, 2010. (p. vi)

121. E. R. van Dam, *Regular graphs with four eigenvalues*, Lin. Alg. Appl. **226–228** (1995) 139–162. (pp. 221, 222)

122. E. R. van Dam, *Graphs with few eigenvalues; an interplay between combinatorics and algebra*, CentER Dissertation 20, Tilburg University, 1996. (pp. 222, 224, 225)

123. E. R. van Dam, *Nonregular graphs with three eigenvalues*, J. Combin. Theory (B) **73** (1998) 101–118. (p. 225)

124. E. R. van Dam, *The spectral excess theorem for distance-regular graphs: a global (over)view*, Electron. J. Combin. **15** (2008) R129. (p. 184)

125. E. R. van Dam & W. H. Haemers, *Graphs with constant μ and $\overline{\mu}$*, Discr. Math. **182** (1998) 293–307. (p. 224)

126. E. R. van Dam & W. H. Haemers, *Spectral characterizations of some distance-regular graphs*, J. Alg. Combin. **15** (2002) 189–202. (pp. 199, 200, 212, 214)

127. E. R. van Dam & W. H. Haemers, *Which graphs are determined by their spectrum?*, Lin. Alg. Appl. **373** (2003) 241–272. (p. 199)

128. E. R. van Dam & W. H. Haemers, *Developments on spectral characterizations of graphs*, Discr. Math. **309** (2009) 576–586. (p. 199)

129. E. R. van Dam, W. H. Haemers & J. H. Koolen, *Cospectral graphs and the generalized adjacency matrix*, Lin. Alg. Appl. **423** (2007) 33–41. (p. 200)

130. E. R. van Dam, W. H. Haemers, J. H. Koolen & E. Spence, *Characterizing distance-regularity of graphs by the spectrum*, J. Combin. Theory (A) **113** (2006) 1805–1820. (pp. 212, 214)

131. E. R. van Dam & J. H. Koolen, *A new family of distance-regular graphs with unbounded diameter*, Invent. Math. **162** (2005) 189–193. (pp. 177, 180)

132. E. R. van Dam & E. Spence, *Small regular graphs with four eigenvalues*, Discr. Math. **189** (1998) 233–257. (pp. 221, 222)

133. R. M. Damerell, *On Moore graphs*, Proc. Cambridge Philos. Soc. **74** (1973) 227–236. (p. 182)

134. K. C. Das, *A characterization of graphs which achieve the upper bound for the largest Laplacian eigenvalue of graphs*, Lin. Alg. Appl. **376** (2004) 173–186. (p. 51)

135. E. B. Davies, G. M. L. Gladwell, J. Leydold & P. F. Stadler, *Discrete nodal domain theorems*, Lin. Alg. Appl. **336** (2001) 51–60. (p. 86)

136. J. Degraer, *Isomorph-free exhaustive generation algorithms for association schemes*, thesis, Ghent University, 2007. (p. 146)

137. J. Degraer & K. Coolsaet, *Classification of some strongly regular subgraphs of the McLaughlin graph*, Discr. Math. **308** (2008) 395–400. (p. 213)

138. Ph. Delsarte, *Weights of linear codes and strongly regular normed spaces*, Discr. Math. **3** (1972) 47–64. (p. 136)

139. Ph. Delsarte, *An algebraic approach to the association schemes of coding theory*, Philips Res. Rep. Suppl. **10** (1973). (pp. 165, 171, 173)

140. Ph. Delsarte, J.-M. Goethals & J. J. Seidel, *Orthogonal matrices with zero diagonal, Part II*, Canad. J. Math. **23** (1971) 816–832. (p. 158)

141. Ph. Delsarte, J.-M. Goethals & J. J. Seidel, *Bounds for systems of lines and Jacobi polynomials*, Philips Res. Rep. **30** (1975) 91*–105*. (pp. 161, 163, 164)

142. Ph. Delsarte, J.-M. Goethals & J. J. Seidel, *Spherical codes and designs*, Geom. Dedicata **6** (1977) 363–388. (p. 164)

143. M. DeVos, L. Goddyn, B. Mohar & R. Šámal, *Cayley sum graphs and eigenvalues of (3,6)-fullerenes*, J. Combin. Theory (B) **99** (2009) 358–369. (pp. 97, 98)

144. P. Diaconis & M. Shahshahani, *Generating a random permutation with random transpositions*, Z. Wahrscheinlichkeitstheorie verw. Gebiete **57** (1981) 159–179. (p. 95)

145. M. Doob, *Graphs with a small number of distinct eigenvalues*, Ann. N.Y. Acad. Sci. **175** (1970) 104–110. (p. 221)

146. M. Doob, *On characterizing certain graphs with four eigenvalues by their spectrum*, Lin. Alg. Appl. **3** (1970) 461–482. (p. 221)

147. M. Doob & W. H. Haemers, *The complement of the path is determined by its spectrum*, Lin. Alg. Appl. **356** (2002) 57–65. (p. 209)

148. Joshua Ducey & Peter Sin, *The elementary divisors of the incidence matrices of skew lines in PG(3,p)*, arXiv:1001.2551v1, 14 Jan 2010. (p. 197)

149. Art M. Duval & Victor Reiner, *Shifted simplicial complexes are Laplacian integral*, Trans. Amer. Math. Soc. **354** (2002) 4313–4344. (pp. 57, 58)

150. M. N. Ellingham, *Basic subgraphs and graph spectra*, Australas. J. Combin. **8** (1993) 247–265. (p. 64)
151. P. Erdős, R. J. McEliece & H. Taylor, *Ramsey bounds for graph products*, Pacific J. Math. **37** (1971) 45–46. (p. 65)
152. S. Fallat & L. Hogben, *The minimum rank of symmetric matrices described by a graph: A survey*, Lin. Alg. Appl. **426** (2007) 558–582. (p. 65)
153. Ky Fan, *On a theorem of Weyl concerning eigenvalues of linear transformations I*, Proc. Nat. Acad. Sci. USA **35** (1949) 652–655. (p. 29)
154. W. Feit & G. Higman, *The nonexistence of certain generalized polygons*, J. Alg. **1** (1964) 114–131. (p. 182)
155. N. C. Fiala & W. H. Haemers, *5-Chromatic strongly regular graphs*, Discr. Math. **306** (2006) 3083–3096. (p. 128)
156. M. Fiedler, *Algebraic connectivity of graphs*, Czech. Math. J. **23** (98) (1973) 298–305. (p. 13)
157. M. Fiedler, *Eigenvectors of acyclic matrices*, Czech. Math. J. **25** (1975) 607–618. (p. 85)
158. M. A. Fiol & E. Garriga, *From local adjacency polynomials to locally pseudo-distance-regular graphs*, J. Combin. Theory (B) **71** (1997) 162–183. (pp. 184, 214)
159. M. A. Fiol, S. Gago & E. Garriga, *A simple proof of the spectral excess theorem for distance-regular graphs*, Lin. Alg. Appl. **432** (2010) 2418–2422. (p. 184)
160. C. Flye Sainte-Marie, *Solution to question nr. 48*, l'Intermédiaire des Mathématiciens **1** (1894) 107–110. (p. 18)
161. D. G. Fon-Der-Flaass, *New prolific constructions of strongly regular graphs*, Adv. Geom. **2** (2002) 301–306. (p. 214)
162. P. W. Fowler, P. E. John & H. Sachs, *(3,6)-cages, hexagonal toroidal cages, and their spectra*, Discrete Mathematical Chemistry (New Brunswick, NJ, 1998), DIMACS Ser. Discr. Math. Theoret. Comp. Sci. 51, AMS, Providence, RI, 2000, pp. 139–174. (p. 98)
163. J. Friedman, *A proof of Alon's second eigenvalue conjecture and related problems*, Memoirs of AMS 195, 2008, 100pp, and arXiv:cs/0405020v1 (118 pages). See also: *A proof of Alon's second eigenvalue conjecture*, pp. 720–724 in: STOC '03: Proc. 35th annual ACM Symp. on Theory of Computing, San Diego, 2003, ACM, New York, 2003. (p. 67)
164. H. Fujii & A. Katsuda, *Isospectral graphs and isoperimetric constants*, Discr. Math. **207** (1999) 33–52. (p. 201)
165. M. Fürer, *Graph isomorphism testing without numerics for graphs of bounded eigenvalue multiplicity*, pp. 624–631 in: Proceedings of the 6th Annual ACM-SIAM Symposium on Discrete Algorithms (SODA), 1995. (p. 63)
166. M. Fürer, *On the power of combinatorial and spectral invariants*, Lin. Alg. Appl. **432** (2010) 2373–2380. (p. 63)
167. F. R. Gantmacher, *Applications of the Theory of Matrices*, Interscience, New York, 1959. (pp. 24, 29)
168. A. L. Gavrilyuk & A. A. Makhnev, *On Krein Graphs without Triangles*, Dokl. Akad. Nauk 403 (2005) 727–730 (Russian) / Doklady Math. 72 (2005) 591–594 (English). (p. 120)
169. S. Geršgorin, *Über die Abgrenzung der Eigenwerte einer Matrix*, Izv. Akad. Nauk. SSSR Otd. Fiz.-Mat. Nauk **7** (1931) 749–754. (p. 31)
170. A. Gewirtz, *Graphs with maximal even girth*, Canad. J. Math. **21** (1969) 915–934. (pp. 144, 146, 213)
171. A. Gewirtz, *The uniqueness of g(2,2,10,56)*, Trans. New York Acad. Sci. **31** (1969) 656–675. (p. 144)
172. C. D. Godsil, *Algebraic Combinatorics*, Chapman and Hall, New York, 1993. (pp. vi, 26, 200, 201)
173. C. D. Godsil, *Spectra of trees*, Annals of Discr. Math. **20** (1984) 151–159. (p. 84)
174. C. D. Godsil, *Periodic graphs*, Electron. J. Combin. **18** (2011) P23. (p. 50)
175. C. D. Godsil & B. D. McKay, *Some computational results on the spectra of graphs*, pp. 73–92 in: *Combinatorial Mathematics IV*, Proc. 4th Australian Conf. on Combin. Math., Adelaide 1975 (L. R. A. Casse & W. D. Wallis, eds.), Lecture Notes in Math. 560, Springer-Verlag, Berlin, 1976. (pp. 202, 205)

176. C. D. Godsil & B. D. McKay, *Constructing cospectral graphs*, Aequat. Math. **25** (1982) 257–268. (pp. 16, 202, 203, 205, 206)
177. C. D. Godsil & G. Royle, *Algebraic Graph Theory*, Springer-Verlag, New York, 2001. (pp. vi, 128)
178. J.-M. Goethals & J. J. Seidel, *Orthogonal matrices with zero diagonal*, Canad. J. Math. **19** (1967) 1001–1010. (p. 159)
179. J.-M. Goethals & J. J. Seidel, *Strongly regular graphs derived from combinatorial designs*, Canad. J. Math. **22** (1970) 597–614. (pp. 147, 148, 161)
180. J.-M. Goethals & J. J. Seidel, *The regular two-graph on 276 vertices*, Discr. Math. **12** (1975) 143–158. (pp. 156, 157, 213)
181. R. L. Graham & H. O. Pollak, *On the addressing problem for loop switching*, Bell System Tech. J. **50** (1971) 2495–2519. (p. 12)
182. M. Grassl, *On SIC-POVMs and MUBs in dimension 6*, quant-ph/0406175. (p. 162)
183. B. J. Green, *Counting sets with small sumset, and the clique number of random Cayley graphs*, Combinatorica **25** (2005) 307–326. (p. 97)
184. M. Grötschel, L. Lovász & A. Schrijver, *The ellipsoid method and its consequences in combinatorial optimization*, Combinatorica **1** (1981) 169–197. (p. 44)
185. R. Grone, *Eigenvalues and degree sequences of graphs*, Lin. Multilin. Alg. **39** (1995) 133–136. (p. 51)
186. R. Grone & R. Merris, *The Laplacian spectrum of a graph II*, SIAM J. Discr. Math. **7** (1994) 221–229. (pp. 51, 53)
187. Robert Grone, Russell Merris & William Watkins, *Laplacian unimodular equivalence of graphs*, pp. 175–180 in: Combinatorial and Graph-Theoretic Problems in Linear Algebra (R. Brualdi, S. Friedland & V. Klee, eds.), IMA Volumes in Mathematics and Its Applications 50, Springer, NY, 1993. (p. 195)
188. D. Grynkiewicz, V. F. Lem & O. Serra, *The connectivity of addition Cayley graphs*, Electr. Notes in Discr. Math. **29** (2007) 135–139. (p. 97)
189. Stephen Guattery & Gary L. Miller, *On the performance of the spectral graph partitioning methods*, pp. 233–242 in: Proc. 2nd ACM-SIAM Symposium on Discrete Algorithms, 1995. (p. 62)
190. Ji-Ming Guo, *A new upper bound for the Laplacian spectral radius of graphs*, Lin. Alg. Appl. **400** (2005) 61–66. (p. 51)
191. Ji-Ming Guo, *On the third largest Laplacian eigenvalue of a graph*, Lin. Multilin. Alg. **55** (2007) 93–102. (p. 52)
192. Ivan Gutman, *The energy of a graph: Old and new results*, pp. 196–211 in: Algebraic Combinatorics and Applications (A. Betten et al., eds.), Springer-Verlag, Berlin, 2001. (pp. 65, 91)
193. Hugo Hadwiger, *Über eine Klassifikation der Streckenkomplexe*, Vierteljschr. Naturforsch. Ges. Zürich **88** (1943) 133–142. (p. 102)
194. W. H. Haemers, *An upper bound for the Shannon capacity of a graph*, Colloq. Math. Soc. János Bolyai 25, Algebraic Methods in Graph Theory (1978) 267–272. (pp. 43, 45, 46)
195. W. H. Haemers, *On some problems of Lovász concerning the Shannon capacity of graphs*, IEEE Trans. Inf. Th. **25** (1979) 231–232. (pp. 43, 45)
196. W. H. Haemers, *Eigenvalue methods*, pp. 15–38 in: Packing and covering in combinatorics (A. Schrijver ed.), Math. Centre Tract 106, Mathematical Centre, Amsterdam, 1979. (p. 41)
197. W. H. Haemers, *Eigenvalue techniques in design and graph theory*, Reidel, Dordrecht, 1980. Thesis (T.H. Eindhoven, 1979) = Math. Centr. Tract 121 (Amsterdam, 1980). (pp. 28, 39, 41, 127, 146)
198. W. H. Haemers, *There exists no (76,21,2,7) strongly regular graph*, pp. 175–176 in: Finite Geometry and Combinatorics (F. De Clerck et al., eds.), London Math. Soc. Lecture Notes Ser. 191, Cambridge Univ. Press, Cambridge, 1993. (p. 145)
199. W. H. Haemers, *Interlacing eigenvalues and graphs*, Lin. Alg. Appl. **226–228** (1995) 593–616. (p. 28)
200. W. H. Haemers, *Distance-regularity and the spectrum of graphs*, Lin. Alg. Appl. **236** (1996) 265–278. (p. 214)

201. W. H. Haemers, *Strongly regular graphs with maximal energy*, Lin. Alg. Appl. **429** (2008) 2719–2723. (p. 161)

202. W. H. Haemers, A. Mohammadian & B. Tayfeh-Rezaie, *On the sum of Laplacian eigenvalues of graphs*, Lin. Alg. Appl. **432** (2010) 2214–2221. (p. 53)

203. W. H. Haemers & F. Ramezani, *Graphs cospectral with Kneser graphs*, pp. 159-164 in: Combinatorics and Graphs, AMS, Contemporary Mathematics 531 (2010). (p. 219)

204. W. H. Haemers & C. Roos, *An inequality for generalized hexagons*, Geom. Dedicata **10** (1981) 219–222. (p. 182)

205. W. H. Haemers & M. S. Shrikhande, *Some remarks on subdesigns of symmetric designs*, J. Stat. Plann. Infer. **3** (1979) 361–366. (p. 79)

206. W. H. Haemers & E. Spence, *Graphs cospectral with distance-regular graphs*, Lin. Multilin. Alg. **39** (1995) 91–107. (pp. 179, 190, 200, 212, 213, 214)

207. W. H. Haemers & E. Spence, *The pseudo-geometric graphs for generalised quadrangles of order (3,t)*, Europ. J. Combin. **22** (2001) 839–845. (p. 145)

208. W. H. Haemers & E. Spence, *Enumeration of cospectral graphs*, Europ. J. Combin. **25** (2004) 199–211. (pp. 204, 205)

209. W. H. Haemers & Q. Xiang, *Strongly regular graphs with parameters* $(4m^4, 2m^4 + m^2, m^4 + m^2, m^4 + m^2)$ *exist for all* $m > 1$, Europ. J. Combin. **31** (2010) 1553–1559. (p. 161)

210. L. Halbeisen & N. Hungerbühler, *Generation of isospectral graphs*, J. Graph Theory **31** (1999) 255–265. (pp. 201, 204)

211. N. Hamada & T. Helleseth, *A characterization of some* $\{3v_2 + v_3, 3v_1 + v_2; 3, 3\}$-*minihypers and some* $[15, 4, 9; 3]$-*codes with* $B_2 = 0$, J. Stat. Plann. Infer. **56** (1996) 129–146. (p. 145)

212. T. H. Haveliwala & S. D. Kamvar, *The Second Eigenvalue of the Google Matrix*, Technical Report 2003-20, Stanford University, 2003. (p. 60)

213. C. Helmberg, B. Mohar, S. Poljak & F. Rendl, *A spectral approach to bandwidth and separator problems in graphs*, Lin. Multilin. Alg. **39** (1995) 73–90. (p. 74)

214. J. van den Heuvel, *Hamilton cycles and eigenvalues of graphs*, Lin. Alg. Appl. **226–228** (1995) 723–730. (p. 72)

215. D. G. Higman, *Partial geometries, generalized quadrangles and strongly regular graphs*, pp. 263–293 in: Atti del Convegno di Geometria Combinatoria e sue Applicazioni (Perugia 1970), Univ. Perugia, 1971. (p. 182)

216. D. G. Higman & C. Sims, *A simple group of order 44,352,000*, Math. Z. **105** (1968) 110–113. (p. 146)

217. R. Hill, *Caps and groups*, pp. 389–394 in: Coll. Intern. Teorie Combin. Acc. Naz. Lincei, Roma 1973, Atti dei convegni Lincei 17, Rome, 1976. (p. 140)

218. A. J. Hoffman, *On the polynomial of a graph*, Amer. Math. Monthly **70** (1963) 30–36. (pp. 14, 38)

219. A. J. Hoffman, *On eigenvalues and colorings of graphs*, pp. 79–91 in: Graph Theory and its Applications (B. Harris, ed.), Acad. Press, New York, 1970. (p. 40)

220. A. J. Hoffman & R. R. Singleton, *On Moore graphs with diameters 2 and 3*, IBM J. Res. Develop. **4** (1960) 497–504. (pp. 119, 144)

221. A. J. Hoffman & J. H. Smith, *On the spectral radii of topologically equivalent graphs*, pp. 273–281 in: Recent Advances in Graph Theory, Proc. Symp. Prague June 1974, Academia Praha, Prague, 1975. (p. 35)

222. S. G. Hoggar, *Two quaternionic 4-polytopes*, pp. 219–230 in: The Geometric Vein, Coxeter Festschrift (C. Davis, ed.), Springer, New York, 1981. (p. 162)

223. S. G. Hoggar, *64 lines from a quaternionic polytope*, Geom. Dedicata **69** (1998) 287–289. (p. 162)

224. H. van der Holst, *A short proof of the planarity characterization of Colin de Verdière*, J. Combin. Theory (B) **65** (1995) 269–272. (p. 103)

225. H. van der Holst, M. Laurent & A. Schrijver, *On a minor-monotone graph invariant*, J. Combin. Theory (B) **65** (1995) 291–304. (p. 103)

226. H. van der Holst, L. Lovász & A. Schrijver, *The Colin de Verdière graph parameter*, pp. 29–85 in: Graph Theory and Combinatorial Biology (L. Lovász et al., eds.), János Bolyai Math. Soc., Budapest, 1999. (pp. 102, 103)

227. R. A. Horn & C. R. Johnson, *Matrix Analysis*, Cambridge Univ. Press, Cambridge, 1985. (pp. 24, 26)

228. T. Huang & C. Liu, *Spectral characterization of some generalized odd graphs*, Graphs Combin. **15** (1999) 195–209. (p. 214)

229. X. Hubaut, *Strongly regular graphs*, Discr. Math. **13** (1975) 357–381. (p. 121)

230. J. E. Humphreys, *Reflection Groups and Coxeter Groups*, Cambridge Univ. Press, Cambridge, 1990. (p. 35)

231. Y. J. Ionin & M. S. Shrikhande, *Combinatorics of Symmetric Designs*, Cambridge Univ. Press, Cambridge, 2006. (p. 161)

232. I. M. Isaacs, *Character Theory of Finite Groups*, Academic Press, New York, 1976. (p. 187)

233. A. A. Ivanov & S. V. Shpectorov, *The association schemes of dual polar spaces of type $^{2}A_{2d-1}(p^{f})$ are characterized by their parameters if $d > 3$*, Lin. Alg. Appl. **114** (1989) 113–139. (p. 132)

234. P. S. Jackson & P. Rowlinson, *On graphs with complete bipartite star complements*, Lin. Alg. Appl. **298** (1999) 9–20. (p. 64)

235. C. R. Johnson & M. Newman, *A note on cospectral graphs*, J. Combin. Theory (B) **28** (1980) 96–103. (p. 199)

236. L. K. Jørgensen & M. Klin, *Switching of edges in strongly regular graphs. I. A family of partial difference sets on 100 vertices*, Electron. J. Combin. **10** (2003) R17. (pp. 146, 149, 161)

237. D. Jungnickel, *On subdesigns of symmetric designs*, Math. Z. **181** (1982) 383–393. (p. 79)

238. W. M. Kantor & R. A. Liebler, *The rank three permutation representations of the finite classical groups*, Trans. Amer. Math. Soc. **271** (1982) 1–71. (p. 121)

239. P. Kaski & P. R. J. Östergård, *There are exactly five biplanes with $k = 11$*, J. Combin. Designs **16** (2008) 117–127. (p. 120)

240. J. B. Kelly, *A characteristic property of quadratic residues*, Proc. Amer. Math. Soc. **5** (1954) 38–46. (p. 119)

241. A. K. Kelmans, *On graphs with randomly deleted edges*, Acta Math. Acad. Sci. Hung. **37** (1981) 77–88. (p. 36)

242. G. Kirchhoff, *Über die Auflösung der Gleichungen, auf welche man bei der Untersuchung der linearen Verteilung galvanischer Ströme geführt wird*, Ann. Phys. Chem. **72** (1847) 497–508. (p. 6)

243. M. Klin & M. Muzychuk, *On graphs with three eigenvalues*, Discr. Math. **189** (1998) 191–207. (p. 225)

244. Jack Koolen, *Euclidean representations and substructures of distance-regular graphs*, Ph.D. thesis, Techn. Univ. Eindhoven, 1994. (p. 185)

245. J. H. Koolen & V. Moulton, *Maximal energy graphs*, Adv. Appl. Math. **26** (2001) 47–52. (p. 65)

246. Yehuda Koren, *On spectral graph drawing*, pp. 496–508 in: COCOON 2003: Computing and Combinatorics, Lecture Notes in Computer Science 2697, Springer, Berlin, 2003. (p. 61)

247. M. Krivelevich & B. Sudakov, *Sparse pseudo-random graphs are Hamiltonian*, J. Graph Th. **42** (2003) 17–33. (p. 72)

248. K. Kuratowski, *Sur le problème des courbes gauches en topologie*, Fund. Math. **15** (1930) 271–283. (p. 102)

249. C. L. M. de Lange, *Some new cyclotomic strongly regular graphs*, J. Alg. Combin. **4** (1995) 329–330. (p. 143)

250. P. W. H. Lemmens & J. J. Seidel, *Equiangular lines*, J. Alg. **24** (1973) 494–512. (pp. 34, 161, 162, 163)

251. Jiong-Sheng Li & Yong-Liang Pan, *A note on the second largest eigenvalue of the Laplacian matrix of a graph*, Lin. Multilin. Alg. **48** (2000) 117–121. (p. 52)

252. Jiong-Sheng Li & Yong-Liang Pan, *De Caen's inequality and bounds on the largest Laplacian eigenvalue of a graph*, Lin. Alg. Appl. **328** (2001) 153–160. (p. 51)

253. Jiong-Sheng Li & Xiao-Dong Zhang, *On the Laplacian eigenvalues of a graph*, Lin. Alg. Appl. **285** (1998) 305–307. (p. 51)

254. M. W. Liebeck, *The affine permutation groups of rank three*, Proc. London Math. Soc. (3) **54** (1987) 477–516. (p. 121)

255. M. W. Liebeck & J. Saxl, *The finite primitive permutation groups of rank three*, Bull. London Math. Soc. **18** (1986) 165–172. (p. 121)

256. J. H. van Lint, *Introduction to Coding Theory*, Springer, New York, 1982. (p. 137)

257. J. H. van Lint & A. Schrijver, *Constructions of strongly regular graphs, two-weight codes and partial geometries by finite fields*, Combinatorica **1** (1981) 63–73. (pp. 133, 147)

258. J. H. van Lint & J. J. Seidel, *Equilateral point sets in elliptic geometry*, Nederl. Akad. Wetensch. Proc. Ser. A **69** = Indag. Math. **28** (1966) 335–348. (pp. 161, 163)

259. P. Lisoněk, *New maximal two-distance sets*, J. Combin. Theory (A) **77** (1997) 318–338. (p. 164)

260. Huiqing Liu, Mei Lu & Feng Tian, *On the Laplacian spectral radius of a graph*, Lin. Alg. Appl. **376** (2004) 135–141. (p. 51)

261. L. Lovász, *On the Shannon capacity of a graph*, IEEE Trans. Inf. Th. **25** (1979) 1–7. (pp. 43, 44, 45)

262. L. Lovász, *Random walks on graphs: a survey*, pp. 1–46 in: Combinatorics, Paul Erdős is Eighty (Volume 2), Keszthely, 1993. (p. 70)

263. L. Lovász & A. Schrijver, *A Borsuk theorem for antipodal links and a spectral characterization of linklessly embeddable graphs*, Proc. Amer. Math. Soc. **126** (1998) 1275–1285. (p. 103)

264. A. Lubotzky, *Cayley graphs: Eigenvalues, expanders and random walks*, pp. 155–189 in: Surveys in combinatorics, 1995 (Stirling), London Math. Soc. Lecture Note Ser. 218, Cambridge Univ. Press, Cambridge, 1995. (p. 201)

265. A. Lubotzky, R. Phillips & P. Sarnak, *Ramanujan Graphs*, Combinatorica **8** (1988) 261–277. (p. 68)

266. Ulrike von Luxburg, *A tutorial on spectral clustering*, Stat. Comput. **17** (2007) 395–416. (p. 62)

267. F. J. MacWilliams & N. J. A. Sloane, *The Theory of Error-Correcting Codes*, North Holland, Amsterdam, 1977. (pp. 136, 191)

268. C. L. Mallows & N. J. A. Sloane, *Two-graphs, switching classes and Euler graphs are equal in number*, SIAM J. Appl. Math. **28** (1975) 876–880. (p. 153)

269. M. Marcus & H. Minc, *A Survey of Matrix Theory and Matrix Inequalities*, Allyn and Bacon, Boston, 1964. (p. 24)

270. M. Marcus & H. Minc, *Introduction to Linear Algebra*, Macmillan, New York, 1965. (p. 166)

271. G. A. Margulis, *Explicit group-theoretical constructions of combinatorial schemes and their application to the design of expanders and concentrators*, Probl. Peredachi Inf. **24** (1988) 51–60. English: Probl. Inf. Transmission **24** (1988) 39–46. (p. 68)

272. R. A. Mathon, *Symmetric conference matrices of order $pq^2 + 1$*, Canad. J. Math. **30** (1978) 321–331. (p. 144)

273. R. A. Mathon, *A note on the graph isomorphism counting problem*, Inf. Proc. Letters **8** (1979) 131–132. (p. 62)

274. B. J. McClelland, *Properties of the latent roots of a matrix: The estimation of π-electron energies*, J. Chem. Phys. **54** (1971) 640–643. (p. 65)

275. R. J. McEliece & H. Rumsey, Jr., *Euler products, cyclotomy and coding*, J. Number Theory **4** (1972) 302–311. (p. 141)

276. B. D. McKay, *On the spectral characterisation of trees*, Ars. Combin. **3** (1979) 219–232. (p. 202)

277. B. D. McKay & E. Spence, *Classification of regular two-graphs on 36 and 38 vertices*, Australas. J. Combin. **24** (2001) 293–300. (pp. 144, 197, 214)

278. R. Merris, *Large families of Laplacian isospectral graphs*, Lin. Multilin. Alg. **43** (1997) 201–205. (p. 201)

279. R. Merris, *A note on Laplacian graph eigenvalues*, Lin. Alg. Appl. **285** (1998) 33–35. (p. 51)

280. B. Mohar, *Isoperimetric numbers of graphs* J. Combin. Theory (B) **47** (1989) 274–291. (p. 70)

281. B. Mohar, *A domain monotonicity theorem for graphs and Hamiltonicity*, Discr. Appl. Math. **36** (1992) 169–177. (p. 72)

282. M. Ram Murty, *Ramanujan graphs*, J. Ramanujan Math. Soc. **18** (2003) 1–20. (p. 95)

283. O. R. Musin, *Spherical two-distance sets*, J. Combin. Theory (A) **116** (2009) 988–995. (p. 164)

284. A. Neumaier, *Strongly regular graphs with smallest eigenvalue* $-m$, Archiv Math. **33** (1979) 392–400. (pp. 121, 124)

285. A. Neumaier, *New inequalities for the parameters of an association scheme*, pp. 365–367 in: Combinatorics and Graph Theory, Proc. Calcutta 1980 (S. B. Rao, ed.), Lecture Notes in Math. 885, Springer-Verlag, Berlin, 1981. (p. 171)

286. A. Neumaier, *The second largest eigenvalue of a tree*, Lin. Alg. Appl. **46** (1982) 9–25. (p. 88)

287. A. Neumaier, *Completely regular two-graphs*, Arch. Math. **38** (1982) 378–384. (p. 157)

288. R. E. A. C. Paley, *On orthogonal matrices*, J. Math. Phys. Mass. Inst. Technol. **12** (1933) 311–320. (p. 159)

289. D. V. Pasechnik, *The triangular extensions of a generalized quadrangle of order (3,3)*, Simon Stevin **2** (1995) 509–518. (p. 157)

290. A. J. L. Paulus, *Conference matrices and graphs of order 26*, Technische Hogeschool Eindhoven, report WSK 73/06, Eindhoven, 1983, 89 pp. (p. 143)

291. S. E. Payne, *An inequality for generalized quadrangles*, Proc. Amer. Math. Soc. **71** (1978) 147–152. (p. 28)

292. M. J. P. Peeters, *Ranks and Structure of Graphs*, Ph.D. thesis, Tilburg University, 1995. (pp. 46, 192)

293. René Peeters, *On the p-ranks of the adjacency matrices of distance-regular graphs*, J. Alg. Combin. **15** (2002) 127–149. (p. 190)

294. J. Petersen, *Sur le théorème de Tait*, L'Intermédiaire des Mathématiciens **5** (1898) 225–227. (p. 143)

295. Tomaž Pisanski & John Shawe-Taylor, *Characterizing graph drawing with eigenvectors*, J. Chem. Inf. Comput. Sci. **40** (2000) 567–571. (p. 61)

296. L. Pyber, *A bound for the diameter of distance-regular graphs*, Combinatorica **19** (1999) 549–553. (p. 181)

297. L. Pyber, *Large connected strongly regular graphs are Hamiltonian*, preprint, 2009. (p. 72)

298. Gregory T. Quenell, *Eigenvalue comparisons in graph theory*, Pacific J. Math. **176** (1996) 443–461. (p. 67)

299. P. Renteln, *On the spectrum of the derangement graph*, Electron. J. Combin. **14** (2007) #R82. (p. 95)

300. N. Robertson & P. D. Seymour, *Graph minors XX: Wagner's conjecture*, J. Combin. Theory (B) **92** (2004) 325–357. (p. 102)

301. N. Robertson, P. D. Seymour & R. Thomas, *A survey of linkless embeddings*, pp. 125–136 in: Graph Structure Theory (N. Robertson & P. Seymour, eds.), Contemporary Math., AMS, 1993. (p. 103)

302. P. Rowlinson, *Certain 3-decompositions of complete graphs, with an application to finite fields*, Proc. Roy. Soc. Edinburgh Sect. A **99** (1985) 277–281. (p. 149)

303. P. Rowlinson, *On the maximal index of graphs with a prescribed number of edges*, Lin. Alg. Appl. **110** (1988) 43–53. (p. 36)

304. P. Rowlinson, *The characteristic polynomials of modified graphs*, Discr. Appl. Math. **67** (1996) 209–219. (p. 201)

305. N. Saxena, S. Severini & I. E. Shparlanski, *Parameters of integral circulant graphs and periodic quantum dynamics*, Internat. J. Quantum Information **5** (2007) 417–430. (p. 50)

306. B. L. Schader, *On tournament matrices*, Lin. Alg. Appl. **162–164** (1992) 335–368. (p. 18)

307. I. Schur, *Über eine Klasse von Mittelbildungen mit Anwendungen auf die Determinantentheorie*, Sitzungsber. Berliner Math. Ges. **22** (1923) 9–20. (p. 28)

308. A. J. Schwenk, *Almost all trees are cospectral*, pp. 275–307 in: New Directions in the Theory of Graphs (F. Harary, ed.), Academic Press, New York, 1973. (p. 201)

309. A. J. Schwenk, *Exactly thirteen connected cubic graphs have integral spectra*, pp. 516–533 in: Theory and Applications of graphs, Proc. Kalamazoo 1976 (Y. Alavi & D. Lick, eds.), Lecture Notes in Math. 642, Springer, Berlin, 1978. (p. 46)

310. A. J. Schwenk & O. P. Lossers, Solution to advanced problem #6434, Amer. Math. Monthly **94** (1987) 885–886. (p. 12)

311. L. L. Scott, Jr., *A condition on Higman's parameters*, Notices Amer. Math. Soc. **20** (1973) A-97, Abstract 701-20-45. (pp. 120, 171)

312. J. Seberry Wallis, *Hadamard matrices*, Part 4 in: W. D. Wallis, A. P. Street & J. Seberry Wallis, *Combinatorics: Room squares, Sum-free sets, Hadamard matrices*, Lecture Notes in Math. 292, Springer-Verlag, Berlin, 1972. (pp. 160, 161)

313. J. Seberry & M. Yamada, *Hadamard matrices, sequences, and block designs*, pp. 431–560 in: Contemporary Design Theory (J. H. Dinitz & D. R. Stinson, eds.), Wiley, New York, 1992. (p. 160)

314. J. J. Seidel, *Strongly regular graphs of L_2-type and of triangular type*, Kon. Nederl. Akad. Wetensch. (A) **70** (1967) 188–196 = Indag. Math. **29** (1967) 188–196. (p. 157)

315. J. J. Seidel, *Strongly regular graphs with $(-1,1,0)$ adjacency matrix having eigenvalue 3*, Lin. Alg. Appl. **1** (1968) 281–298. (pp. 124, 143, 152, 213)

316. J. J. Seidel, *Graphs and two-graphs*, pp. 125–143 in: Proc. 5th Southeast. Conf. Comb., Graph Th., Comp. 1974 (F. Hoffman et al., eds.), Utilitas Mathematica Pub., Winnipeg, 1974. (p. 213)

317. J. J. Seidel, *Strongly regular graphs*, pp. 157–180 in: Surveys in Combinatorics, Proc. 7th British Combinatorial Conf. (B. Bollobás, ed.), London Math. Soc. Lecture Note Ser. 38, Cambridge Univ. Press, Cambridge, 1979. (p. 170)

318. J. J. Seidel, *Discrete non-Euclidean geometry*, pp. 843–920 in: Handbook of Incidence Geometry (F. Buekenhout, ed.), North Holland, 1995. (p. 163)

319. J. J. Seidel & S. V. Tsaranov, *Two-graphs, related groups, and root systems*, Bull. Soc. Math. Belgique **42** (1990) 695–711. (p. 153)

320. E. Seneta, *Non-negative Matrices and Markov Chains*, Springer Series in Statistics, Springer, New York, 1981. (p. 24)

321. Á. Seress, *Large families of cospectral graphs*, Designs, Codes and Cryptography **21** (2000) 205–208. (p. 200)

322. J.-P. Serre, *Répartition asymptotique des valeurs propres de l'opérateur de Hecke T_p*, J. Amer. Math. Soc. **10** (1997) 75–102. (p. 67)

323. C. E. Shannon, *The zero-error capacity of a noisy channel*, IRE Trans. Inf. Theory **2** (1956) 8–19. (p. 42)

324. Jianbo Shi & Jitendra Malik, *Normalized cuts and image segmentation*, pp. 731–737 in: Proceedings of the IEEE Conference on Computer Vision and Pattern Recognition, 1997. (p. 62)

325. S. S. Shrikhande, *The uniqueness of the L_2 association scheme*, Ann. Math. Statist. **30** (1959) 781–798. (pp. 125, 143, 213)

326. Peter Sin, *The p-rank of the incidence matrix of intersecting linear subspaces*, Des. Codes Cryptogr. **31** (2004) 213–220. (p. 197)

327. J. H. Smith, *Some properties of the spectrum of a graph*, pp. 403–406 in: Combinatorial Structures and their Applications, Proc. Conf. Calgary 1969 (R. Guy et al., eds.), Gordon and Breach, New York, 1970. (p. 34)

328. E. Spence, *The strongly regular (40,12,2,4) graphs*, Electron. J. Combin. **7** 2000, R22. (p. 144)

329. E. Spence, http://www.maths.gla.ac.uk/~es/cospecL/cospecL.html and .../cospecSignlessL/cospecSignlessL.html (p. 206)

330. Daniel A. Spielman & Shang-Hua Teng, *Spectral partitioning works: Planar graphs and finite element meshes*, pp. 96–105 in: Proc. 37th Symposium on Foundations of Computer Science (Burlington, 1996), IEEE, 1996. (p. 62)

331. Toshikazu Sunada, *Riemannian coverings and isospectral manifolds*, Ann. Math. **121** (1985) 169–186. (p. 204)

332. J. J. Sylvester, *A demonstration of the theorem that every homogeneous quadratic polynomial is reducible by real orthogonal substitutions to the form of a sum of positive and negative squares*, Philos. Mag. **4** (1852) 138–142. (Reprinted in: Coll. Math. Papers 1, Cambridge Univ. Press, 1904, pp. 378–381.) (p. 29)

333. G. Szegő, *Orthogonal Polynomials*, AMS Colloq. Publ. **23**, 1939 (1st ed.), 1975 (4th ed.). (p. 174)

334. R. M. Tanner, *Explicit concentrators from generalized n-gons*, SIAM J. Algebraic Discr. Methods **5** (1984) 287–293. (pp. 70, 81)

335. D. E. Taylor, *Some Topics in the Theory of Finite Groups*, Ph.D. thesis, Univ. of Oxford, 1971. (p. 156)

336. D. E. Taylor, *Graphs and block designs associated with the three-dimensional unitary groups*, pp. 128–131 in: Combinatorial Math., Proc. 2nd Austr. Conf. (D. A. Holton, ed.), Lecture Notes in Math. 403, Springer-Verlag, Berlin 1974. (p. 147)

337. D. E. Taylor, *Regular 2-graphs*, Proc. London Math. Soc. **35** (1977) 257–274. (pp. 147, 156)

338. W. T. Tutte, *The factorizations of linear graphs*, J. London Math. Soc. **22** (1947) 107–111. (p. 75)

339. W. T. Tutte, *All the king's horses. A guide to reconstruction*, pp. 15–33 in: Graph Theory and Related Topics, Proc. Conf. Waterloo, 1977 (J. A. Bondy & U. S. R. Murty, eds.), Academic Press, New York, 1979. (p. 16)

340. H. Van Maldeghem, *Generalized Polygons*, Birkhäuser, Basel, 1998. (p. 182)

341. R. S. Varga, *Matrix Iterative Analysis*, Prentice-Hall, Englewood Cliffs, N.J., 1962. (p. 24)

342. R. S. Varga, *Geršgorin and his Circles*, Springer, Berlin, 2004. (p. 31)

343. K. Wagner, *Über eine Eigenschaft der ebenen Komplexe*, Math. Ann. **114** (1937) 570–590. (p. 102)

344. Wei Wang & Cheng-xian Xu, *A sufficient condition for a family of graphs being determined by their generalized spectra*, Europ. J. Combin. **27** (2006) 826–840. (p. 217)

345. Wei Wang & Cheng-xian Xu, *Note: On the generalized spectral characterization of graphs having an isolated vertex*, Lin. Alg. Appl. **425** (2007) 210–215. (p. 219)

346. Y. Wang, Y. Fan & Y. Tan, *On graphs with three distinct Laplacian eigenvalues*, Appl. Math. J. Chinese Univ. Ser. B **22** (2007) 478–484. (p. 224)

347. Mamoru Watanabe, *Note on integral trees*, Math. Rep. Toyama Univ. **2** (1979) 95–100. (p. 89)

348. H. Whitney, *Congruent graphs and the connectivity of graphs*, Amer. J. Math. **54** (1932) 150–168. (p. 212)

349. H. A. Wilbrink & A. E. Brouwer, *A (57,14,1) strongly regular graph does not exist*, Indag. Math. **45** (1983) 117–121. (p. 144)

350. H. S. Wilf, *The eigenvalues of a graph and its chromatic number*, J. London Math. Soc. **42** (1967) 330–332. (p. 40)

351. R. C. Wilson & Ping Zhu, *A study of graph spectra for comparing graphs and trees*, Pattern Recognition **41** (2008) 2833–2841. (pp. 62, 206)

352. R. M. Wilson, *The exact bound in the Erdős-Ko-Rado theorem*, Combinatorica **4** (1984) 247–257. (p. 40)

353. R. Woo & A. Neumaier, *On graphs whose spectral radius is bounded by $\frac{3}{2}\sqrt{2}$*, Graphs Combin. **23** (2007) 713–726. (p. 35)

354. Chen Yan, *Properties of spectra of graphs and line graphs*, Appl. Math. J. Chinese Univ. Ser. B **17** (2002) 371–376. (p. 50)

355. Gerhard Zauner, *Quantendesigns, Grundzüge einer nichtkommutativen Designtheorie*, Ph.D. thesis, Vienna, 1999. (p. 162)

356. F. Zhang, *Matrix Theory*, Springer-Verlag, New York, 1999. (p. 28)

357. Xiao-Dong Zhang, *Two sharp upper bounds for the Laplacian eigenvalues*, Lin. Alg. Appl. **376** (2004) 207–213. (p. 51)

358. Xiao-Dong Zhang & Rong Luo, *The spectral radius of triangle-free graphs*, Australas. J. Combin. **26** (2002) 33–39. (p. 50)

Author Index

Ajtai, 70
Alon, 42, 43, 67, 69, 74
Aschbacher, 172
Ayoobi, 227

Babai, 62
Bagchi, 121
Bai, 53
Bang, 181
Bannai, vi, 180–182
Belevitch, 158
Berman, 24
Bigalke, 127
Biggs, vi
Blokhuis, 157, 164, 219
Boppana, 67
Bose, 124, 165, 166
Boulet, 208
Brauer, 187
Bridges, 119, 225
Brooks, 204
Brouwer, vi, 35, 52, 53, 90, 120, 121, 126,
       130, 140, 144, 145, 182, 215,
       219
Brualdi, 36
Burrow, 166
Bussemaker, 46, 144, 211

Calderbank, 136
Cameron, 105, 130, 133, 172, 210
Cauchy, 28
Cayley, 6
Chang, 125, 143
Cherlin, 130
Chesnokov, 207
Chuang, 225

Chung, vi, 69, 72, 97
Chvátal, 71, 127
Cioabă, 35, 77, 127
Cohen, vi, 182, 215
Colin de Verdière, 102, 103
Coolsaet, 144
Courant, 27, 86
Csikvári, 36, 90
Cvetković, vi, 13, 39, 46, 210, 211

van Dam, 135, 177, 180, 199, 200, 221, 222,
       225, 226
Damerell, 182
Davenport, 141
Degraer, 144, 146
Delsarte, v, 136, 163, 165, 170, 171, 173
DeVos, 97
Doob, vi, 13, 210, 221
Dubickas, 181
Duval, 57, 58

Erdős, 127
van Eupen, 140

Fan, 29
Feit, 79, 182
Fiedler, 13, 85
Fiol, 183, 184, 214
Fon-Der-Flaass, 214
Foulser, 121
Friedman, 67
Fürer, 63

Gago, 184
Gantmacher, 24, 29
Garriga, 183, 184, 214

Gavrilyuk, 120
Geršgorin, 31
Gerzon, 162
Gewirtz, 146
Godsil, vi, 26, 201, 202, 205
Goethals, 105, 133, 156, 163, 210
Graham, 12
Gregory, 77
Grigoryev, 62
Grone, 51, 53, 195
Guo, 52

Haemers, 28, 43, 45, 52, 77, 79, 127, 144–
        146, 157, 182, 199, 200, 205,
        207
Hamada, 145
Hasse, 141
Helleseth, 145
Helmberg, 74
Higman, 28, 146, 156, 172, 182
Hilbert, 27
Hill, 140
Hiraki, 181
Hoffman, 14, 35, 36, 38–40, 119, 144
van der Holst, 103
Horn, 24
Hubaut, 121

Isaacs, 187
Ito, vi, 180–182
Ivanov, 132

Johnson, 24, 199, 203
Jung, 127
Jungnickel, 79
Jørgensen, 146

Kallaher, 121
Kantor, 121, 130, 136
Kaski, 120
Kelly, 16, 119
Kelmans, 36
Kharaghani, 233
Kim, 127
Kirchhoff, 6
Klin, 146, 225
Komlós, 70
Koolen, 127, 177, 180, 181, 200
Krivelevich, 72
Kuratowski, 102

de Lange, 143
Laurent, 103
Lemmens, 34, 163

Liebeck, 121
Liebler, 121
van Lint, 121, 133, 137, 163
Lisoněk, 164
Lovász, 43–45, 70
Lubotzky, 68
Luo, 50

MacWilliams, 136, 191
Makhnev, 120
Marcus, 24, 166
Margulis, 68
Mathon, 62, 144
McEliece, 141
McKay, 144, 201, 202, 205
Mena, 119, 225
Merris, 53, 195
Mesner, 126, 166
Milman, 74
Minc, 24, 166
Moulton, 181
Mount, 62
Musin, 164
Muzychuk, 225

Neumaier, vi, 35, 88, 121, 124, 157, 171,
        172, 182, 215
Newman, 199, 203

Omidi, 225, 227
Östergård, 120

Page, 59
Paley, 159
Pasechnik, 157
Paulus, 143
Payne, 28
Peeters, 190
Petersen, 143
Phillips, 68
Pisanski, 61
Plemmons, 24
Pollak, 12
Pyber, 72, 181

Quenell, 67

Reiner, 57, 58
Renteln, 95
Robertson, 119
Roos, 182
Rowlinson, vi, 36
Royle, vi
Rumsey, 141

Sachs, vi, 13
Sarnak, 68
Saxl, 121
Schrijver, 103, 133
Schur, 28
Schwenk, 46, 201
Scott, 120, 171
Seidel, v, vi, 34, 105, 120, 124, 133, 143,
        152, 153, 156, 161, 163, 170–
        172, 210, 211
Seneta, 24
Seress, 200
Serre, 67
Seymour, 42
Shannon, 42
Shawe-Taylor, 61
Shimamoto, 165
Shpectorov, 132
Shrikhande, 79, 125, 143
Shult, 105, 210
Simić, vi
Sims, 28, 146
Sin, 197
Singleton, 119, 144
Sloane, 136, 191
Smith, 34, 35
Spence, 144, 145, 205, 221, 222, 226
Stickelberger, 141
Sudakov, 72
Sunada, 201, 204

Sylvester, 29
Szegő, 174
Szemerédi, 70

Tanner, 70, 81
Tayfeh-Rezaie, 227
Taylor, 156
Tsaranov, 153
Tutte, 75

Ulam, 16

Van Maldeghem, 182
Varga, 24, 31

Wagner, 102
Wang, 217, 219
Watanabe, 89
Watkins, 195
Whitney, 212
Wilbrink, 79, 144
Wilf, 40
Witsenhausen, 12
Woo, 35

Xu, 217, 219

Yan, 50

Zhang, 50

Subject Index

absolute bound, 120, 162, 171
absolute point, 80
access time, 70
addition Cayley graphs, 97
adjacency matrix, 1
algebraic connectivity, 13
antipodal graph, 178
associated two-graph, 153
association scheme, 165
automorphism, 12

bandwidth, 74
Bannai-Ito conjecture, 180
Berlekamp-van Lint-Seidel graph, 140, 148
biaffine plane, 216
biclique, 73
Biggs-Smith graph, 181
bipartite double, 10
bipartite graph, 6, 38
biplane, 122
block design, 78
block graph, 123
Bose-Mesner algebra, 166
Brouwer conjecture, 53
Brouwer-Haemers graph, 132

Cameron graph, 148
cap, 149
Cartesian product, 10
Cauchy-Binet formula, 6
Cayley graph, 11, 93, 94
Cayley sum graph, 97
Chang graphs, 125, 194, 213
characteristic matrix, 24
characteristic polynomial, 2
Cheeger number, 70

chromatic number, 40
3-claw, 105
Clebsch graph, 119, 125
clique, 38, 156
clique number, 38
coalescence, 201
cocktail party graph, 105
coclique, 38
code-clique theorem, 169
cofactor, 5
coherent, 156
coherent triple, 152
Colin de Verdière invariant, 102
collinearity graph, 123, 202
cometric association scheme, 173
companion matrix, 218
complement, 4, 116, 154
complete multipartite graph, 41, 116, 208
completely regular two-graph, 157
cone, 19, 225
conference graphs, 118, 158
conference matrix, 158
conjunction, 10
contraction, 102
core, 159
cospectral, 14–16, 18, 47, 179, 199–207
Courant-Weyl inequalities, 29
cover, 95
covering map, 95
cut, 60
Cvetković bound, 39

de Bruijn cycle, 18
de Bruijn graph, 18
derangement graph, 95
Desargues graph, 47

descendant, 153
design, 78
determinant, 107
diagonally dominant, 30
diameter, 5, 72
dimension, 56
direct product, 10
directed incidence matrix, 1
discriminant, 107
distance-regular graph, 177
dominance order, 58
downshifting, 57
DS graph, 199
dual, 137, 140
dual lattice, 106
Dynkin diagram, 35, 108

edge expansion, 70
effective length, 137
eigenvalue components, 85
elementary divisors, 194
embedding, 101
endpath, 35
energy, 61, 65, 91
equiangular tight frame, 162
equitable partition, 16, 24
equivalent code, 137
Euclidean representation, 65, 105, 183
even lattice, 106
exceptional graph, 111
expander, 70, 97
expander mixing lemma, 69
extended fundamental system, 108

feasible parameters, 143
few eigenvalues, 221
firm generalized n-gon, 129
flag, 78
flag graph, 215
folded 5-cube, 119
folded 7-cube, 42
Fowler conjecture, 98
Frame quotient, 120
$(3,6)$-fullerene, 97
fundamental system, 108

Gauss sum, 141
generalized m-gon, 182
generalized n-gon, 129
generalized adjacency matrix, 199
generalized line graph, 111
generalized quadrangle, 124, 129
geodetic subgraph, 183
geometric srg, 124

Geršgorin circles, 31
Gewirtz graph, 119, 122, 127, 135
GM switching, 16, 202
Godsil's lemma, 84
Google, 59
Gram matrix, 30
graph, 95
graph minor, 102
graphical Hadamard matrix, 160
Grassmann graph, 180
Grone-Merris conjecture, 53
groups (of a design), 122
Guo conjecture, 52

Hadamard matrix, 159
Hadwiger conjecture, 102
Haemers invariant, 45, 101, 175
half case, 118, 120
Hall-Janko graph, 123, 149
halved graph, 130
Hamiltonian, 71, 72, 127
Hamming graph, 178
Hamming scheme, 174
Hermitean linear transformation, 21
Higman-Sims graph, 39, 120
Higman-Sims technique, 28
Hill code, 140
Hoffman bound, 39, 64
Hoffman coloring, 41
Hoffman polynomial, 38, 48
Hoffman-Singleton graph, 74, 119, 120, 127
van der Holst-Laurent-Schrijver invariant,
         103
homomorphism, 95
hypercube, 10
hyperoval, 139

imprimitive drg, 178
imprimitive srg, 117
improper strongly regular graph, 151
incidence graph, 78
incidence matrix, 1
independence number, 38
index, 33, 122
inner distribution, 168
integral graph, 46, 50
integral lattice, 106
integral tree, 89
interlacing, 26
intersection array, 177
intersection matrices, 166
intersection numbers, 165
invariant factors, 194
irreducible lattice, 107

irreducible nonnegative matrix, 22
isoperimetric constant, 70
isospectral graphs, 14

Johnson graph, 179
join, 19, 25
just y-cospectral graphs, 199

k-means, 61
Kelmans operation, 36
Klein graph, 190
Kneser graph, 179
Krein conditions, 120, 171
Krein parameters, 170
Kronecker product, 10
Kuratowski-Wagner theorem, 102

Laplace matrix, 1
Laplace spectrum, 2
lattice, 106
lattice graph, 10, 125
level of a rational matrix, 218
line graph, 9, 105, 130, 202
linear code, 136
linklessly embeddable graph, 101
local graph, 213
loop, 95
Lovász parameter, 44, 65
LP bound, 168

M_{22} graph, 120, 127
Mann's inequality, 81
matrix-tree theorem, 6
McLaughlin graph, 63, 121, 134
Menon design, 159
metric association scheme, 173
minimal polynomial, 188
minimum rank, 65
Moore graph, 129, 172, 182
μ-bound, 121, 124
multiplicity free, 94

net, 123
normalized Laplacian, 62

Odd graph, 179
opposite, 130
orbitals, 121
orthogonal array, 123
orthogonal direct sum, 107
orthogonal vectors, 130
orthogonality relations, 167
orthonormal labeling, 65
outer distribution, 169

outerplanar graph, 101

Paley graph, 68, 115, 159, 191
partial geometry, 124
partial linear space, 202
pentagon, 119
perfect code, 18, 138, 140
perfect dominating set, 18
perfect e-error-correcting code, 18
perfect graph, 44
perfect matching, 75
period of an irreducible matrix, 22
Perron-Frobenius eigenvector, 59
Perron-Frobenius theorem, 22
Petersen graph, 10, 12, 39, 47, 71, 72, 74,
          87, 119, 125, 127
planar graph, 101
point graph, 123, 130, 202
polarity, 80
positive definite matrix, 30
positive semidefinite matrix, 30
primitive association scheme, 167
primitive nonnegative matrix, 22
primitive strongly regular graph, 117
projective code, 137, 139
projective plane, 79
proper vertex coloring, 40
pseudo Latin square graph, 123
pseudogeometric, 124

quasisymmetric design, 121
quotient graph, 96
quotient matrix, 24

Ramanujan graph, 68
Ramsey number, 65
random walk, 69
rank of a permutation group, 121
ratio bound, 39
rationality conditions, 118, 168
Rayleigh quotient, 25
reconstruction conjecture, 16
reduced fundamental system, 108
regular generalized m-gon, 182
regular graph, 4
regular Hadamard matrix, 159
regular tournament, 18
restricted eigenvalue, 117, 151, 223
root lattice, 106, 108
root system, 108
roots, 108

Saltire pair, 205, 207, 219
Schläfli graph, 43, 121, 125, 149

Schur complements, 28
Schur's inequality, 28
Seidel adjacency matrix, 15, 161
Seidel spectrum, 15
Seidel switching, 15, 161
self-dual lattice, 107
Shannon capacity, 43
Shrikhande graph, 125, 194, 213
SICPOVM, 162
sign changes, 174
signless Laplace matrix, 1
simplicial complex, 56
Singleton bound, 139
Smith normal form, 194
spectral center, 89
spectral excess theorem, 184
spectral radius, 33
spectrum, 2
split graph, 54
star complement, 63
star partition, 64
star subset, 63
Steiner system, 78
stepwise matrix, 36
strength, 122
strictly diagonally dominant, 30
strong Arnold hypothesis, 102
strong graph, 151
strong product, 11
strongly connected directed graph, 22
strongly regular graph, 115
subdividing an edge, 35
sum graphs, 97
support, 86, 137
switching class, 15

symmetric design, 79

t-design, 78
ternary Golay code, 133, 140
tesseract, 14
thick generalized n-gon, 129
threshold graph, 53
tight interlacing, 26
toughness, 71
tournament, 18
transitive tournament, 18
transversal design, 122
triangular graph, 10, 116, 125, 179
twisted Grassmann graphs, 180
two-graph, 152
two-weight code, 137
type I lattice, 107
type II lattice, 107

unimodular lattice, 107
unital, 79
universal cover, 95

van Dam-Koolen graphs, 180
vertex connectivity, 126

walk matrix, 217
walk-regular graph, 221
walks, 4
weak generalized quadrangle, 132
weight, 137

y-cospectral graphs, 199

zero-error channel, 42